Alfred Payson Gage

A text-book on the elements of physics : for high schools and academies

Alfred Payson Gage

A text-book on the elements of physics : for high schools and academies

ISBN/EAN: 9783337156657

Printed in Europe, USA, Canada, Australia, Japan

Cover: Foto ©Paul-Georg Meister /pixelio.de

More available books at **www.hansebooks.com**

A TEXT-BOOK

ON THE

ELEMENTS OF PHYSICS

FOR

HIGH SCHOOLS AND ACADEMIES.

BY

ALFRED P. GAGE, A.M.,

INSTRUCTOR IN PHYSICS IN THE ENGLISH HIGH SCHOOL, BOSTON, MASS.

BOSTON:
PUBLISHED BY GINN & COMPANY.
1888.

Entered according to Act of Congress, in the year 1882, by
ALFRED P. GAGE,
in the Office of the Librarian of Congress, at Washington.

TYPOGRAPHY BY J. S. CUSHING & CO., BOSTON.
PRESSWORK BY GINN & CO., BOSTON.

AUTHOR'S PREFACE.

IN his Report for the year 1881, Mr. E. P. Seaver, Superintendent of the Public Schools of Boston, says:—
"It is a cardinal principle in modern pedagogy that the mind gains a real and adequate knowledge of things only in the presence of the things themselves. Hence the first step in all good teaching is an appeal to the observing powers. The objects studied and the studying mind are placed in the most direct relations with one another that circumstances admit. Words and other symbols are not allowed to intervene, tempting the learner to satisfy his mind with ideas obtained at second-hand. One application of this principle is seen in the so-called object-teaching; but the principle is applicable to all teaching, and all methods of teaching based on it are known as objective methods. The theory goes even further, and declares, in general, that no teaching which is not objective in method can properly be called teaching at all. Hence we have this test: Is our teaching objective in method?"

This unequivocal language, from the pen of one of our foremost educators, faithfully and forcibly reflects the sentiment of the age, and leaves nothing further that need be said in advocacy of object or inductive teaching. The question for us to consider is, How shall object-teaching be conducted? Shall the teacher manipulate the apparatus, and the pupil act the part of an admiring spectator? or, Shall the pupil be supplied with such apparatus as he cannot conveniently construct, always of the simplest and least expensive kind, with which he shall be required, under the guidance of his teacher, to interrogate Nature with his own hands? By which method will he acquire the most vigorous growth, and be most likely to catch something of the spirit which animates and encourages the faithful investigator? Can elegantly illustrated works and lucid lectures on anatomy and operative surgery take the place of the dissecting room? Have lecture-room displays proved very effectual in awakening thought and in kindling fires of enthusiasm in the young? Or would a majority of our practical scientists date their first inspiration from more humble beginnings, with such rude utensils, for instance, as the kitchen affords? Is the efficiency of instruction in the natural sciences to be estimated by the amount of costly apparatus kept on show in glass cases, labelled "hands off," or by its rude pine tables and crude apparatus bearing the scars, scratches, and other marks of use? Why should this fundamental study, which logically precedes all other experimental sciences, and ought therefore, beyond all others, to be sound and thorough, be left in the condition of "a mere cram subject"?

Fortunately we are able to appeal to experience in a kindred field for an answer to the first two questions propounded. During the last twenty years there has been almost a universal change from the former method of instruction in Chemistry to the latter, so that to-day our best high schools and academies are provided with chemical laboratories for pupils' work. The result has been that this branch, which was formerly a dull and almost profitless study, has become one of the most interesting and useful in the high school curriculum. Is there any reason why laboratory practice should not do a similar work for Physics? In other words, Do not the same arguments that have been urged for the introduction of chemical laboratories apply with equal propriety and force in advocacy of physical laboratories?

But it is claimed by some that "In Physics the laboratory practice must necessarily be somewhat limited," and the usual, and almost the only reason given, is "on account of the expense." This objection rests upon the flimsiest of foundations. The expense of equipping and maintaining a physical laboratory which will answer the requirements of this book, ought to be considerably less than a similar expense to meet the demands of Eliot and Storer's Elementary Manual of Chemistry. In the English High School, in the city of Boston, the sum of three hundred dollars has furnished a physical laboratory which answers the requirements of a large school. Many and many a school has invested in showy but almost useless apparatus, — for example, in trifling electric playthings, — a sum of money which would go far towards the establishment of a simple working laboratory. But more, much more, depends upon the teacher than the cost of material. "If he has the real scientific spirit, he will do a great deal with small appliances; but if his work is done in a perfunctory manner, then the best equipment in the world will serve him but scantily."

Although this book has been prepared with a view to laboratory work, it may, in common with all text-books, be used as a mere cram-book. It may be advantageously used by those teachers who prefer or are compelled, by a real or a supposed want of time, to perform experiments themselves with elaborate apparatus. Such apparatus, if the teacher possesses it, is best explained to the pupil *viva voce*, and pictures of the apparatus are not needed, while the book will serve an additional and an important purpose of showing how the same results may be obtained in a more simple way. The great central ideas which are kept prominent throughout the book, and which serve to connect the different departments of Physics in one coherent whole, are the doctrines of the conservation of energy and the correlation of forces. So far as practicable, experiments precede the statements of definitions and laws, and the latter are not given until the pupil is prepared, by previous observation and discussion, to frame them for himself. The subjects are so arranged that, in case a year is devoted to this study, *Heat* and *Electricity* may be studied in the winter months, and *Light* in the sunny days of summer.

Many problems are given in connection with the various principles and laws. It is not expected that all pupils will perform all the problems; but the teacher will select judiciously from them. If the minds of the pupils are quite immature, or the time devoted to this study is very limited, it would be advisable to omit some of the more difficult topics; such, for instance, as are treated in §§ 93–97, and others. Most teachers prefer the "too much" to the "too little."

Every teacher has a method of his own. But perhaps the following plan, practised by the author, may be suggestive to some: He divides the experiments into three classes: home, laboratory, and lecture-room experiments. The first class is indicated in the assignment of a lesson. They are such as may be performed with such simple means as every pupil has at his home. The laboratory experiments are conducted as follows: Suppose that the number of pupils engaged at one time is fifteen, about as many as one teacher can care for successfully, and that the number of experiments to be performed during the hour is five, which is about an average number; then, to save a multiplicity of apparatus of the same kind, only three sets of apparatus of a kind are provided for each experiment. As soon as a pupil completes an experiment with one piece of apparatus, he looks about for an idle piece of some other kind; or, finding none, he improves the time in writing notes on his experiments until apparatus is ready for him; in this way each pupil performs five experiments during the hour, and devotes an average time of twelve minutes to each experiment, including the time of writing notes. The third class of experiments includes such as require the use of apparatus that cannot safely be placed in the hands of pupils,—a very limited number,—and those which have been performed by the pupils, and which the teacher may wish to repeat in a more elaborate way.

Laboratory practice and didactic study should go hand in hand, and divide time with one another about equally. In general, let the experiment precede the instruction, the pupils being guided in their investigations in the proper channels by the book and by blackboard directions. Do not teach pupils to swim before entering the water. Supt. Seaver, in another part of his report, exclaims:—

"How many of our text-books begin, not with the. suggestion of concrete illustrations, but with abstract definitions, and still more abstract 'first principles,'—blind guides to the blind teacher, and sources of perplexity to teachers who are not blind," etc.

Why should the pupil so frequently, to his great discouragement, be called upon to break through a wall of such difficulties before coming in contact with Nature?

The author would take this occasion to acknowledge with profound thanks his indebtedness to many distinguished professors of Physics for valuable assistance. Professor C. K. Wead of Michigan University has read the entire work in manuscript, and Dr. C. S. Hastings of Johns Hopkins University has read the larger portion in manuscript and the remainder in proof-sheets; and their many practical suggestions have largely contributed to whatever of success may have been achieved. Prof. T. C. Mendenhall of the Ohio State University has rendered valuable assistance in the preparation of the summary of mechanical formulas and units on page 128, as well as in the revision of the proofs. To Professors A. E. Dolbear, Tufts College; C. R. Cross and S. W. Holman, Mass. Inst. of Technology; C. F. Emerson, Dartmouth College; J. E. Davies, University of Wisconsin; B. C. Jillson, Western University of Pennsylvania; A. C. Perkins, Exeter Academy; J. E. Vose, Cushing Academy, Ashburnham; J. O. Norris, East Boston High School; G. C. Mann, Jamaica Plain High School; and others, who have kindly and patiently read and criticised the proofs as they have passed through the press, our hearty thanks are due.

Under the guidance and counsel of such an array of distinguished instructors, we may well feel a degree of confidence that the teachings of the book are not far wrong. Yet it should be distinctly understood, that for any errors which may have crept into the book, the author holds himself entirely responsible.

CONTENTS.

CHAPTER I.
MATTER AND ITS PROPERTIES.

Introduction. — Molecule. — Constitution of matter. — Physical and chemical changes. — Force. — Three states of matter. — Phenomena of attraction, — adhesion, cohesion, etc. 1

CHAPTER II.
DYNAMICS.

Dynamics of fluids. — Pressure in fluids. — Barometer. — Compressibility and expansibility of fluids. — Transmitted pressure. — Siphon. — Apparatus for raising liquids. — Buoyant force of fluids. — Specific gravity. — Motion. — Laws of motion. — Composition and resolution of forces. — Center of gravity. — Curvilinear motion. — Accelerated and retarded motion. — The pendulum. — Momentum. — Work and energy. — Transformation of energy. — Machines 44

CHAPTER III.
MOLECULAR ENERGY. — HEAT.

Heat defined. — Temperature. — Diffusion of heat. — Effects of heat. — Expansion. — Thermometry. — Laws of gaseous bodies. — Laws of fusion and boiling. — Heat convertible into potential energy, and *vice versa*. — Specific heat. — Thermodynamics. — Steam engine 138

CHAPTER IV.

ELECTRICITY AND MAGNETISM.

Current electricity. — Batteries. — Effects produced by electricity. — Electrical measurements. — Magnets and magnetism. — Laws of currents. — Magneto-electric and current induction. — Thermo-electricity. — Frictional electricity. — Electrical machines. — Applications of electricity 179

CHAPTER V.

SOUND.

Vibration and waves. — Sound-waves. — Velocity of sound. — Reflection and refraction of sound. — Loudness. — Interference. — Forced and sympathetic vibrations. — Pitch. — Vibration of strings. — Overtones and harmonics. — Quality. — Composition of sonorous vibrations. — Sound-receiving instruments. — Musical instruments 272

CHAPTER VI.

RADIANT ENERGY. — LIGHT.

Introduction. — Photometry. — Reflection. — Refraction. — Spectrum analysis. — Color. — Interference. — Refraction and polarization. — Thermal effects of radiation. — Optical instruments 325

APPENDIX 399

ELEMENTS OF PHYSICS.

ELEMENTS OF PHYSICS.

CHAPTER I.
MATTER AND ITS PROPERTIES.

I. INTRODUCTION.

§ 1. Experimentation. — An experiment is a question put to Nature. We receive the answer by means of a *phenomenon*, — that is, a change which we observe, sometimes by the sight or hearing, sometimes by other senses. In every experiment, certain facts or conditions are always known; and the inquiry consists in ascertaining the facts or conditions that follow as a consequence. The following experiments and discussions will illustrate: —

§ 2. Things known and things to be ascertained. — We are certain that we cannot make our right hand occupy the same space with our left hand at the same time. All experience teaches us that *no two portions of matter can occupy the same space at the same time.* This property which matter possesses of excluding other matter from its own space, is called *impenetrability*. It is peculiar to matter; nothing else possesses it. These facts being known, let us proceed to put certain interrogatories to Nature. Is air matter? Is a vessel full of air a vessel full of nothing? Is it "empty"? *Can matter exist in an invisible state?*

Experiment 1. Float a cork on a surface of water, cover it with a tumbler or tall glass jar, and thrust the glass vessel, mouth downward,

MATTER AND ITS PROPERTIES.

into the water. In case a tall jar (Fig. 1) is used, the experiment may be made more attractive by placing on the cork a lighted candle. State *how* the experiment answers each of the above questions, and what evidence it furnishes that air is matter; or, at least, that air is like matter.

Experiment 2. Hold a test-tube for a minute over the mouth of a bottle containing ammonia water. Hold another tube over a bottle containing hydrochloric acid. The tubes become filled with gases that rise from the bottles, yet nothing can be seen in either tube. Place the mouth of the first tube over the mouth of the second, and invert. Straightway a white cloud appears in the tubes. Soon a white, flaky solid collects on the bottom of the lower tube. Surely, out of nothing we cannot create something. Which one of the above questions does this experiment answer? How does the experiment answer it?

Again, we are quite familiar with the fact that matter exerts a downward pressure on things upon which it rests; and that matter, in a liquid state at least, exerts pressure in other directions than downward, as, for instance, against the sides of the containing vessel. *Does air exert pressure?*

Experiment 3. Thrust a tumbler, mouth downward, into water, and slowly invert. You see bubbles escape from the mouth. What is this that displaces the water, and forms the bubbles? When the tumbler becomes filled with water, once more invert, keeping its mouth under the surface of the water, and raise it nearly out of the water, as in Figure 2. The water does not fall out of the tumbler, but remains in it, entirely filling it. Hence, there is some pressure exerted on the free surface of the water; otherwise, the level would be the same in the two communicating vessels. This pressure on the surface of the water can only be produced by *some body* resting thereupon. But there is no body, except the air, that rests upon it. What conclusion do you draw from this?

Fig. 2.

Experiment 4. Pass a glass tube through the stopper of a bottle (Fig. 3). Attach a rubber tube to the glass tube. Exhaust the air by "suction" from the bottle; pinch the rubber tube in the middle, insert the open end into a basin of water, and then release the tube. What causes the water to enter the bottle? Why does not the water fill the bottle?

Fig. 3.

Finally, we know that matter has weight, and nothing else has it. *Has air weight?*

Experiment 5. Exhaust the air by means of an air-pump from a hollow globe (Fig. 4). Having turned the stop-cock to prevent the entrance of air, carefully balance the globe on a scale-beam, as in Figure 5. Afterwards turn the stop-cock, and admit the air. The globe is no longer balanced. Once more apply weights till it is balanced.

The experiments with *air* teach us that *it is matter, since, like matter, it can exclude other matter from the space it occupies, it exerts pressure, and has weight*, while all the above experiments draw from nature one reply, MATTER CAN EXIST IN AN INVISIBLE STATE.

§ 3. Minuteness of particles of matter. — Physiologists teach us, that, in order to smell any substance, we must take into our nostrils, as we breathe, small particles of that substance which are floating in the air. The air, for several meters around, is sometimes filled with fragrance from a rose. You cannot see anything in the air, but it is, nevertheless, filled with a very fine dust that floats away from the rose. The odor of rosemary at sea renders the shores of Spain distinguishable long before they are in sight. A grain of musk will scent a room for many

Fig. 4.

Fig. 5.

years, by constantly sending forth into the air a dust of musk. Though the number of particles that escape must be countless, yet they are so small that the original grain does not lose perceptibly in weight.

The microscope enables us to see, in a single drop of stagnant water, a world of living creatures, swimming with as much liberty as whales in a sea. The larger prey upon the smaller, and the smaller find their food in the still smaller, and so on, till the power of the microscope fails us. The whale and the minnow do not differ more in size than do some of these animalcules, the largest of which are hardly visible to the naked eye. But as the smallest of these perform very complex operations in collecting and assimilating food, we must conclude that they are composed not only of many particles, but of many kinds of matter. These minute living forms that people the microscopic world are exceedingly large, in comparison with the inconceivably minute particles called *molecules*, which physicists now " measure without seeing."

§ 4. **The molecule.**—**Experiment 1.** Examine carefully a drop of water with the naked eye, or with a microscope. So far as you are

Fig. 6.

able to see, the space occupied by the drop is entirely filled with water. Fill a test-tube with water (Fig. 6). Insert a cork stopper, pierced with a glass tube; heat over a lamp-flame, and note the phenomena produced. The water expands, and rises in the smaller tube; still the test-tube seems to be full of water. Place it in ice-water, and the water contracts.

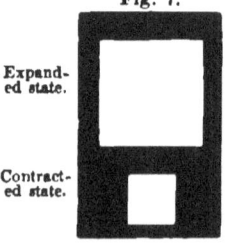

Fig. 7.

Expanded state.

Contracted state.

This change of volume can be explained only on one of two suppositions: the space occupied by the water may, as it appears, be full of water, which the heat causes to expand, and occupy a greater space, as represented graphically in Figure 7; or the body of water may consist of a definite number of distinct particles called *molecules* (as represented in Figure 8), separated from one another by spaces so small as not to be perceptible, even with the aid of a microscope. Expansion, in this case, is accounted for by a simple separation of molecules to greater distances. *There is no increase in the number of molecules, no increase in their size, only an enlargement of space between them.* Which of these suppositions is the more probable?

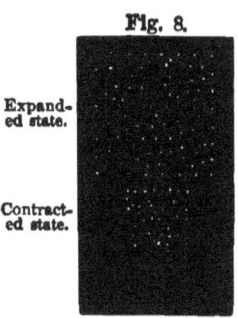

Fig. 8.

Expanded state.

Contracted state.

Experiment 2. Place a tumbler full of cold water in a warm place, and in about an hour examine it. You find many small bubbles of air clinging to all parts of the interior surface of the glass. Is it probable that outside air has descended into the liquid?

Experiment 3. Place a tumbler half full of water under a glass receiver of an air-pump (page 54), and exhaust the air. When a very good vacuum has been obtained, bubbles of air will be seen to form at all points in the liquid, and to rise and burst near the surface.

Evidently the air was previously in the same space occupied by the water. This seems to contradict the first of the above suppositions; for, according to that, the space occupied by the water is *full* of water, leaving no room for other matter. But according to the second supposition, the space is not *filled* with water; there is still room for particles of other matter in the spaces among the molecules of water. Now, as we cannot conceive of two portions of matter occupying the same space at the same time (*e.g.*, where air is, water cannot be), we conclude that the glass "full of water" is not full of water. In a similar manner, it may be shown that no visible body completely fills the space enclosed by its surface, but that there are spaces in every body that may receive foreign matter. If there are spaces, then the bodies of matter that our eyes are permitted to see are not continuous, as space is continuous. But every visible body is an aggregation of a countless number of separate and individual bodies called *molecules*.

Perform, at your homes, the two following experiments:

Experiment 4. Pulverize one-half of a teaspoonful of starch, and boil it in two tablespoonfuls of water, stirring it meantime. What phenomena occur? What do they teach? What becomes of the water?

Experiment 5. Fill a bowl half full with peas or beans. Just cover them with tepid water, and set away for the night. Examine in the morning. What phenomena do you observe? Explain each.

Strictly speaking, are bodies of matter impenetrable? What only is impenetrable? When you drive a nail into wood, do you make the two bodies occupy the same space at the same time? Do the wood and the iron occupy the same space? How only can you explain this phenomenon, consistently with the principles of impenetrability of matter?

§ 5. **Theory of the constitution of matter.** — For reasons which appear above, together with many others that will appear as our knowledge of matter is extended, physicists have generally adopted the following theory of the constitution of matter. *Every visible body of matter is composed of exceedingly small particles, called molecules; in other words, every body is the sum of its molecules. No two molecules of matter in the universe are in contact with each other. Every molecule of a body is separated from its neighbors, on all sides, by inconceivably small spaces. Every molecule is in quivering motion in its little space, moving back and forth between its neighbors, and rebounding from them. When we heat a body we simply cause the molecules to move more rapidly through their spaces; so they strike harder blows on their neighbors, and usually push them away a very little; hence, the size of the body increases.*

This theory seems, at first, little more than an extravagant guess. But if it shall be found that this theory, and this theory alone, will enable us to account for most of the known phenomena of matter, then we shall be content to adopt it till a better can be produced.

§ 6. **Porosity.** — If the molecules of a body are nowhere in absolute contact, it follows that there are unoccupied spaces among them which may be occupied by molecules of other substances. These spaces are called *pores*. Water disappears in cloth and beans. It is said to penetrate them; but it really enters the vacant spaces or pores between the molecules of these substances. All matter is porous; thus water may be forced through solid cast-iron, and dense gold will absorb the liquid mercury much as chalk will water. The term *pore*, in physics, is restricted to the invisible spaces that separate molecules. The cavities that may be seen in a sponge are not pores, but holes; they are no more entitled to be called pores, than the cells of a honeycomb, or the rooms of a house, are entitled to be called, respectively, the pores of the honeycomb or of the house.

Small as animalcules are, they are coarse lumps in comparison with the size of the molecule. By means of delicate calculations, the physicist has succeeded in ascertaining approximately the probable size of the molecule. If a drop of water could be magnified to the size of the earth, it is thought that its molecules would appear smaller than an apple. In other words, the molecule, in size, is to a drop of water what an apple is to the earth. If we should attempt to count the number of molecules in a pin's head, counting at the rate of ten million in a second, we should require 250,000 years.

§ 7. **Density.** — Cut several blocks of wood, apple, putty, lead, etc., of just the same size, and weigh them. Do they have the same weight? Can you explain the difference by a difference of porosity?

Again, if you can try the experiment illustrated in Figs. 4 and 5, using various gases, you will find that the weights of the same volumes of different gases are different. But the chemist has reasons for believing that there is the same number of molecules in the globe whatever be the gas, if the pressure and the temperature are the same. We see then that some bodies have more matter in a given volume than others, either because the molecules are closer together, or because the molecules are different; we call them more *dense*. By the *mass of a body* we understand *the quantity of matter in it;* and by its *density*, the *mass in the unit volume of it*. For example, the density of cast-iron is about 450 pounds per cubic foot.

§ 8. Simple and compound substances. — Place a small quantity of sugar on a hot stove. In a few minutes it changes to a black mass. This black substance is found to be charcoal, or carbon, as chemists call it. Evidently the sugar must have contained carbon, for the carbon came from the sugar. Chemists are also able to obtain water from sugar. The heat, in our experiment, expels the water in the form of steam, and leaves the carbon. Carbon can be extracted from sugar in another way. Prepare a very thick syrup, by dissolving sugar in hot water, and pour upon the syrup two or three times its bulk of sulphuric acid. You will quickly obtain a bulky, spongy mass of carbon.

By suitable processes, there may be obtained from marble three substances, each one of which is entirely unlike marble. One of the substances is carbon; another is a metal called calcium, which looks very much like silver; the third is a gas called oxygen, which, when set free from its prison-house in the solid, expands to many times the size of the marble from which it was liberated.

If we should grind a small piece of marble for many hours in a mortar, we should reduce the marble to a very fine powder, but should fall very far short of reducing it to its molecules. Still, each little particle of the powder is as truly marble as the original lump. If we should continue the division, in our imaginations, till the marble were reduced to molecules, we should expect to find all the molecules just alike. Now, since our smallest piece, our molecule, our unit of marble, is *marble*, and since marble is composed of the three substances, carbon, calcium, and oxygen, we conclude that our molecule itself must be capable of division. No one has been able to separate any one of these substances into other substances. No one has taken away from calcium anything but calcium, or extracted from carbon, or from oxygen, anything but carbon and oxygen.

Those substances that have resisted all efforts to break them up into other substances are called *simple substances* or elements. Those substances that may be broken up into other substances are called *compound substances*. Of the large number of substances known to man, only 71 are elements. All other substances are compounds of two or more of these 71 elements.

A molecule of any substance, simple or compound, is that minute mass of the substance which cannot be divided without destroying its properties.

§ 9. **Physical and chemical changes.** — When sugar is ground to a powder, the particles are simply torn apart, but do not lose their characteristics. The powder is just as sweet as the lump. Such a division is called a *physical division*. Generally, *any change in a substance that does not cause it to lose its identity, in other words, to cease to be that substance, is called a physical change*. When sufficient heat is applied to sugar, the molecules themselves are divided; and when a molecule of sugar is divided, the result is not two parts of a molecule of sugar, but the two substances, carbon and water. The sweetness is destroyed; sugar no longer exists; other substances have taken its place. The molecule of sugar is no more like the substances into which it has been separated, than a word is like the letters that compose it. Such a division is called a *chemical division*. Generally, *any change in a substance that causes it to lose its identity, or cease to be that substance, is called a chemical change.*

Ice, heated, melts to water; water, heated, becomes steam; steam, cooled, condenses to water; water, cooled, becomes solid. During these changes the substance, the molecule, has not changed. There has been only a change *among* the molecules, in distance and arrangement. What kind of change is this? But if the steam is subjected to a very intense heat, the result is that it becomes converted into a

10 MATTER AND ITS PROPERTIES.

mixed gas, consisting of two gases, oxygen and hydrogen. This gas is not condensable at any ordinary temperature. Unlike steam, it burns and even explodes. What kind of separation is this? What has been separated?

Blackboard crayons are prepared by subjecting the dust of plaster of Paris to great pressure, which causes the particles to unite and form the crayon. What kind of change is this? What kind of union? In the experiment (page 2) with the ammonia and hydrochloric-acid gases, the two gases disappear, and a solid is left in their place. What kind of change is this: chemical or physical? Is it union or separation?

§ 10. Annihilation and creation of matter impossible.
— Experiment 1. Prepare a saturated solution of calcium chloride. Mix with an equal bulk of water and weigh the solution. Prepare a dilute solution of sulphuric acid (1 to 4), and pour an equal weight of the last solution on the first, all at once, and shake gently. Instantly the mixed liquid becomes a solid. The solid formed is commonly called plaster of Paris. It is an entirely different substance from either of the two liquids used. What kind of change is this? A new substance has been formed. Has matter been created? Weigh the resulting solid; its weight equals the sum of the weights of the two liquids. The conclusion is, that no matter has been created, none lost.

Solids may be converted into liquids or gases; gases may be converted into liquids or solids; substances may completely lose their characteristics : but *man has not discovered the means by which a single molecule of matter can be created out of nothing, or by which a single molecule of matter can be reduced to nothing.* Matter cannot be created, cannot be annihilated; it is a constant quantity. The discovery of this fact laid the foundation of the science of Chemistry.

This statement may not seem to accord with many occurrences of every-day experience. Wood, coal, and other substances burn; matter disappears, and very little is left that can be seen. But does matter pass out of existence when it disappears in burning, or does it assume the invisible state known by the name of gas?

Experiment 2. Hold a cold, dry tumbler over a candle-flame. The bright glass instantly becomes dimmed; and, on close examination, you find the glass bedewed with fine drops of a liquid. This liquid is water.

ANNIHILATION AND CREATION OF MATTER. 11

You may think it strange that water is formed in the hot flame; yet this simple experiment shows that this is really the case. If water is formed during the burning, what is the reason we do not see it? Simply because it rises in the form of steam, which is an invisible gas. The visible cloud, often called steam, which is formed in front of the nozzle of a tea-kettle, is not steam, but fine drops of water floating in the air, — a sort of water-dust. All clouds are of the same nature. A cloud always stands over Niagara Falls, even on the clearest days. The water of the river falls a distance of 150 feet, and, striking a bed of rocks below, some of it is dashed into fragments, or dust, which rises in a cloud.

Experiment 3. Introduce a candle-flame into a clean glass bottle; after it has burned a few minutes the flame goes out. Why does it go out? See whether the air in the bottle is the same as it was before. Pour a wineglass full of lime-water into the bottle, cover tightly, and shake. Also pour lime-water into a bottle filled with air. The former becomes white and cloudy, the latter remains clear. It is apparent that some new substance has been formed during the burning, which, unlike air, can turn the lime-water white. This new substance is likewise an invisible gas.

So that, before we can decide whether or not matter is annihilated while burning, it is necessary to collect carefully, not only the ashes, but all the invisible gases that are formed. This is a somewhat troublesome experiment; but it has been frequently performed, and it is found that their collective weight is quite equal to the weight which the candle loses.

Water does not pass out of existence when it "dries up"; nor are raindrops and dewdrops created out of nothing. Matter is everywhere undergoing great and various changes, both chemical and physical. Nature is ever arraying herself in new forms. The sun warms the tropical ocean, converting the liquid into vapor; the vapor rises in the air, is recondensed on mountain hights, and returns in rivers to the ocean whence it came. Geology teaches us that continents and oceans, and even the "everlasting hills," have a birth and decay, as well as whole tribes of animals and vegetables. Although we may be counted among the living ten years hence, our bodies will, ere that, have crumbled into dust; and the matter that will then compose our bodies is to-day to be found mainly in the earth upon which we tread. Change is stamped upon all matter; nothing is exempt. Only the *quantity* of matter remains unchanged.

12 MATTER AND ITS PROPERTIES.

§ 11. Force.—**Experiment 1.** From a piece of cardboard suspend, by means of silk threads, six pith-balls, so that they may be about 2^{cm}[1] apart. Procure a clean, dry glass tube, about 40^{cm} long and 3^{cm} in diameter. Rub a portion of this tube briskly with a silk handkerchief, and hold it about 2^{cm} below the balls. The balls seem to become suddenly possessed of life. They gather about the rod, and strive to reach it. If we cut one of the threads, the ball will fly straight to the rod, and cling to it for a time. The means by which the rod pulls the balls is invisible. Yet evidence is positive that the rod has an influence on the balls, — that it *pulls* them. Slip a piece of glass between the rod and the balls; still the influence is felt by the balls. The glass does not sever the invisible bonds that connect the balls with the rod.

Fig. 9.

Now slowly bring the rod near the balls, till they touch. They at first cling to the rod; but soon the rod, as if displeased with their company, begins to push them away. Withdraw the rod; the balls do not hang by parallel threads as before, but appear to be pushing one another apart. Gradually bring the palm of the hand up beneath the balls, but without touching them. The balls gradually yield to the pull of the hand, and come together. Remove the hand, and they again fly apart. Matter does not seem to be the dead, inert thing which it is often called; it can *push* and *pull*.

Experiment 2. Raise one of these balls with the fingers, and then withdraw the fingers. Something from below seems to reach up, and pull the ball down again. The same happens with each one of the balls; every ball is pulled by something below. What is it that pulls the balls? Carry the balls into another room, the same thing occurs. Carry them to any part of the earth, the same thing occurs. It must be that it is the earth itself that pulls the balls. The earth pulls the fruit and the leaf from the tree to itself; it pulls all objects to itself; and more, — it holds them there. Attempt to raise anything from the ground, and you feel the earth's pull resisting you.

Attempt to break a string, or crush a piece of chalk, and you find

---

[1] Tables of the Metric Measures may be found in the Appendix, Section A.

MOLAR AND MOLECULAR FORCES. 13

that, notwithstanding the molecules of these bodies do not touch one another, they possess a force which tends to keep them together, and to resist your attempt to separate them.

§ 12. **Force defined.** — This tendency to push and to pull, which matter possesses, is called *force*. We do not know why separate portions of matter tend to approach one another, or to separate from one another. We do not know the nature of force; we cannot see it or grasp it; we simply know that there must be a *cause* for certain *effects* produced. The familiar effects produced are motion and rest. For example, we see a body move; we know that there is a cause: that cause we attribute to force. When a body in motion comes to rest, we look for a cause, and that cause we attribute to force. It is difficult to define force; probably the most comprehensive definition that has been given is the following: *Force is that which can produce, change, or destroy motion.*

All force exhibits itself in pushes or pulls. All motion is produced by pushes or pulls, or by a combination of both. A pulling force is called an *attractive* force, or simply *attraction*. A pushing force is called a *repellent* force, or *repulsion*.

§ 13. **Attraction and repulsion mutual.** — Experiment. Suspend a wooden lath in a sling. Rub one end of a glass rod with silk, and bring that end of the rod near to one end of the lath. The lath is attracted by the rod and moves toward it. Now place the rod in the sling, and bring the lath near to its excited end. The lath draws the rod to itself. We conclude that the pulling force belongs to both — that both are concerned in the pulling. In the experiment with the pith-balls (§ 11, Exp. 1), they seem to be mutually pushing each other. *All attraction and repulsion between different portions of matter are mutual.*

§ 14. **Molar and molecular forces.** — The glass rod does not seem to possess any attractive force, until it is rubbed with the handkerchief. The pith-balls do not repel one another until they have first touched the glass rod. After a time, the rod and the balls lose both their attractive and repellent forces. Or, if we pass the hand several times over the part of the rod that has been rubbed, and over the balls, they quickly surrender their forces. These forces are temporary. They are called *electric forces*, and their cause *electricity*. The attractive force that draws the balls to the earth existed before the experiment. No manipulation can destroy it or increase it; it is eternal and unchangeable, and exists between all portions of matter. This force is called the *force of gravity*, and the phenomenon is called *gravitation*.

We have seen the effects of attractive and repellent forces, reaching across sensible distances. Have we any evidence that these forces exist among portions of matter, at insensible distances, *i.e.*, at distances too short to be perceived by our senses? Stretch a piece of rubber; you realize that there is a force resisting you. You reason that if the supposition be true, that the grains or molecules that compose the piece of rubber do not touch each other, then there must be a powerful attractive force reaching across the spaces between the molecules, to prevent their separation. After stretching the rubber, let go one end. It springs back to its original form. What is the cause? Compress the rubber; its volume is diminished. (Does this confirm our supposition respecting the granular structure of matter?) Remove the pressure; the rubber springs back to its original form. What is the cause?

Every body of matter, with the possible exception of the molecule, whether solid, liquid, or gaseous, may be forced into a smaller volume by pressure, — in other words, *matter is compressible*. When pressure is removed, the body expands into nearly or quite its original volume. This shows two things: first, that *the matter of which a body is formed does not really fill all the space which the body appears to occupy;* and, second, that *in the body is a force, which, acting from within outward, resists outward pressure tending to compress it, and expands the body to its original volume when pressure is removed.* This is, of course, a repellent force, and is exerted among molecules, tending to push them farther apart.

But it has previously been shown that there is also an attractive force existing between the molecules. Now what is the effect, when two forces act on a body in opposite directions? Let two boys, at opposite ends of a table, push the table. If both push with equal force, the table does not move; it is as if no one pushed it. But if one boy pushes a little harder than the other, then the table moves in the direction in which the greater force is applied. Now we have the key to the solution of a difficulty, which always arises in the mind of a beginner in science, when he first hears the startling statement that the molecules of bodies, of his own body even, do not touch one another. If faith were of quick growth, he would shudder at the thought of falling to pieces, or of being wafted away by the winds as so much dust.

The ancients, perceiving that matter must be built up of small parts, overcame this difficulty by supposing that the minute particles have hooks or claws by which they grasp one another. Our knowledge of the operation of forces enables us to dispense with hooks and claws, much to the advantage of science. We see that the molecules of a body are kept from falling apart, or from separation, by a universal attractive force; they are also kept from falling together, or from permanent contact, by an ever-existing repellent force. These forces act at insensible distances between molecules, and hence are called *molecular forces*. When forces act between bodies at sensible distances they are called *molar forces*. Give illustrations (1) of molar forces; (2) of molecular forces.

II. THREE STATES OF MATTER.

§ 15. **Matter** presents itself in three different states: *solid*, *liquid*, and *gaseous*, — fairly represented by earth, water, and air. Because these forms are so common and abundant, some ancient philosophers held that all solid matter is formed of earth, all liquids of water, and all gases of air. On this account they called them, together with fire, elements or primary matter. They cannot now be so regarded from a chemical point of view, because each of them has been separated into still more simple substances; nor from a physical standpoint, because, as will soon be shown, most substances may exist in any one of these states.

§ 16. **Characteristics of each of these states.**
Experiment 1. Provide two vessels, a cubical dish and a goblet, each having a capacity of about 200^{ccm}. Also provide 200^{ccm} of sand, 200^{ccm} of water, and a cubical block of wood containing 200^{ccm}. Grasp the block, and place it in the cubical vessel. Attempt to do the same thing with the water. Why can you not grasp the water? Pour a portion of the water into the cubical vessel. When you move a portion of the block, the whole block moves. When you pour a portion of the water into the cubical vessel, the whole does not necessarily go. Why is this? Why is it that we can dip a cupful of water out of a pailful, without raising the whole? Pour all the water into the goblet. The water adapts itself to the shape of the goblet, and the vessel is filled. Attempt to place the block of wood in the goblet. What difference in phenomena do you observe? Why this difference? Pour the sand from vessel to vessel. It adapts itself to the shape of each vessel. Why? Drop the block of wood on a table. Pour water on the table. How does a liquid behave when there is no vessel to confine it?

Experiment 2. Throw small particles of sawdust into the goblet of water; you can thus render perceptible any motion of the water in the goblet, just as, by throwing blocks of wood on the smooth surface of a river, you can discover the motion of the river. Notice the ease with which the particles move about, rise, and sink. As they become quiet, slightly jar the vessel, or tap it with the end of a pencil, and notice the ease with which disturbance is produced throughout the liquid. Now rap the side of the block with a hammer, and observe how immovable are the particles of wood.

Our experiments teach us that *the molecules of solids are not easily moved out of their places;* consequently, *solid masses form such a firmly connected whole that their shape is not easily changed, and a movement of one part causes a movement of the whole.* On the other hand, *the molecules of liquids have scarcely any fixedness of position, but easily slip between and around one*

THREE STATES OF MATTER. 17

another; consequently, liquid bodies easily mold themselves to the shape of the vessel that contains them, are *poured* from vessel to vessel, and are easily separated into parts.

But what shall we say of the sand, which, like water, adapts itself to the shape of the containing vessel, and can be poured? Is sand a liquid? and are powders liquids? No, powders are a collection of small *lumps* of solid matter. When powders are poured, lumps of matter roll around one another, as when potatoes are poured from basket to basket. When liquids are poured, *molecules* glide past one another.

It is not so easy to study the characteristics of gases, because we cannot usually see them. But we may be aided by a device similar to that employed to make the movement of water visible.

Experiment 3. Darken a room, and admit, through a small crack or hole, a beam of direct sunlight. You see particles of dust dancing in the path of the light; the motion never ceases. See how easily the motion is quickened by gently waving the hand at some distance from the beam of light.

Experiment 4. Place under the receiver of an air-pump a partially inflated balloon, Fig. 32, page 53 (or a Seven-in-one apparatus with the piston near the closed end of the cylinder, and stop-cock closed), and exhaust the air. The tendency of gases to expand becomes evident.

In gases, fixedness of position of the molecules is entirely wanting, and freedom of motion among themselves is almost perfect. They appear to be in a continual state of repulsion, and consequently have a tendency to expand to greater and greater volumes. They expand indefinitely, unless confined by pressure, while liquids and solids tend to preserve a uniformity of volume.

Liquids do not rise above what is called their surface, and we may have a vessel half full of a liquid; but *gases have no definite surface*, and there is no such thing as a vessel half full of gas. On the other hand, *if gases are subjected to pressure, their volume may be indefinitely diminished ;* for instance, the air that now fills a quart vessel may be compressed into a pint vessel, or even into less space, if sufficient force is used. *The com-*

pression of liquids is barely perceptible, even when the pressure is very great.

§ 17. Philosophy of the three states of matter. — We conclude from the difficulty which we experience in separating the parts of a solid body, that the molecular attractive force in solids is very great. From the ease with which we usually separate the parts of a body of liquid, we might conclude that this force in liquids is very weak. But before arriving at any conclusion, it is necessary to consider how the difficulty of separation of the parts of a liquid is to be measured. It is very easy to tear off a portion of a sheet of tinfoil, but we should not surely regard this as an evidence that the molecules of tin have but little attraction for each other, for in tearing such a body we only apply the force to a comparatively few molecules at a time. We can form a just estimate of the strength of molecular attraction only by attempting to separate the foil into two portions by such means as that the separation may take place no sooner at one point than at another. So, too, it is very easy to separate a drop of water into two portions, but this is no measure of the attractive forces unless we take precautions that we do not apply the separating force successively to different molecules. If we succeed in preventing such a successive action, and there are certain methods of doing this more or less perfectly, we should find the process much more difficult, — more so indeed, than to produce a similar change in many solids.[1]

There is, however, a difference in the molecular action in solids and liquids; such that, in the latter state, the molecular forces offer no resistance to a *shaping* force, while in the former state, change of shape can only be brought about by the application of considerable force.

In a gas, on the contrary, there is little attraction between the molecules; but as they are constantly hitting one another, and thereby tending to drive one another apart, it requires an external force to keep them together.

[1] The cohesive force of water is at least 132 lbs. per square inch. — MAXWELL.

NOTE. In gases, the molecules are thought to be in motion like gnats in the air; in liquids, like men moving through a crowd; in solids, the motion of each molecule is like that of a man in a dense crowd where it is almost or quite impossible to leave his neighbors, yet he may turn around, and have some motion from side to side.

Practically, the condition of any portion of matter depends upon its temperature and pressure. (See p. 160.) Just as at ordinary pressures water is a solid, a liquid, or a gas, according to its temperature, so any substance may be made to assume any one of these forms unless a change of temperature occasions a chemical change.

There are certain apparent exceptions to the last statement; for example, charcoal, though it has been vaporized, has never been obtained in a liquid state, simply because sufficient pressure has never been used. Ice will change to a vapor, but cannot be melted unless the pressure exceeds six grams per square centimeter. For a similar reason, iodine and camphor vaporize, but do not melt. Alcohol has never been solidified, or frozen.[1] It has been rendered thick and pasty,— a semi-solid condition, — showing that it only requires a little lower temperature than any to which it has been exposed, to complete the solidification.

As regards the temperature at which different substances assume the different states, there is great diversity. Oxygen and nitrogen gases, or air,— which is a mixture of the two,— liquefy and solidify only at extremely low temperatures; and then, only when the attractive force is aided by tremendous pressure. On the other hand, certain substances, as quartz and lime, are liquefied only by the most intense heat generated by an electric current. The facts, summed up, are as follows: *no one of the three states of matter, solid, liquid, or gaseous, is peculiar to any substance; the state that a substance assumes depends solely on its temperature and pressure;* so that every solid may be regarded as simply matter in a frozen state, every liquid as matter in a melted state, and every gas as matter in a state of vapor.

[1] Since this statement was written alcohol has been frozen at about −130° C.

Every liquid has been solidified and volatilized, and every gas has been liquefied and solidified. Air was one of the last of the gases to surrender its reputation of being a "permanent gas." Not till the year 1878 was it reduced to lumps. We may predict the future of our globe. If its heat increases sufficiently, the whole world will become a thin gas. If its heat diminishes indefinitely, all earth and air will become a solid mass.

III. PHENOMENA OF ATTRACTION.

ACCORDING to the circumstances under which attraction acts, we have the various phenomena called *gravitation*, *cohesion*, *adhesion*, *capillarity*, *chemism*, and *magnetism*. Sometimes these terms are used as names of the unknown forces that cause the phenomena.

§ 18. **Gravitation.** — That attraction which is exerted on all matter, at all distances, is called *gravitation*. Gravitation is universal, that is, every molecule of matter attracts every other molecule of matter in the universe. The whole force with which two bodies attract one another is the sum of the attractions of their molecules, and depends upon the number of molecules the two bodies collectively contain, and the mass of each molecule. The whole attraction between an apple and the earth is equal to the sum of the attractions between every molecule in the apple and every molecule in the earth.

§ 19. **Weight.** — It is scarcely necessary to state, that what is understood by the *weight* of a body is the mutual attraction between it and the earth. The term *mass* is equivalent to the expression *quantity of matter*. It follows, then, that *weight* is proportional to *mass*. Why do we weigh articles of trade, such as sugar and tea?

§ 20. **Does the apple attract the earth with as much force as the earth attracts the apple?** — Let us examine this question. First assume that the molecules of the apple and the earth have equal masses, *i.e.*, are homogeneous; then the attraction of any

LAW OF GRAVITATION. 21

molecule in the apple for any molecule in the earth is equal to the attraction of any molecule in the earth for any molecule in the apple. That is, if the earth and the apple consisted each of a single like molecule, their attraction for each other would be equal. Now suppose that the apple contains two and the earth five such molecules. Let the force with which one molecule attracts another be represented by n. Now, each molecule of the apple attracts the five molecules in the earth with a force of $5n$; the two molecules in the apple would attract the earth with a force of $10n$. On the other hand, each molecule of the earth attracts the molecules of the apple with a force of $2n$, and the five molecules in the earth would attract the apple with a force of $10n$. It is obvious that the same course of reasoning will apply in case the attraction is between two molecules whose masses differ, and consequently between all bodies of whatever mass or substance. *Hence a body of small mass attracts a body of large mass as strongly as the latter attracts the former.*

If the apple attracts the earth as strongly as the earth attracts the apple, why does not the earth rise to meet the apple? Let us examine a similar case. Suppose that a man in a boat pulls on a rope attached to a ship. His pulling draws the boat to the ship, but the ship does not appear to move. But if five hundred men, in as many boats, pulled together, the ship would be seen to move. Did the one man produce no motion? If so, then would the five hundred men produce no motion, since five hundred times nothing is nothing? Yes, the apple moves the earth as surely as the earth moves the apple; but the apple has *more to move*, and, consequently, it moves the earth a distance as many times less than it is moved by the earth, as the quantity of matter in the earth is times the quantity of matter in the apple. The respective distances the two bodies move vary inversely as their masses.

§ 21. **The force of gravity varies with the distance from the center.** — Observations made in various ways show that the force of gravity varies over the surface of the earth. It can be proved by geometrical methods that a sphere or a spheroid acts upon a molecule without it as though all its attractive force were concentrated at its center. Now it is found that the nearer an object without the earth's surface is to the center of the earth the greater is the force of gravity. The polar diameter of the earth is about 26 miles less than its equatorial diameter, and, consequently, the distance from the center to the surface at the poles is 13 miles less than to the surface at the equator. This considerable difference in distance from the center occasions an appreciable difference between the weight of a body (having any considerable mass) at the equator and at the poles; and, since the distance of the surface from the center constantly increases as we go from the poles toward the equator, the weight of all objects transported from the poles toward the equator constantly diminishes.

It is obvious that any object raised above the earth's surface, as in a balloon, must weigh less than at the surface of the earth. But the hights with which we commonly have to deal in our experiments are so small in comparison with the earth's radius, that the differences in weight due to differences in hight at a given place can scarcely be detected by most delicate tests.

The statement that "weight is proportional to mass" (§ 19) must, therefore, be restricted to a comparison of *masses at the same place and at the same altitude* only. The propriety of making a distinction between the terms *mass* and *weight* is now apparent, as the former implies that which does not change when a body is transferred from place to place, while the latter may change.

If the earth were of uniform density, bodies carried below its surface would lose in weight as the distance below the surface increases. At one-fourth the distance to the center there would be a loss of one-fourth the weight. At one-half the distance the weight would be one-half; and at the center nothing. Is weight an essential property of matter? State certain conditions on which a body would have no weight.

The terms *up* and *down* are derived from the attraction between the earth and terrestrial objects. *Down* is toward the center of the earth, or it is the direction in which a body falls or tends to move in consequence of gravitation. *Up* is the opposite direction. It is apparent that the up and down of one place cannot correspond with the up and down of any other place.

COHESION. 23

QUESTIONS.

1. If an iron pound-weight and a pound of sugar were balanced with ordinary scales at the equator, and transported to one of the poles of the earth, would they cease to balance each other?
2. If the same quantity of sugar be suspended from a spring-balance at the pole, will this instrument indicate just a pound, more or less?
3. Imagine yourself at the center of the earth. In what direction must you turn your face in order to look up?
4. Imagine a person at one of the poles, and another at the equator, to be looking down upon you at the center of the earth. Would they both look in the same direction?
5. Draw a circle to represent the earth, and two lines to represent the direction in which the two persons would look.
6. What is the origin of "water-power"?
7. What is the cause of tides?
8. Which is more difficult, to ascend or descend a hill, and why?
9. The earth has about 81 times as much matter in it as the moon. At which body would you weigh more?
10. Is there a place between the two bodies at which you would weigh nothing? If so, why?
11. How far does the earth's attraction extend?
12. Which would you prefer, a pound of gold weighed with a spring-balance at the surface of the earth, or a pound weighed 3,000,000^m below the surface?

§ 22. **Cohesion.**—That attraction which holds the molecules of the same substance together, so as to form larger bodies, is called *cohesion*. It is the force that prevents our bodies, and all bodies, from falling down into a mass of dust. It is that force which resists a force tending to break or crush a body. It is greatest in solids, usually less in liquids, and nothing in gases. It acts only at insensible distances, and is strictly a molecular force. When once the cohesion is overcome, it is difficult to force the molecules near enough to one another for this force to become effective again. Broken pieces of glass and crockery cannot be so nicely readjusted that they will hold together. Yet two polished surfaces of glass, placed in contact, will cohere quite strongly. Or if the glass is heated till it is soft, or in a semifluid condition, then, by pressure, the molecules at the two surfaces will flow around one another, pack themselves closely together, and the two bodies will become firmly united. This process is called *welding*. In this manner iron is welded.

Cohesive force varies greatly in different substances, according to the variation in the nature, form, and arrangement of the molecules of which they are composed. These modifications of the force of attraction give rise to certain *conditions of matter*, designated as *crystalline, amorphous, hard, flexible, elastic, brittle, viscous, malleable, ductile,* and *tenacious*.

§ 23. **Crystalline and amorphous conditions of matter.** —If our vision could be rendered keen enough to enable us to see and examine the molecular structure of different substances, to look into their bodies, as we look into the starry heavens, and observe the positions, the spaces, and the arrangement of that unexplored world, there would undoubtedly be unfolded to us wonders and beauties of which we have never dreamed. We should probably behold an endless variety of arrangement among the molecules. We might learn why it is that the molecule of the diamond, of graphite, and of charcoal being the same (*i.e.*, the same substance), we get, possibly by different arrangement and different behavior of molecular forces, the hard, transparent, and brilliant diamond in the one case, the soft, opaque, metallic-looking graphite in another, and finally the porous, black, and shapeless charcoal.

Obtain a piece of mica, or Iceland spar, and a piece of chalk, and attempt to cut them in two, by applying the knife in different directions. You find that you can easily cleave the mica in one direction, and obtain a smooth, shining surface. This is called its *plane of cleavage*. Cut it in any other direction, and you get rough and ragged surfaces. The spar may be cleft easily and smoothly in three directions. But the chalk may be cleft in one direction as well as another, and in no direction can a smooth surface be obtained. We learn by these trials that matter may have method in its arrangement, or possess definite structure.

When matter exhibits structure or method in its molecular arrangement, it is said to be *crystalline*. Examples of crystalline arrangement are mica, Iceland spar, and carbon in the form of diamond. When its molecular arrangement is methodless or structureless, it is said to be *amorphous*. Examples of amorphous matter are chalk, glue, glass, and carbon in the form of charcoal.

Experiment 1. Pulverize 20^g of alum, and dissolve in 50^{ccm} of hot water; suspend a thread in the solution, and put it away where it can quietly and slowly cool. After it has become cold, you will find attached to the thread beautiful transparent bodies of regular shape. The process by which matter in solidifying assumes a structural condition is called *crystallization*, and bodies which have acquired regular shape by this process are called *crystals*. Obtain crystals of saltpetre, blue vitriol, and potassium bichromate, by dissolving as much as possible of these substances in hot water, and allowing the solutions to cool, *always slowly and quietly.*

Experiment 2. Thoroughly clean a piece of window-glass, and pour upon it a hot, concentrated solution (see § 36) of ammonium chloride or saltpetre. Allow the liquid to drain off, hold it up to the sunlight, and you will see beautiful crystals rapidly springing into existence, spreading and branching like vegetable growth.

Very interesting illustrations of crystallization are those delicate lacelike figures which follow the touch of frost on the window-pane. Figure 10 represents a few of more than a thousand forms of snowflakes that have been discovered, resulting from a variety of arrangement of the water molecules.

Nature teems with crystals. Nearly every kind of matter, in passing from the liquid state (whether molten or in solution) to the solid state, tends to assume symmetrical forms. *Crystallization is the rule ; amorphism, the exception.* Break open a sugar-loaf, and you will find the surface fracture composed of small, shining, crystalline surfaces. You can scarcely pick up a stone and break it, without finding the same crystalline fracture. Every piece of ice is a mass of crystals, so closely packed together that the individuals are not distinguishable.

§ **24. Change of volume by crystallization.** — This tendency of matter to structural arrangement is not only very interesting, but very important in the arts. It is very natural to

Fig. 10.

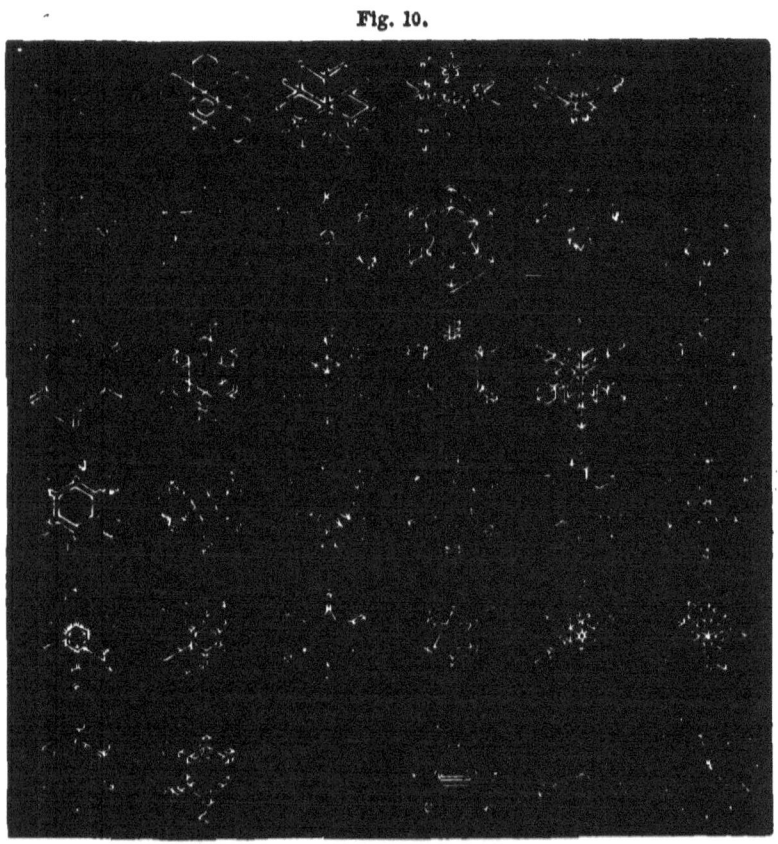

suppose that the new arrangement of molecules, when passing from the liquid to the solid state, should occasion either an increase or diminution in volume. We are not surprised when we find that water, in freezing, disregards the law of contraction by cold, and that the molecules are not found so closely packed together, in the new and structural state, as when under the influence of cohesion alone.

The force exerted by the molecules in changing positions is so enormous as to burst the strongest vessels. Hence our service-pipes are burst when water is allowed to freeze in them. Huge rocks are dislodged from their resting-places in the native quarry on the mountain-side by water getting into the crevices, freezing, expanding year after year, and pushing the rocks from their support. Cast-iron and many alloys, such as typemetal, expand on solidifying. Such metals may be cast in molds, since, in expanding, they fill all the minute cavities of the mold. Most metals contract on solidifying. Hence gold, silver, and copper coins require to be stamped. Cast-iron, when broken, exhibits a crystalline fracture. Wrought-iron, when subjected to long-continued jarring, — for instance, the axles of car-wheels, and iron cannon, — becomes very brittle, and, when broken, exhibits a very marked crystalline fracture which it would not have shown before long use. It is probable that the molecules of iron, when shaken up by the jarring, are free to arrange themselves in their peculiar method, and that, in this new arrangement, the cohesive force is weakened.

§ 25. **What is the cause of this almost universal tendency of matter to crystallize?** — We have no absolute knowledge of the doings in the molecular world. But we have very satisfactory methods of judging. Analogy is the light by which we must frequently explore inaccessible space. We determine the laws that govern large, tangible masses, and from these we infer the laws that govern small, intangible bodies, or molecules. Let us adopt this method in attempting to unravel the mystery before us.

Experiment 1. Take two cambric needles, and draw each several times, from the eye to the point, over the same end of a magnet. Now suspend each needle by a thread, so that it will be balanced in a horizontal position. Bring the eye of one near the point of the other. When brought near enough, they attract each other. Bring the point of one near the point of the other; they repel one another. Bring the eye of one near the eye of the other; they repel one another. We thus discover that the relation of these two needles to one another is such, that if unlike ends are brought together they attract one another, but if like ends are brought together they repel one another. The opposite character which the ends of the needle exhibit is called *polarity*.

Now break one of the needles into two pieces, and experiment as before. The two pieces exhibit the same polarity that the two unbroken needles did. Break them into still smaller pieces, and the smallest piece that you can obtain possesses polarity, as certainly as the original needle. Imagine the work of division to be continued till the molecule is reached. Is it too much to assume that the molecule may possess polarity?

Experiment 2. Next, place a magnet beneath a sheet of paper, and sift iron filings over it. The instant they strike the paper they arrange themselves in lines around the magnet (see Fig. 162, page 221). Gently tap the paper, and they arrange themselves still more definitely. This reminds us of the effect of jarring on the car-axle and cannon, where molecules, once set in motion, tend to arrange themselves according to some guiding principle. Next, lay the magnet on a bed of iron filings (see page 214), and then raise it. We find the filings clinging most abundantly to the ends, diminishing in number toward the middle.

We pass readily from these facts to conclusions respecting the molecular arrangement in the crystal. Only grant the supposition that the molecule is endowed with something similar to polarity, and we can picture to ourselves the molecules, like the iron filings, wheeling into line in obedience to their polar forces. Crystals are more easily cleft in some directions than in others: may not this be accounted for by supposing that, like the magnet, the attraction on some sides of the molecule is greater than on others?

§ 26. Hardness. — Name some metal that you can scratch with a finger-nail. See if you can scratch a piece of copper with a piece of lead, and *vice versa*. Get as many specimens as possible of the following substances: talc, chalk, glass, quartz, iron, silver, lead, copper, rock-salt, and marble. Ascertain which of them will scratch glass, and which are scratched by glass. What term do we employ in speaking of those substances that are easily scratched? To those that are scratched with difficulty? Which is the softest metal that you have tried?

FLEXIBILITY.

The hardest? Which is the softer metal, iron or lead? Which is the more dense metal? Does hardness depend upon density? What force must be overcome in order to scratch a substance? When will one substance scratch another?

To enable us to express degrees of hardness, the following table of reference is generally adopted: —

MOHR'S SCALE OF HARDNESS.

1. Talc.
2. Gypsum (or Rock-Salt).
3. Calcite.
4. Fluor-Spar.
5. Apatite.
6. Orthoclase (Feldspar).
7. Quartz.
8. Topaz.
9. Corundum.
10. Diamond.

By comparing a given substance with the substances in the table, its degree of hardness can be expressed approximately by one of the numbers used in the table. If the hardness of a substance is indicated by the number 4, what would you understand by it?

§ 27. **Flexibility.** — Such substances as may be bent, or admit of a hinge-like movement among their molecules, are called *flexible*. What difference have you noticed in different jack-knife blades? How can you tell a soft blade from a hard blade?

Fig. 11.

If you bend a stick, as in Figure 11, it is apparent that the molecules on the upper side must be separated from each other a little farther than usual, and that they must have slightly rolled round one another, while those on the under side must be crowded together more closely than usual. On the other hand, the molecules in a glass rod have fixed relative positions which will permit very little disturbance.

§ 28. **Elasticity.** — Obtain thin strips of the following substances: rubber, wood, ivory, whalebone, steel, brass, copper, iron, zinc, and lead. Stretch the piece of rubber. What change in its molecular condition must occur when it is stretched? What molecular force causes it to contract when the stretching force is removed? Compress the rubber. What change of molecular condition takes place in compression? What force causes it to expand when the pressure is removed? Bend each one of the above strips. Note which completely unbends when the force is removed. Arrange the names of these substances in the order of the rapidity and completeness with which they unbend.

What change takes place among the molecules on the concave side of the bent strips? What, among the molecules on the convex side? What two forces are concerned in the unbending? Twist the cord of a window-tassel. What causes it to untwist? The property which matter possesses of recovering its former shape and volume, after having yielded to some force, is called *elasticity*. To what forces is elasticity due? Does all matter possess this property in the same degree? Does the rubber possess the same ability to unbend, as to contract after being stretched? In what four ways have you tested the elasticity of substances? Does a substance possess equal power of recovering its form after yielding to each of these four methods of applying force? Why are pens made of steel? What moves the machinery of a watch? What is the cause of the softness of a hair mattress or feather-bed?

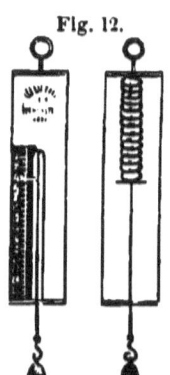

Fig. 12.

A common spring-balance used for weighing consists of a steel spring wound into a coil. The weight of the body to be weighed straightens or draws out the spring. A pointer moving over a plate which is divided into equal parts shows how much the spring has been drawn out. But the entire virtue of this apparatus consists in the elasticity of the spring, or its power to recover its original form after being drawn out. Give other illustrations of the application of elasticity to practical purposes.

Any alteration in the form of a body due to the application of a force is called a *strain*, and the force by which the strain is produced is called the *stress*. A body which, having experienced a strain due to a certain stress, completely recovers its original condition when the stress is removed, is said to be *perfectly elastic*. Liquids and gases are perfectly elastic (see § 48). Solids are perfectly elastic up to a certain limit, which varies greatly in different substances. If the stress exceeds a certain limit, the form of the solid becomes permanently altered, and the state of the body, when the permanent alteration is about to take place, is called the *limit of perfect elasticity*. In soft or plastic bodies this limit is soon reached. What is the result of overloading carriage springs?

§ 29. **Brittleness.**—Apply sharp blows with a hammer to each of the substances whose hardness you have tested (§ 26), and ascertain which are the most easily broken or pulverized. Observe that some substances suffer a permanent change in form when subjected to a stress which exceeds their limit of elasticity, while others break before there is any permanent alteration in form. The latter are said to be *brittle*.

§ 30. **Viscosity.**—Support in a horizontal position, at one of its extremities, a stick of sealing-wax, and suspend from its free extremity a small weight, and let it remain in this condition several days, or perhaps weeks. At the end of the time the stick will be found permanently bent. Had an attempt been made to bend the stick quickly, it would have been found quite brittle. A body which, subjected to a stress for a considerable time, suffers a permanent change in form, is said to be *viscous*. Hardness is not opposed to viscosity. A lump of pitch may be quite hard, and yet in the course of time it will flatten itself out by its own weight, and flow down hill like a stream of syrup. Liquids like molasses and honey are said to be viscous, in distinction from limpid liquids like water and alcohol.

§ 31. **Malleability and ductility.**—Some substances possess, in the solid state, a certain amount of *fluidity*; that is, their molecules may be displaced without overcoming their cohesion. Place a piece of lead on an anvil, and hammer it. It spreads out under the hammer into sheets, without being broken, though it is evident that the molecules have moved about among one another, and assumed entirely different relative positions. Heat a piece of soft glass tube in a gas-flame, and, although the glass does not become a liquid, it behaves very much like a liquid, and can be drawn out into very fine threads. When a solid possesses sufficient fluidity to admit of being drawn out into threads, it is said to be *ductile*.[1] When it will admit of being hammered or rolled into sheets, it is said to be *malleable*.

As might be expected, *those substances that are ductile are also malleable.* But the same substance does not usually possess the two properties in an equal degree. Platinum is the most ductile metal. It can be drawn into wire finer than a spider's thread. It is the seventh metal in the rank of malleability. Gold is the most malleable metal. It can be hammered into leaves so thin, that it would require 300,000 to make a book one inch thick. It ranks next to platinum in ductility. Iron, at a red heat, is very malleable and ductile. What metals can be drawn into wires? What metals can be rolled or hammered into sheets?

§ 32. **Tenacity.**—In order that a substance may be ductile, it is evident that it must possess a strong cohesive force, so as to prevent rupture. The power that matter possesses of resisting rupture, by a pulling force, is called *tenacity*.[2] *A body may be tenacious without being ductile, but it cannot be ductile without being tenacious.* It is remarkable that the tenacity of most metals is increased by being drawn out into wires. It would seem, that, in the new arrangement which the molecules assume, the cohesive force is stronger than in the old. Hence cables made of iron wire twisted together, so as to form an iron

---

[1] Ductile, *draw-able.*   [2] Malleable, as it were *mallet-able.*
[3] Tenacity, property of *holding.*

ADHESION. 33

rope, are stronger than iron chains of equal weight and length, and are much used instead of chains, where great strength is required.

§ 33. **Adhesion.**—Grasp with your finger a piece of gold-leaf, and, honest as you may be, it will stick to your fingers; it will not drop off, it cannot be shaken off, and to attempt to pull it off is to increase the difficulty. Dust and dirt stick to clothing. Thrust your hand into water, and it comes out wet. You can climb a pole, because your hands stick to the pole; but if the pole is greased, climbing is not so easy. We could not pick anything up, or hold anything in our hands, were it not that these things stick to the hands.

Every minute's experience teaches us that not only is there an attractive force between molecules of the same kind of matter, but there is also an attractive force between molecules of unlike matter. That force which causes unlike substances to cling together, is called *adhesion*. Is adhesion a molar or a molecular force? How does it differ from cohesion? Why do not gold watches, and other articles of gold jewelry, appear to stick to the fingers? What keeps nails, driven into wood, in their places? What would happen if all adhesion between the different parts of the building you are in should be suddenly destroyed? When a liquid sticks to a solid, what term do we usually employ in describing the phenomenon?

Fig. 13.

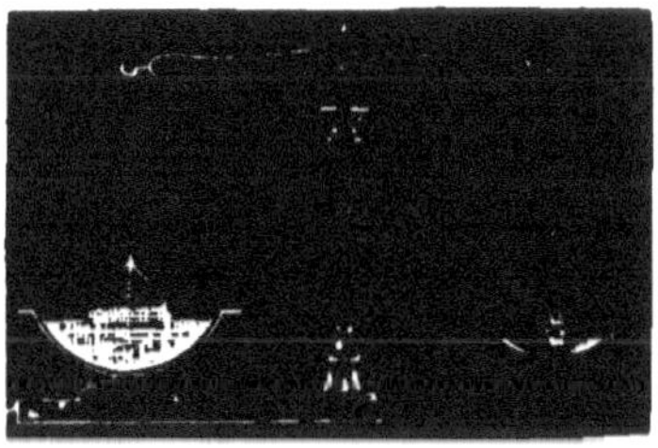

Experiment 1. Suspend a plate of glass, about 8^{cm} square, from one arm of a scale-beam, attaching the threads to the plate with sealing-wax. Balance it, and place a dish of water under the glass, so that its under surface will just touch the surface of the water. You may now add several grams' weight to the other side of the beam without destroying the balance. Finally, the glass is apparently pulled away from the water. But on examination you will find it wet, so that you

have really succeeded, not in separating the glass from the water, but water from water. Then the weight that you were obliged to add does not measure the adhesive force between the glass and the water; it merely measures the amount of force necessary to tear the liquid apart. The same force was not sufficient to tear the liquid from the solid, hence we infer that *the adhesion between a solid and a liquid may be greater than the cohesion in the liquid.*

Glass is wet by water, but is not wet by mercury. Is there no adhesion between mercury and glass?

Experiment 2. Substitute mercury for water in the last experiment. As soon as the glass touches the mercury a slight adhesion occurs, which can be measured by the weight required to be placed in the opposite scale-pan in order to separate them.

It is probable that *there is some adhesion between all substances when brought in contact. If a liquid adheres to a solid more firmly than the molecules of the liquid cohere, then will the solid be wet by the liquid.* If a solid is not wet by a liquid, it is not because adhesion is wanting, but because cohesion in the liquid is stronger. That gases adhere to solids is proved by the phenomena of absorption described in § 37.

QUESTIONS.

1. Why will not water wet articles that have been greased?
2. Why is it difficult to lift a board out of water?
3. Why does water run down the side of a tumbler when it is inclined, instead of falling vertically? Suggest some method of preventing it.
4. In what does the value of cement, glue, and mucilage consist?
5. What enables you to leave a mark with a pencil or crayon?

§ 34. Capillarity. — Examine the surface of water in a goblet. You find the surface level, as in A (Fig. 14), except around the edge next the glass, where the water is curved upward so as to resemble the interior surface of a watch crystal. Mercury placed in a goblet (B) has its edge turned downward, resembling the exterior surface of a watch crystal. This seems to indicate

CAPILLARITY. 85

a repulsion between mercury and glass. But a previous experiment (page 34) has shown that, instead of repulsion, there is a slight adhesion between these substances.

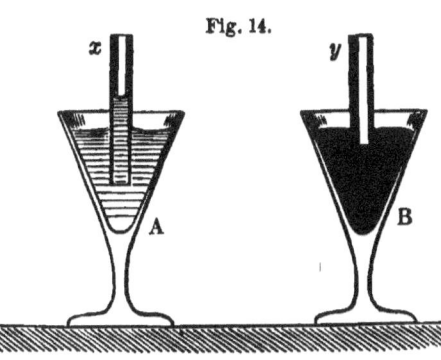

Fig. 14.

Pour any liquid on a level surface which it does not wet, — e. g., water on paraffine or wax, or mercury on glass. It spreads itself over the surface, but the edges are everywhere rounded or turned down like the edges of mercury in a goblet. Surely these rounded edges are not caused by the repulsion of the sides of a vessel. The edges of all liquids will be turned down unless the adhesion between them and the sides of the vessels exceeds the cohesion in the liquid. The glass does not cause the turning down of the surface of mercury in the goblet, — its tendency is rather to prevent it.

Thrust vertically two plates of glass into water, and gradually bring the surfaces near each other. Soon the water rises between the plates, and rises higher as the plates are brought nearer. Thrust a glass tube of very fine bore into water; the attraction within it, on all sides, will raise the water to twice the hight it would reach when between two plates whose distance apart is equal to the diameter of the bore of the tube. Thrust a tube of the same bore into alcohol; this liquid rises in the tube, but not so high as water. The surfaces of both the water and the alcohol are concave. If the tube is placed in mercury, the opposite phenomena occur: the mercury is depressed, and its surface is convex.[1] Both ascension and

[1] The scope of this book will not admit of an explanation of the phenomena of capillarity. The student can find a lucid treatment of this subject in Maxwell's "Theory of Heat," pp. 260-274; also under "Capillary action," Encyclopædia Britannica.

depression diminish as the temperature increases, being greatest at the freezing point of the given liquid, and least at its boiling point. (Regarding heat as a repellent force, can you give any reason why the ascension should be less at high than at low temperatures?) Inasmuch as the phenomena are best shown in tubes having bores of the size of hairs, they are in such cases called *capillary*[1] *phenomena*, and the tubes are called *capillary tubes*.

The phenomena of capillary action are well shown by placing

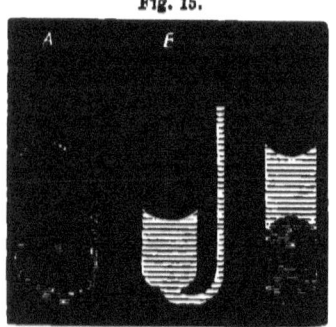

Fig. 15.

various liquids in U-shaped glass tubes, having one arm reduced to a capillary size, as A and B in Figure 15. Mercury poured into A assumes convex surfaces in both arms, but does not rise so high in the small arm as it stands in the large arm. Pour water into B, and all the phenomena are reversed. C is a glass tube containing water and mercury, and showing the shapes that the surfaces of the two liquids take.

Generalizing the above facts, we have the *four laws of capillary action:* —

I. *Liquids rise in tubes when they wet them, and are depressed when they do not.*
II. *The ascension or depression varies inversely as*[2] *the diameter of the bore.*
III. *The ascension and depression vary with*[2] *the nature of the substances employed.*
IV. *The ascension or depression varies inversely with the temperature.*

Illustrations of capillary action are abundant. It feeds the lamp-flame with oil. It wets the whole towel, if one end is left for a time in a basin of water. It draws water into wood, and causes it to swell with a force sufficient to split rocks, and to raise large weights. How does a little water in a wooden tub prevent its falling to pieces?

---

[1] Capillary, *hair-like*. [2] Observe that throughout this treatise the word *as* expresses an exact proportion. When there is not an exact proportion, the word *with* is used.

SOLUTION OF SOLIDS. 87

§ 35. Other molecular phenomena.—Besides the phenomena we have just studied, there are a great many others depending in part on molecular attraction, but much more on the molecular motions, of which we learned in § 5, page 6. Many of them are quite familiar and important; but the explanation, even when it can be given, is usually complicated and incomplete. The principal names given these phenomena are *solution*, *absorption*, and *diffusion*.

§ 36. Solution of solids — *depends mainly on molecular attraction.* Hold a lump of sugar so that it will just touch the surface of water. Soon water is drawn up into the pores of the lump by capillary action, and the whole lump, including the part not submerged, becomes moist. Next you discover that the lump becomes smaller, and slowly disappears in the water.

When a solid becomes diffused through a liquid, it is said to be *dissolved*. The dissolving liquid is called a *solvent*, and the resulting liquid is called a *solution*. *A liquid will dissolve a solid, only when the adhesion between them is greater than the cohesion in the solid.* A liquid always dissolves a solid more rapidly at first, less rapidly as the adhesion becomes more nearly satisfied; and when it is completely satisfied, or is balanced by the cohesion in the solid, the liquid will dissolve no more of the solid; and the solution is said to be *saturated*. When a solution will take much more of a solid, it is said to be *dilute;* and *concentrated*, when it will take little or no more.

If the solid be first pulverized, the liquid has more surface on which to act, and the solid is dissolved much more rapidly. *Heat generally weakens cohesion more than it weakens adhesion;* hence, with few exceptions, hot liquids dissolve solids more rapidly and in greater quantities than cold liquids. Boiling water dissolves three times as much alum as cold water; consequently, when a hot saturated solution of alum is allowed to cool, at least two-thirds of the alum must be restored to the solid state (see Exp. 1, page 25), while one-third, or the amount that the cold liquid is capable of dissolving, remains in solution. The remaining solution is called the *mother-liquor*. Lime, and a few other substances, are dissolved better in cold water. Crystals of such substances are only obtained by gradual evaporation of the solvent.

Water is the great solvent. When we speak of the solubility of a substance, water is always understood to be the solvent, unless some other liquid is specified. Why is it fortunate that water is so good a solvent? Name substances that water does not dissolve. Of the many substances insoluble in water, some, as phosphorus, gums, and resin, find a solvent in alcohol; sulphur, in bi-sulphide of carbon; lead, in mercury; and fats, in ether or benzine. Would you wash varnished furniture with alcohol? How are grease-spots removed from clothing?

§ 37. **Absorption of gases by solids** — *depends mainly on molecular attraction, and is generally superficial.* Certain solids possess so strong an attraction for gases that they not only draw the gases into the small cavities or holes within them, but greatly condense them there. It should be carefully noted that the attraction in this case is generally between the gases and the *surfaces of cavities*, and is hence called *superficial*, in distinction from *intermolecular attraction*, which is the name given to the phenomenon when gases are taken into the *pores* of a body.

Freshly-burned charcoal placed in dry air, may, in a few days, have its weight increased one-fiftieth in consequence of the air that it absorbs. (Has air weight?) The attraction of charcoal for noxious gases is especially great, making it very efficient in cleansing the air in hospitals, and in removing noxious odors from putrid animal and vegetable matter by absorbing the foul gases that are generated. It does not check decay, but rather hastens it. A rat, which had been buried in charcoal dust, was uncovered at the end of a month; nothing visible was left but the hair and bones, yet no bad odor was perceptible. Why do farmers mix muck with manures?

§ 38. Absorption of gases by liquids — *depends on molecular attraction and motion, and is intermolecular.* Water, at a temperature of 0° Cen., is capable of condensing in its *pores* six hundred times its own bulk of ammonia gas. Water thus charged with this gas is called "ammonia water." The amount of gas that a liquid will absorb is increased by pressure. "Soda water" is simply water saturated with carbonic-acid gas under great pressure; it contains no soda. When the pressure is removed, a large part of the gas escapes, causing effervescence.

§ 39. Free diffusion of liquids — *depends mainly on motion.* — **Experiment 1.** Into a test-tube containing 20^{ccm} of water, pour about 2^{ccm} of olive-oil, and shake. By shaking, the oil becomes divided into small particles, which give the water an opaque, milky-white appearance, but it is not separated into its molecules. After standing for a few minutes, the oil almost completely separates from the water, and rises to the top.

Experiment 2. Partially fill a glass jar (Fig. 16) with water. Then introduce beneath the water, by means of a long tunnel, a concentrated solution of sulphate of copper. The lighter liquid rests upon the heavier, and the line of separation between the two liquids is at first distinctly marked. But in the course of days or weeks this line will gradually become obliterated, the heavier blue liquid will gradually rise, and the lighter colorless liquid will descend, till they become thoroughly mixed.

Fig. 16.

Experiment 3. Take about 1^{ccm} of bisulphide of carbon, color it by dropping into it a small particle of iodine, and pour this colored solution into a test-tube nearly filled with water. The colored liquid, being heavier than the water, sinks directly to the bottom, and shows no tendency to mix with the water. But, in the course of time, you discover that the colored liquid diminishes in quantity, and finally disappears. The peculiar odor of this substance which pervades the air in the vicinity shows that a considerable portion has evaporated. But it must have worked its way gradually through the water above it.

If, during the operation of diffusion in the last two experiments, you examine the liquid with a microscope, you will not be able to trace any currents; hence the motion of liquids in diffusion is not in mass, but by molecules, — a kind of intermolecular motion. We learn that some liquids, even when stirred together, will not remain mixed; while others, whose densities are very different, when merely placed in contact with each other, slowly mix of themselves.

§ 40. Diffusion of liquids through porous partitions. — Osmose. — Dialysis. — *Very complex.* — Experiment. Cut off the bottom of a conical-shaped bottle[1] (or, better, use a glass tunnel or lamp-chimney); fit to the neck of the bottle a cork, having a glass tube passing through it (Fig. 17). Tie tightly over the bottom a piece of gold-beater's skin or parchment paper. Fill the bottle with a concentrated solution of sulphate of copper, and press the cork into the bottle so that the liquid will stand a little way up the tube, say at a.

Fig. 17.

Now suspend the apparatus in a vessel of water, so that the bottom may be covered. In less than an hour it will be found that the liquid has risen in the tube, showing that water must have passed through the septum,[2] and mixed with the solution. Examine the water in the outer vessel, and you will find that it is slightly tinged with the blue vitriol, showing that some of the solution has also passed through the septum. But the liquid has risen in the tube, showing that more of the water than of the solution has passed through the septum.

When liquids or gases force their way through porous septa, and mix with each other, the diffusion is called *osmose*.[3] To distinguish the two opposite currents, the flow of the liquid or gas towards that which increases in volume

[1] See Appendix, Section B.   [2] Septum, *partition.*   [3] Osmose, *impulse.*

is called *endosmose*,[1] and the opposite current is called *exosmose*.[2]

It is found that crystallizable substances are the best subjects of osmose, while those which are usually amorphous, such as gelatine and gummy substances, are very little inclined to osmose. Those substances that pass readily through septa are called *crystalloids*;[3] those that do not are called *colloids*.[4]

The principle of unequal diffusibility of liquids through septa finds important application in chemical and pharmaceutical laboratories. For example, from a rod (Fig. 18) is suspended a glass vessel having a bottom of parchment paper. Such a vessel is called a *dialyzer*. In the dialyzer is placed, for instance, the liquid contents of the stomach or intestines of a dead animal, suspected of containing some poison, and the vessel is floated in a vessel of water. If either arsenic or strychnine is present it will separate from the albuminous matter in the food, and pass through the septum into the water. The process of separating mixed liquids by osmose is called *dialysis*.

Fig. 18.

§ 41. **Free diffusion of gases** — *depends almost wholly on molecular motion.* — **Experiment.** Fill a test-tube with oxygen gas, and thrust in a lighted splinter; the splinter burns much more rapidly than in the air. Fill another tube with hydrogen gas, and keep the tube inverted (for, this gas being about sixteen times lighter than air, there will be no danger of its falling out). Thrust in a lighted splinter; the gas takes fire, and burns with a pale flame at the mouth of the tube. Next fill one tube with oxygen and the other with hydrogen gas, and place the mouth of the latter over the mouth of the former, as in Figure 19. In about a minute apply a lighted splinter to the mouth of each

[1] Endosmose, *inward impulse.*
[2] Exosmose, *outward impulse.*
[3] Crystalloid, *like crystal.*
[4] Colloid, *like gum.*

tube (let the mouth of each tube be freely open to prevent accident); a slight explosion takes place in each instance. It is apparent that although the oxygen gas is sixteen times heavier than the hydrogen, some of it has risen into the upper tube, while some of the lighter hydrogen has descended into the lower tube, and the two gases have become diffused.

Many pairs of liquids do not diffuse into each other, but *every gas diffuses into every other gas*, and it is impossible to prevent two gases from mixing when placed in contact. (It is thought best to introduce the subject of diffusion of liquids and gases in this place, though it has little or no connection with the subject of adhesion. The explanation of diffusion must be deferred to its proper place in the chapter on Heat, page 158.)

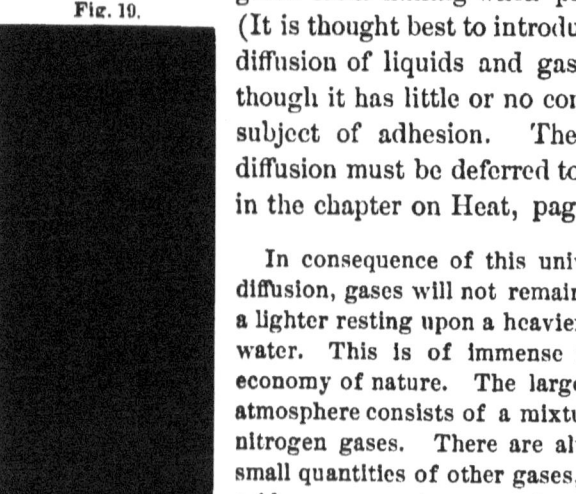

Fig. 19.

In consequence of this universal tendency to diffusion, gases will not remain separated, — *i.e.*, a lighter resting upon a heavier, as oil rests upon water. This is of immense importance in the economy of nature. The largest portion of our atmosphere consists of a mixture of oxygen and nitrogen gases. There are always present also small quantities of other gases, such as carbonic-acid gas, ammonia gas, and various other gases, which are generated by the decomposition of organic matter. These gases, obedient to gravity alone, would arrange themselves according to their weight, — carbonic-acid gas at the bottom, or next the earth, followed respectively by oxygen, nitrogen, ammonia, and other gases. Neither animal nor vegetable life could exist in this state of things. But, in consequence of their diffusibility, they are found intimately mixed, and in the same relative proportions, whether in the valley or on the highest mountain peak.

§ 42. Diffusion of gases through porous partitions — *depends on the size of molecules, size of pores, and on molecular motion; very complex.*

DIFFUSION OF GASES.

Experiment. Take a thin, unglazed earthen cup, such as is used in Bunsen's battery (page 190), and plug up the open end with a cork through which extends a glass tube. Place the exposed end of the tube in a cup of colored water. Lower a glass jar, filled with hydrogen or coal-gas, over the porous cup, as in Figure 20. Instantly air is forced down through the tube, and escapes in bubbles from the colored liquid. The gas in the larger vessel forces its way through the pores of the cup, diffuses itself in the air contained in it, and causes an unusual pressure on the colored liquid, as is evinced by the air that is forced out through it. In a minute remove the glass jar. The hydrogen now escapes through the sides of the cup, and mixes with the air on the outside; a partial vacuum is formed in the cup, and water rises in the tube. In both cases air passed through the sides of the porous cup, but the influx and efflux of hydrogen was much more rapid.

Fig. 20.

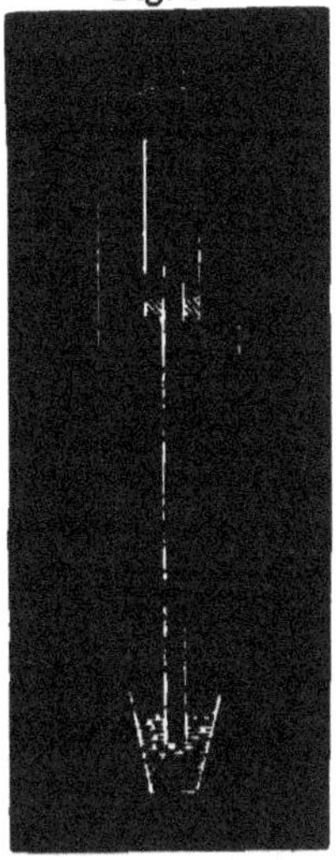

An interesting modification of this apparatus is the *diffusion fountain* (Fig. 21). By passing the glass tube of the porous cup through the cork of a tightly-stopped vessel, and having another glass tube pass through another perforation in the same cork, water is forced out in a jet several feet in hight, when the hydrogen jar is held over the porous cup.

Fig. 21.

Children well understand that toy balloons, which are made of collodion and filled with coal-gas, collapse in a few hours after they are inflated. This is caused by the escape of the gas by osmose. Nature furnishes an illustration of osmose of gases in respiration. In the lungs the blood is separated from the air by the thin, membranous walls of the veins. Carbonic-acid gas escapes from the blood through these septa, and oxygen gas enters the blood through the same septa.

CHAPTER II.

DYNAMICS.

IV. DYNAMICS OF FLUIDS.

§ 43. Equilibrium, pressure, and tension. — That branch of science which treats of force and the motions it produces is called *dynamics*. It has been shown that force may act on a body to produce motion or rest; also that two or more forces may so act on a body as to neutralize each other's effect. In the latter case, the body continues in the same condition, either of motion or rest, as if it were independent of the action of the forces, and is said to be in *equilibrium*,[1] and the forces acting on it are also said to be in equilibrium. Inasmuch as no body is ever free from the action of force, it must be that *a body at rest is in a state of equilibrium*.

If any portion of a force is not effective in producing motion, — *i.e.*, if part or all of it is exerted against other forces, — there may result what is called a *pressure* on the body; as when we push on a wall or on a heavy sled moving over the ice, or a book presses the table. The same force which causes a body to fall when unsupported, causes it to press on any obstacle which prevents it from falling. Or, if the force is exerted on a body in which the molecular attraction is strong, — *i.e.*, on a solid, — we may have a pull or *tension*, as when we hang in a swing, or hang a stone from a rubber band. If the body under the influence of a force maintains a uniform velocity, we may *measure the force by the pressure* (*or tension*) *exerted*, ·or may *measure the pressure by the amount of the force*, whichever may be more convenient. The case of uniform velocity includes the case of rest.

[1] Equilibrium, *equal balance*.

PRESSURE IN FLUIDS. 45

§ 44. Pressure in fluids. — It will be seen that, with the exception of the phenomena of capillarity and those occasioned by difference in compressibility and expansibility, liquids and gases are governed by the same laws. We shall, therefore, treat them together, in so far as they are alike, under the common term of *fluid*.

It should be borne in mind that we are placed on the borders of two oceans. A watery ocean borders our land; an aerial ocean, which is called the atmosphere, surrounds us. Every molecule, in both the gaseous and liquid oceans, is drawn toward the earth's centre by gravity. This gives to both fluids a downward pressure upon everything upon which they rest.

The gravitating power of liquids is everywhere apparent, as in the fall of drops of rain, the descent of mountain streams, the power of falling water to propel machinery, and the weight of water in a bucket. But to prove the downward pressure of air requires special experiments. If we lower a pail into a well, it fills with water, but we do not perceive that it becomes heavier thereby; the downward pressure is not felt. But when we raise a pailful out of the water, it suddenly becomes heavy. If we could raise a pailful of air out of the ocean of air, might not the weight of the air become perceptible? If we dive to the bottom of a pond of water, we do not feel the weight of the pond resting upon us. We do not feel the weight of the atmospheric ocean resting upon us; but we should remember that our situation with reference to the air is like that of a diver with reference to water.

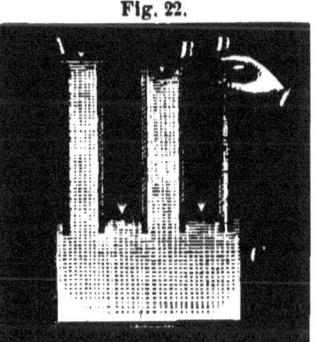

Fig. 22.

Experiment 1. Fill two glass jars (Fig. 22) with water, A having a glass bottom, B a bottom provided by tying a piece of sheet-rubber tightly over the rim. Invert both in a larger vessel of water, C. The water in A does not feel the downward pressure of the air directly above it, the pressure being sustained by the rigid glass bottom. But it indirectly feels the pressure of the air on the surface of the water in the open vessel, and it is this pressure that sustains the water in the jar. But the rubber bottom of the jar B yields somewhat to the downward pressure of the air, and is forced inward, until it is balanced by the upward pressure of the water, plus the tension of the rubber.

Take a glass tube D, 1^m long, having a bore of 1^{cm} diameter. Covering one end with a finger, fill with water, and invert in C. You feel the weight of the air pressing your finger against the tube. Remove the finger and the water in the tube at once sinks to the level of the water in the vessel C, because the downward pressure of the air on the column of water, plus the weight of the column of water, is greater than the upward pressure. In every instance we find that the downward pressure of air gives rise to an upward pressure in the liquid. In this respect fluids differ widely from solids, whose molecules are so firmly held together that, when one part is pushed in any direction, that part drags the rest with it.

We have accounted for water being sustained in the vessels A, B, and D, by an upward pressure produced by the downward pressure of the air. Does this downward pressure create an upward pressure in the air itself, so that, if the vessels are lifted out of the water, the water will not fall out?

Fig. 23.

Experiment 2. Keeping the finger pressed on the end of D, raise it ,slowly and vertically out of the water. The water does not fall out. Why? Slip a thin glass plate, or a piece of thick pasteboard, under the mouth of A, and, pressing it against the mouth, raise the vessel carefully out of the water, and remove the hand from the plate. The water does not fall out, nor does the plate fall. Why?

Experiment 3. Force a tin pail (Fig. 23), having a hole in its bottom, as far as possible into water, without allowing water to enter at the top. A stream of water spurts through the hole. Why? Why does it require so much effort to force the pail down into the water? Does downward pressure cause a lateral pressure?

Experiment 4. Make holes, at different depths, in the side of a vessel (Fig. 24) containing water. Water issues in streams, with considerable force, from the orifices. Why?

Experiment 5. Bind a piece of thin sheet-rubber tightly over a wide-mouthed bottle, and place it in water in different positions. In whatever position the bottle is placed, the rubber is pressed inward. What lesson does this teach?

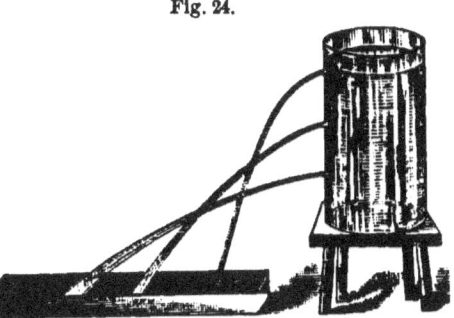

Fig. 24.

Experiment 6. The Magdeburg hemispheres (Fig. 25) are two hemispherical cups, having their edges made smooth so as to be "air-tight" when placed in contact. Each cup is provided with a handle. One of the handles consists of two parts, a stem and a ring, the two parts being connected by a screw. The stem has a bore passing through it, and a stop-cock, which regulates the passage of air through the bore. Place the lips of the cups in contact, remove the ring, screw the stem to the plate of an air-pump, and exhaust the air from the sphere; then close the stop-cock, and replace the ring. Now two boys grasping the rings, and holding the sphere in any position they choose, can only with great difficulty pull them apart. Why?

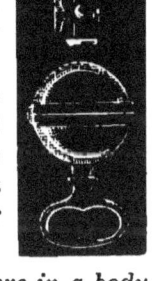

Fig. 25.

Boys amuse themselves by lifting bricks (Fig. 26) with a circular piece of leather, moistened and pressed against the surface of the brick, so as to exclude the air. The pressure of air against the leather binds it to the brick in whatever position placed.

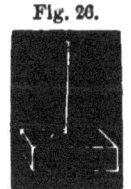

Fig. 26.

We conclude that *gravity causes pressure in a body of fluid in all directions.*

§ 45. Pressure increases with the depth. — In the experiment with the vessel with apertures in its side (Fig. 24), we find that the deeper the orifice, the greater the velocity of the stream. And in the experiment with the wide-mouthed bottle covered with rubber, we find that, at the same depth, the rubber is pressed inward equally in all directions, but, as it is carried to greater depths, the pressure is increased.

Experiment. Take a glass tube bent in the form represented by a, Figure 27; place mercury in the lower part of the tube, so as to fill the short arm, and gradually lower the tube into a deep vessel of water. The downward pressure of the water will force the mercury up the long arm to a hight proportional to the depth of the tube in the water.

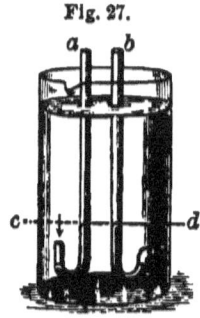

Fig. 27.

§ 46. Pressure at any point in a fluid equal in all directions.—Experiment 1. Introduce another tube, containing mercury, of the form represented by b, Figure 27; lower both tubes so that the orifices in the water shall be at the same level, and it will be found that the downward pressure in a and the lateral pressure in b will force the mercury to the same level, cd.

Experiment 2. Cover one end of a lamp-chimney (Fig. 28) with a circular piece of leather, and suspend from the hand by means of a string attached to the center of the leather and passing through the chimney. Hold the leather firmly against the bottom of the chimney, and lower the covered end a little way into a vessel of water. You may now drop the string, and the upward pressure of the water will keep the leather in place. Pour water slowly into the chimney, and, when the water in the chimney nearly reaches the level of the water outside, the leather will fall. The upward pressure of the water in the vessel against the leather is just balanced by the downward pressure of the water in the chimney and the weight of the leather. Why does not a pailful of water in a well seem heavy?

Fig. 28.

The results of experiments thus far show that, *at every point in a body of fluid, gravity causes pressure to be exerted equally in all directions, and that in liquids the pressure increases as the depth increases.*

Have we any means of ascertaining the pressure at any point in the atmosphere?

Experiment 3. Prepare a U-shaped glass tube closed at one end (Fig. 29), 80^{cm} in hight from the center of the bend, and with a bore of 1^{qcm} section. Fill the closed arm with mercury and invert. The mer-

PRESSURE AT ANY POINT IN A FLUID.   49

cury in the closed arm will sink about 2^{cm} to A, and will rise 2^{cm} in the open arm to C; but the surface A is 76^{cm} higher than the surface C. This can be accounted for only by the atmospheric pressure. The column of mercury BA, containing 76^{ccm}, is an exact counterpoise for a column of air of the same diameter extending from C to the upper limit of the atmospheric ocean, — an unknown hight.

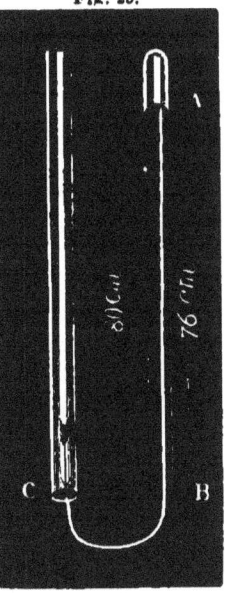

Fig. 29.

The weight of the 76^{ccm} of mercury in the column BA is 1033.3^g exactly, but, for convenience, may be said to be about 1^k. Hence the weight of a column of air of 1^{qcm} section, extending from the surface of the sea to the upper limit of the atmosphere, is about 1^k. But gravity causes equal pressure in all directions. Hence, *at the level of the sea, all bodies are pressed upon in all directions by the atmosphere, with a force of about 1^k per square centimeter, about 15 pounds (exactly 14.7 lbs.) per square inch, or about one ton per square foot.* Fluid pressure is generally expressed in atmospheres. *An atmosphere* (when the term is used to denote pressure) *is the pressure of 1^k per square centimeter.*

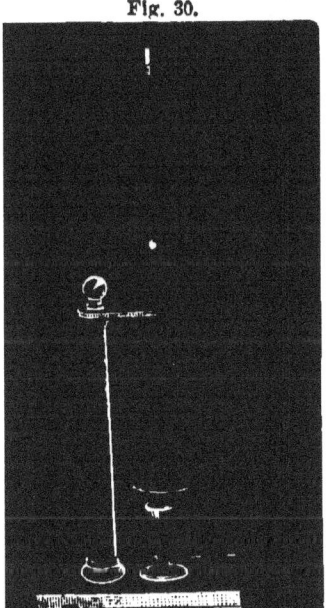

Fig. 30.

A man of average size sustains an external pressure of about fifteen tons. If the area of the bottom of an "empty" pail is one square foot, the downward pressure on its bottom is a little more than one ton; how can any person carry such a pail? and why is its bottom not forced out?

DYNAMICS.

§ 47. Barometer. — Figure 30 represents another form of apparatus, which is more commonly used for ascertaining atmospheric pressure. It consists of a straight tube about 85^{cm} long, closed at one end, and filled with mercury. When this tube is inverted, the open end having been covered with a finger and plunged into an open cup of mercury, and the finger withdrawn, the mercury in the tube will sink till it balances the atmospheric pressure. This experiment was devised by Torricelli, an Italian. The apparatus is called a *barometer*.[1] The empty space above the mercury in the tube is called a *Torricellian vacuum*. The history of this experiment is very interesting and important, inasmuch as it was the first demonstration of the pressure of the atmosphere. (See Whewell's History of Inductive Sciences, Vol. I., page 345.)

Fig. 31.

The hight of the barometric column is subject to fluctuations;

[1] Barometer, *weight measurer*.

this shows that the atmospheric pressure is subject to variations from various causes. The barometer is always a faithful monitor of all changes in atmospheric pressure. It is also serviceable as a weather indicator. Not that any particular point at which mercury may stand foretells any particular kind of weather, but *any sudden change in the barometer indicates a change in the weather.* A rapid fall of mercury generally forebodes a storm, while a rising column indicates clearing weather.

If the barometer is carried up a mountain, it is found that the mercury constantly falls as the ascent increases. This shows that the pressure is greater near the bottom of the aerial ocean than near its top. It is found that the pressure increases very rapidly near the bottom, as may be understood by studying Figure 31. The shading shows the variation in density of the air. The figures in the left margin show the hight of the atmosphere, in miles; those on the right the corresponding hight of the mercury, in inches. The average hight of the mercurial column, at the level of the sea, is about 76^{cm} (30 inches).

It will be seen that the density at a hight of 3 miles is but little more than $\frac{1}{2}$ the density at the sea-level; at 6 miles, $\frac{1}{4}$; at 9 miles, $\frac{1}{8}$; at 15 miles, $\frac{1}{30}$; at 35 miles it is calculated to be only $\frac{1}{30000}$, so that the greatest part of the atmosphere must be within that distance of the surface of the earth. On the other hand, if an opening could be made in the earth, 35 miles in depth below the sea-level, it is calculated that the density of the air at the bottom would be 1,000 times greater than at the sea level, so that water would float in it. Air has been compressed to this density.

To what hight the atmosphere extends is unknown. It is variously estimated at from 50 to 200 miles. If the aerial ocean were of uniform density, and of the same density that it is at the sea-level, its depth would be a little short of five miles. Certain peaks of the Himalayas would rise above it. It may be readily seen that hights of mountains may be measured approximately by the aid of a barometer.

QUESTIONS.

1. A person on the top of Mt. Blanc would take in what portion of the air, on expanding his lungs to a certain extent, that he would at the bottom?

2. How would this affect breathing, considering that a person requires a definite amount of air in a given time, in order to sustain life?

3. A person ascending 6 miles in a balloon leaves what proportional part of the whole mass of air below him?

4. When the barometric column stands at 492^{mm}, what is the atmospheric pressure in grams per square centimeter?

5. A barometer carried into a mine stands at 982^{mm}; what is the atmospheric pressure in the mine?

§ 48. Compressibility and expansibility of gases. — The increase of pressure attending the increase in depth, in both liquids and gases, is readily explained by the fact that the lower layers of fluids sustain the weight of all the layers above. Consequently, if the body of fluid is of uniform density, as is very nearly the case in liquids, the pressure will increase in nearly the same ratio as the depth increases. But the aerial ocean is far from being of uniform density, in consequence of the extreme compressibility of gaseous matter. The contrast between water and air, in this respect, may be seen in the fact that water, subjected to a pressure of one atmosphere, contracts .0000457 of its volume; under the same circumstances, air contracts one-half. For most practical purposes, we may regard the density of water at all depths as uniform, while it is far otherwise in large masses of gases.

The pressure at different depths in liquids may be illustrated by piling several bricks one on another, when the pressures that different bricks sustain vary directly with their depths below the upper surface of the pile. On the other hand, pressure of gases at different depths may be illustrated by piling fleeces of wool one on another. Since the volume of each successive fleece varies with the weight it bears, the pressures which different fleeces sustain are not proportional to their respective depths

COMPRESSIBILITY AND EXPANSIBILITY OF GASES. 53

below the upper surface of the pile. At twice the depth, there would be much more than twice the pressure, because the lower point would sustain more than twice the number of fleeces.

Closely allied to compressibility is the elasticity of gases, or their power to recover their former volume after compression. *The elasticity of all fluids is perfect.* By this is meant, that the force exerted in expansion is always equal to the force used in compression; and that, however much a fluid is compressed, it will always completely regain its former bulk when the pressure is removed. Liquids are perfectly elastic; but, inasmuch as they are perceptibly compressed only under tremendous pressure, they are regarded as practically incompressible, and so it is rarely necessary to consider their elasticity. It has already been stated (page 17) that matter in a gaseous state expands indefinitely, unless restrained by external force. The atmosphere is confined to the earth by the force of gravity.

Experiment. Partially fill an india-rubber balloon with air, and tightly close it. What is the external force that prevents the air in the balloon from expanding and completely inflating the balloon? Place it under the glass receiver of an air-pump (Fig. 32), and exhaust the air; the balloon becomes completely distended, and possibly bursts. Before it is placed under the receiver, the balloon sustains a pressure of 15 pounds on every square inch. What prevents a collapse under this pressure? Inasmuch as the balloon shows no signs of distention, or collapse, until placed under the receiver, it would seem that this great outward pressure is exactly balanced by the tension of the air within.

Fig. 32.

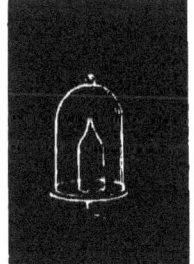

Fig. 33.

Glass-blowers prepare thin glass bottles (Fig. 33) for the purpose of illustrating the tension of air. Containing air of ordinary density, they are sealed and placed under the receiver of an air-pump; the surrounding air (in other words, the outside pressure) is removed, and the enclosed air then bursts the bottles, throwing fragments of glass in all directions.

At every point, then, in a body of air, forces are acting outwards. The air is somewhat like a spring coiled up, and ready to relax itself, when opportunity is given. Since this elastic force at the bottom of the column exactly balances the force of gravity acting on the whole column, *i.e.*, equals the weight of the whole column, it follows that, *at the sea-level, the elastic force of air is ordinarily 1^k per square centimeter*.

§ 49. **Air-pump.** — The air-pump, as its name implies, is used to withdraw air from a closed vessel. Figure 34 will serve to illustrate its operation. R is a glass receiver from which air is to be exhausted. B is a hollow cylinder of brass, called the pump-barrel. A plug P, called a piston, is fitted to the interior of the barrel, and can be moved up and down by the handle H; *s* and *t* are valves. A valve acts on the principle of a door intended to open or close a passage. If you walk against a door on one side, it opens and allows you to pass; but if you walk against it on the other side, it closes the passage, and stops your progress. Suppose the piston to be in the act of descending. The compression of the air in B closes the valve *t*, and opens the valve *s*, and the enclosed air escapes. After the piston reaches the bottom of the barrel, it begins its ascent; when the air above the piston, in attempting to rush down

Fig. 34.

to fill the vacuum that is formed between the bottom of the barrel and the piston, closes the valve s. But as soon as a vacuum is formed above t, and the downward pressure on the valve removed, the air in R expands, opens the valve t, and fills the space in B that would otherwise be a vacuum. But, as the air in R expands, it becomes rarefied; and, as there is less air, so there is less tension. The external pressure of the air on R, being no longer balanced by the tension of the air within, presses the receiver firmly upon the plate L. Each repetition of a double stroke of the piston removes a portion of the air remaining in R. The air is removed from R by its own expansion. However far the process of exhaustion may be carried, the receiver will always be filled with air, although it may be exceedingly rarefied. The operation of exhaustion is practically ended when the tension of the air in R becomes too feeble to lift the valve t.

D is another receiver, opening into the tube T, that connects the receiver with the barrel. Inside the receiver is placed a barometer. It is apparent that air is exhausted from D as well as from R; and, as the pressure is removed from the surface of the mercury in the cup, the barometric column falls; so that the barometer serves as a gauge to indicate the approximation to a vacuum. For instance, when the mercury has fallen 380^{mm} (15 inches), one-half of the air has been removed.

QUESTIONS.

1. Why is it difficult for a person to lift the receiver from the pump after the air is exhausted from it?
2. Why is it easily raised before the air is exhausted?
3. Suppose that the air in the pump-barrel, when the piston is raised, is one-eighth of all the air in the pump, including the air in the receivers; what portion of the air is removed by the first double stroke?
4. What portion of the original amount of air is removed at the second double stroke?
5. Which double stroke removes the most air?
6. If there were no force required to lift the valve t, why could not a perfect vacuum be obtained?

7. It is a very good pump that reduces the hight of the mercurial column to 3^{mm}. What portion of the air has been removed in that case?

An *absolute vacuum* has never been attained. The difficulty may be readily understood. According to the most recent calculations, the number of molecules contained in a cubic centimeter of air of ordinary density is something like 21,000,000,000,000,000,000 (twenty-one million trillion); consequently, when it is reduced to one-millionth its usual density, 21,000,000,000,000 (twenty-one trillion) molecules are still left. The exhaustion may be carried much farther than by purely mechanical means, by heating a piece of charcoal in the receiver while the pumping is going on. Heat expels the air in its pores. After the pumping has ceased, the charcoal is allowed to cool, when it condenses a large portion of the remaining air in its pores. (See § 37, page 38.)

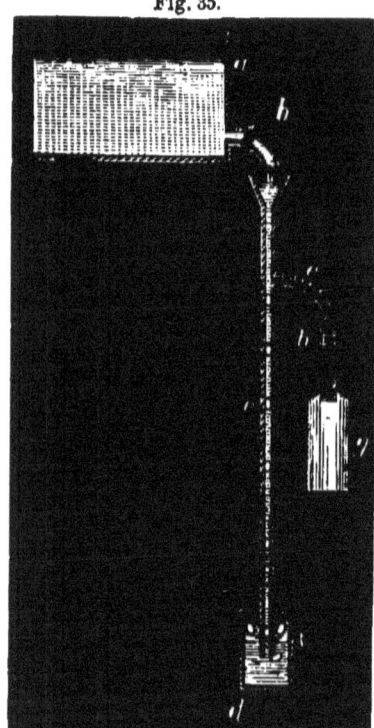

Fig. 35.

A very cheap and efficient substitute for an air-pump for many purposes may be arranged as in Figure 35, in which a is an elevated tank of water having a faucet b by which the rapidity of the flow of water may be regulated. The tube c should be as long as the hight of the room will admit, and its lower end should dip into a cup of water d. To the end of the branch-pipe e there may be connected, by means of rubber tubing h, a glass tube leading to a vessel g, from which air is to be exhausted. Water falling freely through a vertical tube exerts no lateral pressure; consequently there is no tendency to enter the branch e. As the water in falling increases in velocity, it tends to separate, leaving between the cylinders of water vacuous spaces. The lower end of the pipe c being immersed in water, air cannot enter there; but the air in the receiver g expands and rushes through the tube e, to fill these vacua, and thus exhaustion is effected. In Sprengel's air-pump mercury is substituted for water, and air is reduced by it to less than one-millionth its usual density.

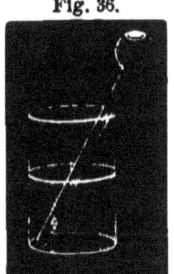

Fig. 36.

Experiment 1. Take a glass tube (Fig. 36), having a bulb blown at one end. Nearly fill it with water, so that when inverted there will be only a bubble of air in the bulb. Insert the open end in a glass of water, place under a receiver, and exhaust. Nearly all the water will leave the bulb and tube. Why? What will happen when air is admitted to the receiver?

Experiment 2. Through a cork of a tightly-stopped bottle pass one arm of a U-shaped glass tube C (Fig. 37). Introduce the other arm into the empty vessel B. Place the whole under a glass receiver, and exhaust the air. What phenomena will occur? What will happen when air is admitted to the receiver?

Fig. 37.

§ 50. *or Boyle's* Mariotte's Law. — The experiment illustrated by Figure 32 showed that the volume of a given body of gas depends upon the pressure to which it is subjected. To find more exactly the relation between these quantities, proceed as follows: —

Experiment 1. Take a bent glass tube (Fig. 38), the short arm being closed, and the long arm, which should be at least 85^{cm} long, being open at the top. Pour mercury into the tube till the surfaces in the two arms stand at zero. Now the surface in the long arm supports the weight of an atmosphere. Therefore the tension of the air enclosed in the short arm, which exactly balances it, must be about 15 pounds to the square inch. Next pour mercury into the long arm till the surface in the short arm reaches 5, or till the volume of air enclosed is reduced one-half, when it will be found that the hight of the column A C is just equal to the hight of the barometric column at the time the experiment is performed. It now appears that the tension of the air in A B balances the atmospheric pressure, *plus* a column of mercury A C, which is equal to another atmosphere; ∴ the tension of the air in A B = two atmospheres. But the air has been compressed into half the space it formerly occupied, and is, consequently, twice as dense. If the length and strength of the tube would admit of a column of mercury above the surface in the short arm equal to twice A C, the air would be compressed into one-third its original bulk; and, inasmuch as it would balance a pressure of three atmospheres, its tension would be increased threefold.

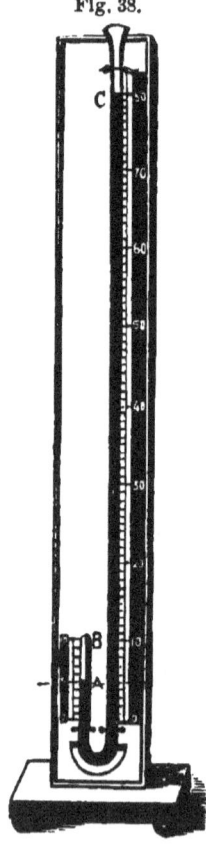

Fig. 38.

Fig. 39.

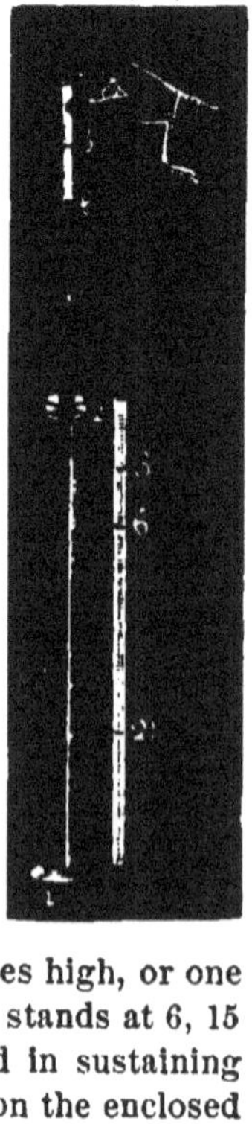

Experiment 2. Next take a glass tube (Fig. 39) open at both ends, and about 24 inches long. Tie three strings around the tube, — one 3 inches from the top, another 6 inches, and the third 21 inches. Nearly fill a glass jar, B, 25 inches high with mercury. Lower the tube into the mercury till it reaches the string at 3. Press a finger firmly over the upper end, and raise the tube till the string at 21 is on a level with the surface of the mercury in the jar. The mercury in the tube will stand at 6. At first the air enclosed in the tube between 3 and the finger withstands an upward pressure of the mercury sufficient to sustain a column of mercury 30 inches high, or one atmosphere. When the tube is raised and the mercury stands at 6, 15 inches high, one-half of that upward pressure is exerted in sustaining the 15 inches of mercury, and the other half is exerted on the enclosed

air. But the pressure on the air is reduced one-half, while the volume is doubled. The results of the two sets of experiments may be tabulated as follows: —

Pressure	$\frac{1}{3}, \frac{1}{2}, 1, 2, 3, 4$, &c.
Volume	$3, 2, 1, \frac{1}{2}, \frac{1}{3}, \frac{1}{4}$, &c.
Density	$\frac{1}{3}, \frac{1}{2}, 1, 2, 3, 4$, &c.
Elastic force	$\frac{1}{3}, \frac{1}{2}, 1, 2, 3, 4$, &c.

From these results we learn that, at twice the pressure there is half the volume, while the density and elastic force are doubled. At half the pressure the volume is doubled, and the density and elastic force are reduced one-half. Hence the law: *The volume of a body of gas varies inversely as the pressure, density, or elastic force.* This is sometimes called Mariotte's, and sometimes Boyle's, law, from the names of the two men who discovered it at about the same time. This law is true for all gases within certain limits, but under extreme pressure the reduction in volume is greater than indicated by it. The greatest deviation from it occurs with those gases that are most easily liquefied.

QUESTIONS.

1. Into the neck of a bottle partly filled with water (Fig. 40), insert a cork very tightly, through which passes a glass tube nearly to the bottom of the bottle. Blow forcibly into the bottle. On removing the mouth, water will flow through the tube in a stream. Why?

Fig. 40.

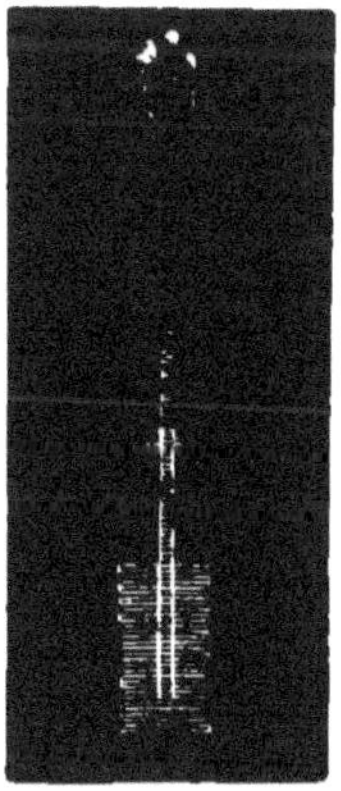

2. How can an ounce of air, in a closed fragile vessel, sustain the outside pressure of the atmosphere, amounting to several tons?

Fig. 41.

3. What drives the pellets from a pop-gun?

4. Figure 41 represents a dropping-bottle, much used in chemical laboratories. Why do bubbles of air force their way down into the liquid?

5. Stop the upper orifice, and the liquid will quickly cease to drop. Why?

6. The inconvenience arising, in many culinary and laboratory operations, from water "boiling away," may be remedied as represented in Figure 42. A bottle filled with water is so suspended that its mouth is just below the surface of the boiling liquid. As the water evaporates, and its surface falls below the mouth of the bottle, an air-bubble enters the bottle, expands, and pushes out enough water to cover once more the mouth of the bottle. Why does not the air push out all the water from the bottle?

Fig. 42.

7. Figure 43 represents a weight-lifter. Into a hollow cylinder s is fitted air-tight a piston t. The cylinder is connected with an air-pump by a rubber tube u. When air is exhausted the piston rises, lifting the heavy weight attached to it. Why?

8. If the area of the lower surface of the piston is 20^{qcm}, how heavy a weight ought to be lifted when the air is one-half exhausted?

9. Suppose you tightly stopper a bottle at the top of Mont Blanc, carry it to the sea-level, insert the mouth of the bottle in water, and withdraw the stopper; what would happen?

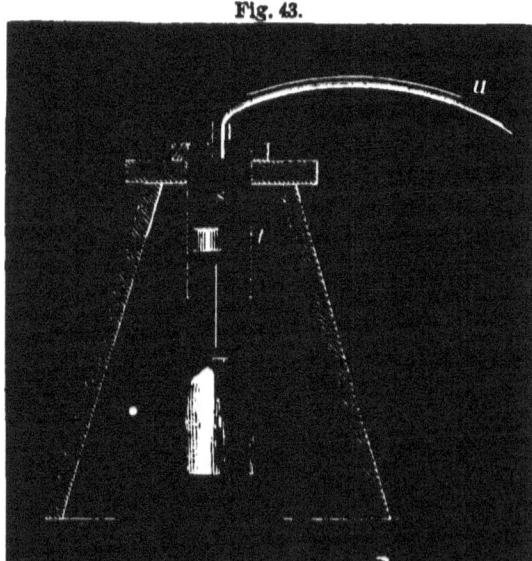

Fig. 43.

10. Show that the labor of working the kind of air-pump described (§ 49) increases as the exhaustion progresses.

§ 51. Condenser. — In the experiment with the bottle (Fig. 40), air was condensed in the mouth by muscular contraction, and forced into the bottle. An apparatus A (Fig. 44), intended to condense air in a closed vessel, is called a *condenser*. Its construction is like that of the barrel of the air-pump, except that the position of the valves is reversed. (Compare with Fig. 34.) What differences do you notice in respect to the valves? What happens to the valves when the piston in the condenser is forced down? If the condenser is connected with a closed vessel B, how much air would be forced into it at one down stroke? What prevents the air from escaping during an up stroke? If, after air is condensed in B, the cylinder C is connected with it by a screw, and the stop-cock t is suddenly turned, what would happen to the bullet s? What name would you give to such an apparatus?

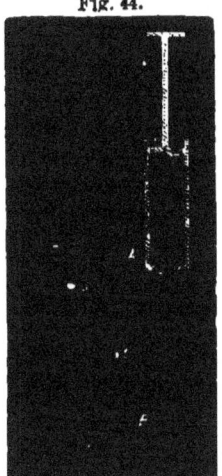

Fig. 44.

The Western Union Telegraph Company, in New York City, employs atmospheric pressure in forwarding messages to its central office from the various telegraph stations in that city. Tubes of uniform size, free from sudden curvatures, and laid under ground, connect the branch offices with headquarters. Rolls of paper, or letters to be despatched, are deposited in a cylindrical box c (Fig. 45), which fits the interior of the tube. The box being dropped into the end of the tube at a, and the air being exhausted from the tube at the end b, by means of an air-pump worked by steam, air rushes in at a and pushes the box through the tube with a force of several pounds for every square inch of the end of the box. The operation is still further facilitated by the aid of a condensing-pump worked by steam at the end a.

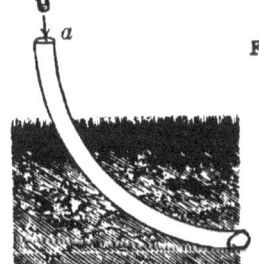
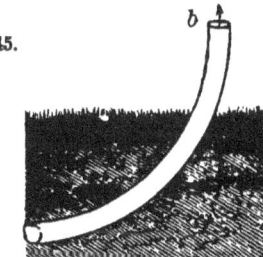

Fig. 45.

§ 52. Pressure transmitted undiminished in all directions. — Fill the globe G (Fig. 46), and about one-fifth the cylinder C, with water. The water in the tubes a, b, c, and d, will rise

62 DYNAMICS.

to the same level with the water in the cylinder C. Now force the piston P into the cylinder, and the downward pressure will cause jets of water to issue from each of the tubes. But the streams from the tubes *a*, *b*, and *c*, rise to exactly the same hight that the stream from the tube *d* does, although the liquid in the latter tube receives the direct action of the downward force. It thus appears that the pressure is not felt alone by that portion of the liquid that lies in the path of the force, but is felt equally in all parts and in all directions.

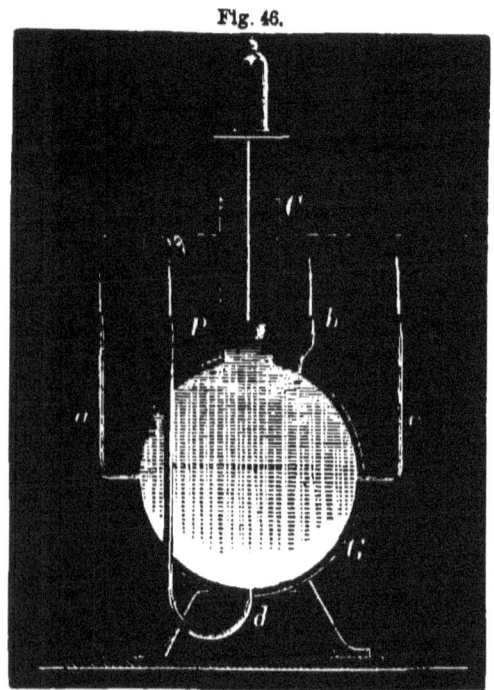

Fig. 46.

If the globe is filled with air, and subjected to pressure as above, currents of air will issue from the several tubes with equal force. This property of transmitting pressure equally in all directions, which is peculiar to fluids, is due to their *mobility* and *perfect elasticity*.

Figure 47 represents a number of elastic hoops enclosed in the vessel A B C D. A weight, placed on *a*, communicates to it a downward pressure. It is evident, that not only is the pressure communicated to the hoops below it in succession, and finally to the bottom of the box, but there is also a lateral pressure due to the elastic property of the hoops. The hoop *c*, receiving pressure from *b*, above, reacts, exerting an upward pressure; it also presses laterally upon the side A, and the hoop

n, and downward upon d; d and n in turn transmit pressure to their adjacent hoops, and thus every hoop receives and transmits, upward, downward, and laterally, a force equal to the downward pressure of the weight W. Hence that portion of the bottom immediately under the weight receives no greater pressure from W than an equal area of any other part of the bottom, or than an equal area of either of the sides, A and B, or the top C. This operation illustrates, somewhat imperfectly, the method by which elastic fluids transmit pressure undiminished in all directions.

Fig. 47.

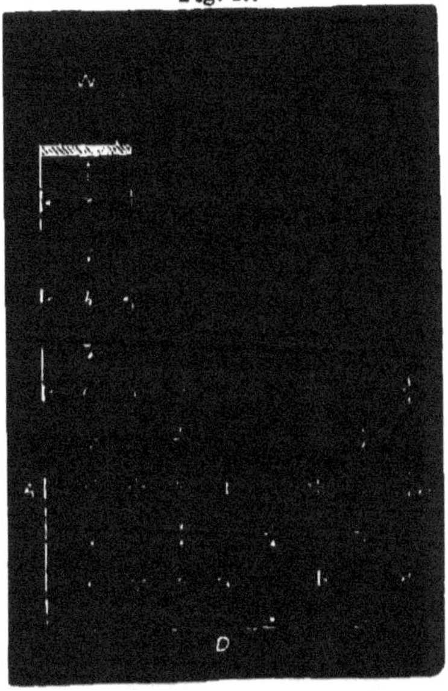

If we take a quantity of water in a vessel A. (Fig. 48), shut in by two pistons, a and b, whose areas are respectively 16^{qcm} and 4^{qcm}, and place a 10-gram weight on the platform d, and an equal weight on the platform c, it will be found that the latter is not sufficient to balance the former, but that it will require a 40-gram weight placed on c to preserve equilibrium. But the area of the piston b is 4^{qcm}, while the piston a contains four such areas; hence it follows that a pressure of 10^g is transmitted to each of the 4^{qcm} of a, and just supports the 40-gram weight. Had the area of the piston b been 1^{qcm}, then the 10-gram weight placed on it would require a 160-gram weight placed on a to balance it; that is, a pressure of 10^g would be exerted on every square centimeter of a.

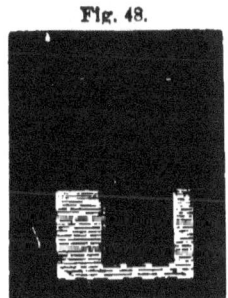

Fig. 48.

Obviously this form of apparatus cannot be made to work well on account of the friction of the pistons; but we may substitute for the pistons and weights columns of liquids. For instance, let the connecting tube and the lower part of the barrels be filled with mercury; the two free surfaces will be at the same level. Now if 10^g of any liquid, e.g. water, is poured into b, the level of the mercury will be changed; and, to bring it back to its original level, 40^g of some liquid must be poured into a.

We conclude, therefore, that *a pressure exerted on a given area of a fluid enclosed in a vessel is transmitted to every equal area of the interior of the vessel; and that the whole pressure that may be exerted upon the vessel may be increased in proportion as the area of the part subjected to external pressure is decreased.*

§ 53. **Hydrostatic bellows.** — This principle is well illustrated by means of the *hydrostatic bellows.* Two boards, b and c (Fig. 49), each having an area of (say) 400^{qcm}, are so connected, by leather attached to their edges, as to form an airtight vessel called the bellows. A glass tube a, having a bore of 1^{qcm} section, communicates with the interior of the bellows. Let water be poured into the tube a till the board b is raised a few centimeters. The water will stand at the same hight in the tube and bellows. Now, if 50^g of water be poured into the tube, it will require a weight of 20,000^g to be placed upon b to prevent its rising. Any weight less than that will be raised by the 50^g of water. If, instead of water being introduced into the bellows, a person stand on b, and blow into the tube, he can easily raise himself by the force of his breath.

Fig. 49.

§ 54. **Hydrostatic press.** — Closely allied to the bellows is the *hydrostatic press*, sometimes called *Bramah's press* from the name of the inventor. You see two pistons, t and s, Figure 50.

The area of the lower surface of t is (say) one hundred times that of the lower surface of s. As the piston s is raised and depressed, water is pumped up from the cistern A, forced into the cylinder x, and exerts an upward pressure against the piston t one hundred times greater than the downward pressure exerted upon s. Thus, if a pressure of one hundred pounds is applied at s, the cotton bales will be subjected to a pressure of five tons.

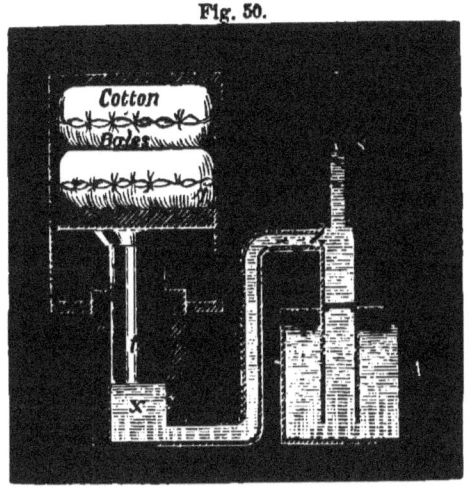

Fig. 50.

The pressure that may be exerted by these presses is enormous. The hand of a child can break a strong iron bar. But observe that, although the pressure exerted is very great, the upward movement of the piston t is very slow. In order that the piston t may rise 1^{cm}, the piston s must descend 100^{cm}. The disadvantage arising from slowness of operation is little thought of, however, when we consider the great advantage accruing from the fact that one man can produce as great a pressure with the press as a hundred men can exert without it.

The press is used for compressing cotton, hay, etc., into bales, and for extracting oil from seeds. The modern engineer finds it a most efficient machine, whenever great weights are to be moved through short distances, as in launching the Great Eastern steamship.

§ 55. **Pressure in fluids due to gravity.**—Having considered the transmission to the walls of the containing vessel, of external pressure applied to any portion of a surface of a liquid, we will examine the effects of pressure due to the weight of the liquids themselves. Suppose that we have three vessels filled with water, A, B, and C (Fig. 51), of equal depth, and having bottoms of equal areas. It is plain that the bottom of vessel A sustains a pressure equal to the weight of the column of water $abcd$, or just the weight of the water in the vessel. The pressure on hj, a portion of the bottom of vessel B, is equal to the weight of a column of water $ghji$. But this pressure is transmitted undiminished to the surface fh; consequently, the pressure on fh is equal to the weight of a column of water of the size of $efhg$, and the pressure on jl is equal to the weight of a column $ijlk$. Hence the pressure on the whole bottom fl is equal to the pressure of a column of water $eflk$, or the same as the pressure on the bottom of vessel A. But the weight of the water in B is less than the weight of the water in A. Hence, (1) *the pressure on the bottom of a vessel may be greater than the weight of the water in the vessel.*

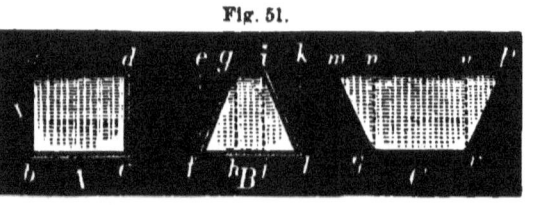

Fig. 51.

In vessel C, the side mq sustains the downward pressure of the body of water mqn; and the side pr sustains the pressure of the body orp; while the bottom qr sustains only the pressure of the column $nqro$, which is equal to the pressure on the bottoms of each of the vessels, A and B. Hence, (2) *the pressure on the bottom of a vessel may be less than the weight of the water in the vessel.*

We conclude, therefore, that (3) *the pressure on the bottom of a vessel depends on the depth and area of the bottom and the density of the liquid, and is independent of the shape of the vessel and the quantity of liquid.* — The important fact that the pressure on the bottom does not depend on the shape of the vessel is often called the *hydrostatic paradox*, because, though true, it seems at first absurd.

PRESSURE IN FLUIDS DUE TO GRAVITY. 67

Experiment. The last conclusion may be verified with apparatus like that represented in Figure 52. Vessels A, B, and C have different capacities, but equal depths, and the disk d is to serve successively for the bottom of each. Each vessel, when in use, is supported by the tripod e. The disk is supported and pressed up strongly against the bottom of the vessel by means of a string passing up through the vessel, and attached to a spring-balance. Let water be poured into vessel C, and regulate at pleasure the amount of downward pressure necessary to push the bottom off and allow the water to escape. Note the depth of water when the bottom is forced off, and mark the level of the surface with the pointer f. Also note the pressure indicated by the index of the balance. Substitute vessel A for vessel C. Pour the water caught in the basin g, in the last experiment, into vessel A, till it reaches the pointer f, when the bottom will be forced off at the same depth as before, as shown by the pointer, and by the same pressure, as shown by the spring-balance. But much less water is required than was used with the vessel C. The experiment, repeated with vessel B, will give the same results with the use of a still less quantity of water.

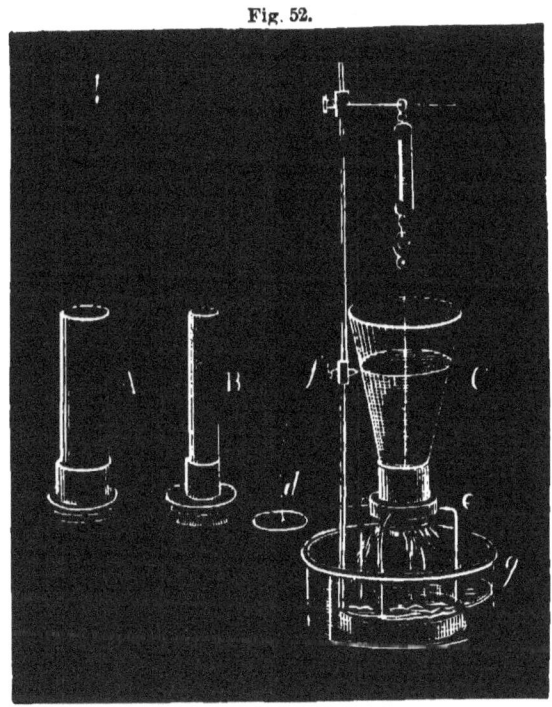

Fig. 52.

(4) *The pressure due to gravity on any portion of the bottom of a vessel is equal to the weight of a column of that liquid whose base is the area of that portion of the bottom pressed upon, and whose hight is the greatest depth of the water in the vessel.* Thus,

suppose the area of hj, of the bottom of vessel B (Fig. 51), is 100^{qcm}, and the depth gh is 9^{cm}; then the column $ghji$ contains 900^{ccm}. And, since the weight of one cubic centimeter of water is one gram, the weight of the column is 900^{g}, which is the pressure on the surface hj; and the pressure on each of the equal surfaces fh and jl being the same as on hj, the pressure on the entire bottom is 2700^{g}.

Evidently the lateral pressure at any point of the side of a vessel depends upon the depth of that point; and, as depth at different points of a side varies, hence, (5) *to find the pressure upon any portion of a side of a vessel, we find the weight of a column of water whose base is the area of that portion of the side, and whose hight is the average depth of that portion.* Thus, we compute the pressure on the side ab of vessel A (Fig. 51), by multiplying the area of the side 90^{qcm} (dimensions, 9×10^{cm}), by the depth to the middle point x, $4\frac{1}{2}^{cm}$, and this by the weight of 1^{ccm} of water, which gives 405^{g} for the pressure on the side ab.

QUESTIONS AND PROBLEMS.

1. It is apparent that a dam (Fig. 53), to be equally capable of resisting pressure in all its parts, should be made thicker towards the bottom. How rapidly should its thickness increase?

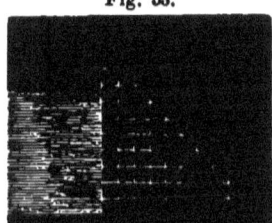
Fig. 53.

2. At high tide, suppose the flood-gate of a dock to be closed, leaving the surface of water on the inside and outside of the gate at the same level. From which does the gate sustain the greater pressure, the water in the dock, or the ocean of water outside? Why?

3. The interior dimensions of the rectangular vessel (Fig. 54) are 25^{cm} in length, 20^{cm} in width, and 15^{cm} in depth. The vessel is full of water. Compute the total pressure on each of the six sides.

4. Suppose that the plug n (Fig. 54), the area of whose end is 4^{qcm}, is pressed down upon the surface of the water with the force of 100^{g}; what additional pressure will each side of the vessel sustain?

THE SURFACE OF A LIQUID AT REST IS LEVEL. 69

5. How great will be the whole pressure that each side sustains, due to the weight of the liquid and the external pressure?

6. Suppose mercury, which is 13.6 times heavier than water, to be employed instead of water, what would be the answers to the three preceding questions?

Fig. 54.

7. Into the top of a keg filled with water, a brass tube 10^m long is inserted, a transverse section of whose bore is 1^{qcm}. The depth of the water in the cask is 30^{cm}, and the area of the bottom of the cask is 40^{qcm}. (a) Compute the pressure on the bottom of the keg. (b) Compute the pressure on the bottom of the cask if the tube is filled with water. (c) What is the weight of the water in the tube that causes this extra pressure?

8. What crushing-force on each side would an empty cubical box, the area of one of whose sides is 1^{qm}, sustain, if lowered 1^{km} into the sea?

9. What crushing-force on each side would this box sustain from the atmospheric pressure at the sea-level, if the air were completely exhausted therefrom?

10. Suppose the top of the vessel (Fig. 54) to be the weak part of the vessel, not able to sustain more than 50^g pressure on 10^{qcm}, what pressure applied to the plug will burst the vessel?

§ 56. **The surface of a liquid at rest is level.** — By jolting a vessel the surface of a liquid in it may be made to assume the form seen in Figure 55. Can it retain this form? Take two molecules of the liquid at the points a and b, on the same horizontal level. The downward pressure upon a is the weight of a column of molecules ac, and the downward pressure upon b is the weight of the column bd. Now, since the pressure at a given depth is equal in all directions, bd and ac represent

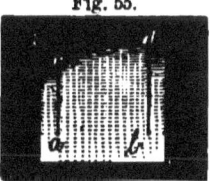

Fig. 55.

the lateral pressures at the points b and a respectively. But bd is greater than ac; hence, the molecules a and b, and those lying in a straight line between them, are acted upon by two unequal forces in opposite directions. There will, therefore, be a movement of molecules in the direction of the greater force toward *a*, till there is equilibrium of forces, which will only occur when the points *a* and *b* are equally distant from the surface; or, in other words, *there will be no rest till all points in the surface are on the same horizontal level.*

This fact is commonly expressed thus: "Water always seeks its lowest level." In accordance with this principle, water flows down an inclined plane, and will not remain heaped up. An illustration of the application of this principle, on a large scale, is found in the method of supplying cities with water. Figure 56 represents a modern aqueduct, through which water is conveyed from an elevated pond or river *a*, beneath a river *b*, over a hill *c*, through a valley *d*, to a reservoir *e*, in a city, from which water is distributed by service-pipes to the dwell-

Fig. 56.

ings. The pipe is tapped at different points, and fountains rise theoretically to the level of the water in the pond, but practically not so high, on account of the resistance of the air and the check which the ascending stream receives from the falling drops. Where should the pipes be made stronger, on a hill or in a valley? Where will water issue from faucets with greater force, in a chamber or in a basement? How high may water be drawn from the pipe in the house *f*?

§ 57. **Artesian wells, etc.**— In most places, the crust of the earth is composed of distinct layers of earth and rock of various kinds. These layers frequently assume concave shapes, so as to resemble cups placed one within another. Figure 57 represents a vertical section exposing a few of the surface-layers of the earth's crust: *a* is a stratum of loose sand or gravel; *b*, a clay-bed; *c*, a stratum of slate; *d*, a stratum of limestone; the whole resting on a bed of granite *e*. If you hollow out a lump of clay, and pour water into the cavity, you will find that the water will percolate through the clay very slowly. Water that falls in rain passes readily through the gravel *a*, till it reaches the clay-bed *b*, where it collects. Hence a *well*, sunk to the clay-bed, will

PARADOX. 71

fill with water as high as the water stands above the clay. Water also works its way from elevated places down between the strata of rocks. If a hole is bored through the slate c, water will rise above the surface of the ground in a fountain, in attempting to reach the level of its source on the hill; and if bored still lower, through the stratum d, a still higher fountain may result. Such borings are called *Artesian wells*. Water frequently forces its way through fissures in the rocky strata to the surface, as at t, and gives rise to *springs*.

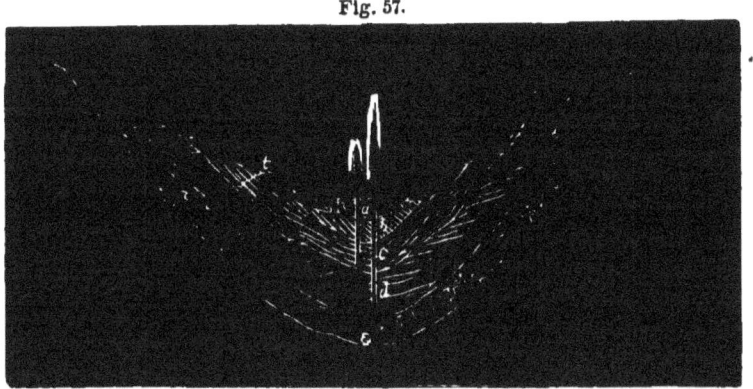

Fig. 57.

§ 58. "Any quantity of liquid, however small, may balance any quantity of liquid, however large." — If you lead a pipe through a dike by the seashore, and curve it upward, the water will rise no higher than the sea-level, even though the pipe should end in a quill.

Notwithstanding that every-day experience teaches that "liquids seek a level," it may seem strange that the large quantity of water in a teapot is balanced by the small quantity in the nozzle. Why, for instance, should the liquid in the small arm B balance the liquid in the large arm A, of the vessel in Figure 58? Imagine the liquid in A to be divided into columns a, b, c, and d, each equal to the column e. It is clear that the downward pressure of any one of the columns a, b, c, d, or e, will balance the downward pressure of any one of the other columns, and that there is no reason why e should rise above any one, or all, of the others.

§ **59. Siphon.** — A siphon is an instrument used for transferring a liquid from one vessel to another through the agency of atmospheric pressure. It consists of a tube of any material (rubber is often most convenient), bent into a shape somewhat like an inverted U. To set it in operation, fill the tube with a liquid, stop each end with a finger or cork, insert one end in the liquid to be transferred, bring the other end below the level of the surface of the liquid, remove the stoppers, and the liquid will immediately flow. Why? The force that raises the liquid in the short arm of the siphon A (Fig. 59) is the pressure of the atmosphere less the downward pressure of a column of water dc. The excess tends to carry the water over through the bend. On the other hand, the upward pressure at b, tending to carry the water back into the vessel, is an atmosphere less the weight of a column of water ba. But the former excess is greater than the latter by the weight of a column eb; consequently the liquid moves in the direction of the greater force towards b, with a velocity dependent on the distance eb. When the distance eb becomes zero, as in B, the flow ceases, and the liquid stands in the tube.

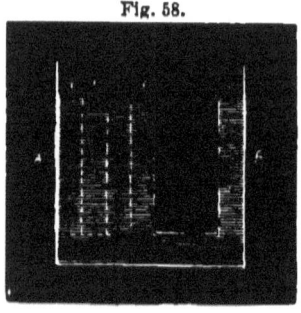

Fig. 58.

If one of the vessels is raised a little, as in C, the liquid will flow from the raised vessel, till the surfaces in the two vessels are on the same level. The remaining diagrams in this cut represent some of the great variety of uses to which the siphon may be put. D, E, and F are different forms of siphon fountains. In D, the siphon tube is filled by blowing in the tube f. Explain the remainder of the operation. A siphon of the form G is always ready for use. It is only necessary to dip one end into the liquid to be transferred. Why does the liquid not flow out of this tube in its present condition? H illustrates the method by which a heavy liquid may be removed from beneath a lighter liquid. By means of a siphon a liquid may be removed from a

SIPHONS.

vessel in a clear state, without disturbing sediment at the bottom. I is a *Tantalus cup*. A liquid will not flow from this cup till the top of the bend of the tube is covered. It will then continue to flow as long as the end of the tube is in the liquid. The siphon J may be filled with a liquid that is not safe or

Fig. 59.

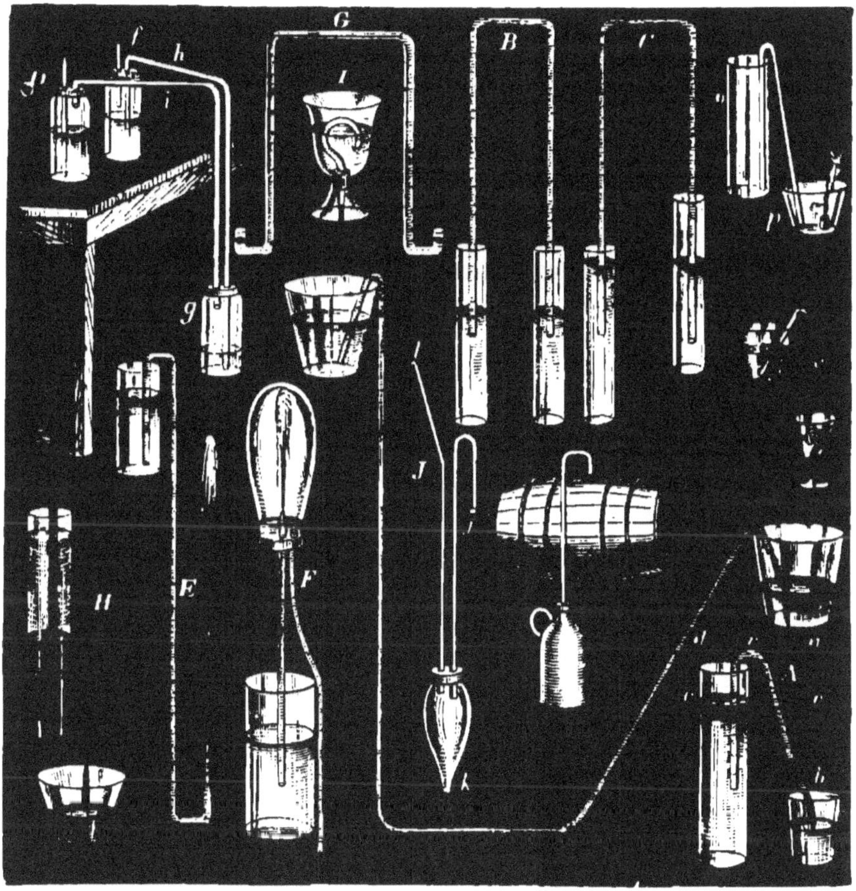

pleasant to handle, by placing the end j in the liquid, stopping the end k, and sucking the air out at the end l till the lower end is filled with the liquid.

Gases heavier than air may be siphoned like liquids. Vessel o

contains carbonic-acid gas. As the gas is siphoned into the vessel p, it extinguishes a candle-flame. Gases lighter than air are siphoned by inverting both the vessels and the siphon.

QUESTIONS.

1. What is the greatest hight to which the bend r (in A, Fig. 59) can be carried, and allow water to flow?
2. What would be the greatest hight if mercury were used?
3. Suppose the bend r is 15^m above the liquid; what theoretically ought to happen when the end b is unstopped?
4. What would happen if the long arm were cut off at e?
5. What would happen if it were cut off between e and a?
6. What would happen if the siphon were lifted out of the liquid?
7. What would be the effect of lengthening the long arm?
8. Must the two arms of a siphon be of unequal length?
9. How far can a liquid be carried by a siphon?
10. Will a siphon work in a vacuum?
11. Imagine that some such condition of things as is represented by the apparatus K (Fig. 59) exists in the earth, and that the siphon a has a smaller bore than the siphon c; can you account for intermittent springs which flow and cease to flow at nearly equal intervals of time?

§ 60. Apparatus for raising liquids. — The siphon can only be used for transferring liquids over hights to a lower level. Liquids cannot be transferred to a higher level by atmospheric pressure alone. In fact, atmospheric pressure is only a convenience, and *never does work*. If the piston a of a *syringe* (Fig. 60) is raised, the air is rarefied below it, and the atmospheric pressure will force water up into the syringe; but to raise the piston, against the atmospheric pressure tending to force it downward, requires as much muscular energy as would be required to raise the same quantity of water to the same hight as that to which it is raised in the syringe.

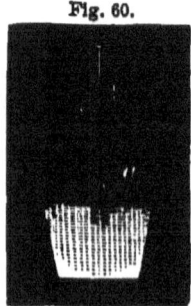
Fig. 60.

The common *lifting-pump* is constructed like the barrel of an air-pump. Figure 61 represents the piston in the act of

APPARATUS FOR RAISING LIQUIDS.

rising. As the air is rarefied below it, water rises by atmospheric pressure, and opens the lower valve. The weight of the water above the piston closes the upper valve, and the water is discharged from the spout. When the piston is pressed down, the lower valve closes, the upper valve opens, and the water between the bottom of the barrel and the piston passes through the upper valve above the piston. How high can the bottom of the barrel be above the surface of the liquid, if the liquid to be pumped is water? How high if it is mercury?

Fig. 61.

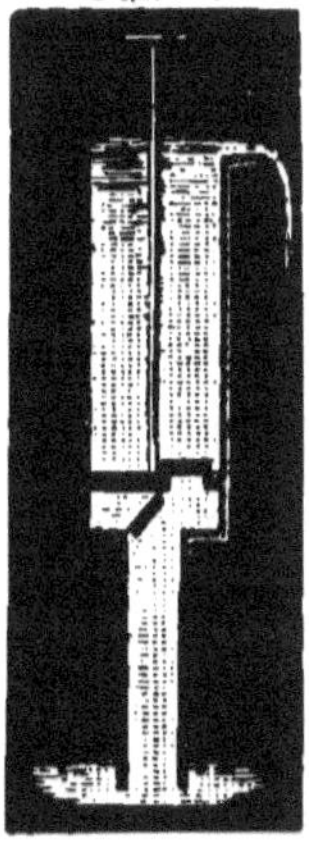

The liquid is sometimes said to be raised in a lifting-pump by the "force of suction." Is there such a force?

Fig. 62.
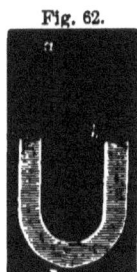

Experiment. Bend a glass tube into a U shape, with unequal arms, as in Figure 62. Fill the tube with a liquid to the level cb. Close the end b with a finger, and try to suck the liquid out of the tube. You find it impossible. Remove the finger from b, and you can suck the liquid out with ease. Why?

Fig. 63.
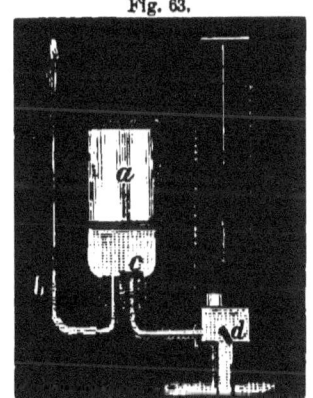

The piston of a *force-pump* (Fig. 63) has no valve, but a branch pipe leads from the lower part of the barrel to an air-condensing chamber a, at the bottom of which is a valve c, opening upward. As the piston is raised, water is forced up through the valve d, while water in a is prevented from returning by the valve c. When the piston is forced down, the valve d closes, the valve c opens, and water is forced into the chamber a, condensing the air above the water. The elasticity of the condensed air forces the water out of the hose b in a continuous stream.

V. BUOYANT FORCE OF FLUIDS.

Experiment 1. Gradually lower a large stone, by a string tied to it, into a bucket of water, and notice that its weight gradually becomes less till it is completely submerged. Slowly raise it out of the water, and note the change in weight as it emerges from the water. Suspend the stone from a spring balance, weigh it in air and then in water, and ascertain its loss of weight in the latter. Repeat the experiment with pieces of iron, wood, and other substances. Inflate a bladder, and force it beneath a surface of water. Fill a thin rubber balloon with coal-gas, and it will rise to the top of the room.

In all these experiments it seems as if something in the fluid, underneath the articles submerged, were pressing up against them. This lifting-force is called the *buoyant force* of fluids. Every body immersed appears to lose part of its weight; some bodies appear to lose all their weight. Do bodies really lose any portion of their weight when immersed in a liquid?

Experiment 2. Place a beaker of water on a scale-pan of a balance-beam, and weigh. Weigh a stone first in the air, then in water, and ascertain the apparent loss of weight. Then suspend the stone from a support (Fig. 64), and weigh the beaker of water with the stone immersed, and it will be found that the beaker and its contents gain in weight precisely as much as the stone loses. That is, *the water supports what is not supported by the string*, and *no weight is really lost*. Repeat the experiment with a block of wood.

Fig. 64.

Experiment 3. Make a saturated solution of salt in water. Weigh the same stone in air, fresh water, and salt water. The apparent loss of weight is greater in salt than in fresh water. Throw a piece of iron into mercury. It floats on the mercury like cork on water. Fill a vessel with carbonic-acid gas; blow a soap-bubble, and drop it into the vessel. It will not sink in the vessel, but rolls over the side and falls to the floor. It appears that some fluids have greater buoyant force than others. The water of the Dead Sea, in Palestine, is so salt (*i.e.*, so heavy) that a person could not possibly sink in it.

WHY A SOLID IS BUOYED UP BY A FLUID, ETC. 77

§ 61. Why a solid is buoyed up by a fluid, and with how great a force it is buoyed up. — Suppose *dcba* (Fig. 65) to be a cubical block of marble immersed in a liquid. It is obvious that the downward pressure upon the surface *da* is equal to the weight of the column of liquid *edao*. The upward pressure on the surface *cb* is equal to the weight of a column of liquid *ecbo*. The difference between the upward pressure against *cb* and the downward pressure on *da*, is the weight of a column of liquid *ecbo* less the weight of a column of liquid *edao*, which is a column of liquid *dcba* (*ecbo* − *edao* = *dcba*). But a column of liquid *dcba* has precisely the volume of the solid submerged. Therefore, *a solid is buoyed up by a fluid in consequence of the unequal pressures upon its top and bottom at their different depths*, and *the amount of the buoyancy is the weight of a body of that fluid equal in volume to the solid immersed*. The last proposition is generally stated as follows: *A solid loses in weight as much as the weight of the fluid it displaces.*

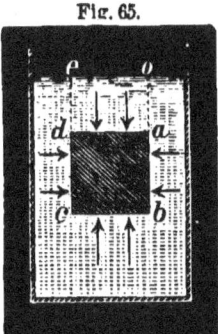

Fig. 65.

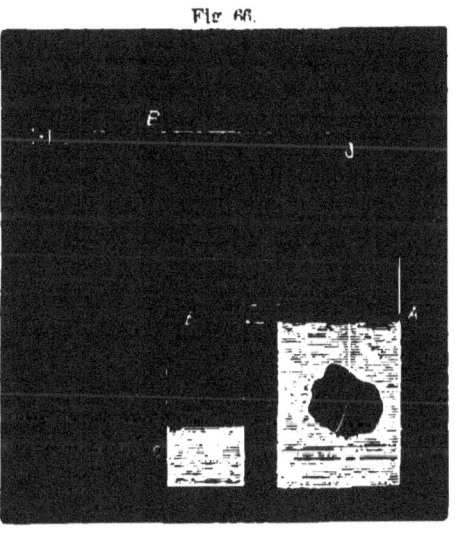

Fig. 66.

Experiment 4. The last statement may be verified with apparatus like that shown in Figure 66. Fill the vessel A till the liquid overflows at E. After the overflow ceases, place a vessel *c* under the nozzle. Suspend a stone from the balance-beam B, and weigh it in air, and then carefully lower it into the liquid, when some of the liquid will flow into the vessel *c*. The vessel *c* having been weighed when empty, weigh it again with its liquid contents, and it will be found that its increase in weight is just equal to the loss of weight of the stone.

Experiment 5. Next suspend a block of wood that will float in the liquid, and weigh it in air. Then float it upon the liquid, and weigh the liquid displaced as before, and it will be found that the weight of the liquid displaced is just equal to the weight of the block in air.

Hence, *a floating mass displaces its own weight of liquid;* in other words, *a floating mass will sink till it displaces an equal weight of the liquid, or till it reaches a depth where the buoyant force is equal to its own weight.*

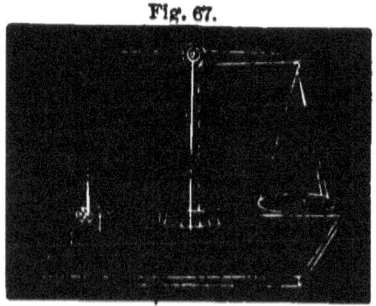

Fig. 67.

Experiment 6. Next, partially fill with water a glass (Fig. 67), graduated in cubic centimeters and fractions of the same. Note the level of the water. Drop one of the solids into the water, and note again the level of the water. The difference between the two levels is the number of cubic centimeters of water that the solid displaces. But one cubic centimeter of water weighs one gram. Hence, *the number of cubic centimeters displaced is equal to the weight in grams of the water displaced, and this is the loss in weight the solid sustains in water.*

There is an adage that "a pound of feathers weighs more than a pound of gold." Is there truth in the statement?

Experiment 7. Instead of feathers, we will employ a hollow globe *a* (Fig. 68); in place of the "pound of gold," we will use a counterpoise *b*, of any metal whose weight is just equal to the weight of the globe. Then, when the globe and counterpoise are suspended from the opposite arms of the balance-beam *c*, the beam will be horizontal. Now place the whole on the plate of an air-pump, cover with a receiver, and exhaust the air. As soon as the exhaustion commences, the globe begins to descend, and at the end of the operation the beam is completely tilted. Although the globe and counterpoise were both buoyed up by the air, it becomes evident, when this support is removed, that the globe was buoyed up more than the counterpoise, as we might expect from the fact that it displaces more air.

A pound of feathers displaces more air than a pound of gold, and is therefore buoyed up more by the air; consequently the pound of gold, which balances a pound of feathers in the air, does not balance them in a vacuum. We learn from this experiment that *bodies weigh less in air than in a vacuum*, and that *we never learn the true weight of a body, except when weighed in a vacuum*.

It has been stated (page 51) that the density of the atmosphere is greatest at the surface of the earth. A body free to move cannot displace more than its own weight of a fluid; therefore a balloon, which is a large bag filled with a gas about fourteen times lighter than air at the sea-level, will rise till the balloon, plus the weight of the car and cargo, equals the weight of the air displaced. The aeronaut, wishing to ascend still higher, throws out a portion of his cargo; wishing to descend, he allows some of the gas to escape at the top of the balloon by means of a valve, which he controls by means of a cord passing through the balloon to the car.

QUESTIONS.

1. Why is it difficult to stand in water reaching the neck?
2. Why can a person raise a stone under water, which he cannot lift when out of water?
3. A piece of cork weighs 50^g; what weight of water does it displace when floating?
4. What weight of mercury will a piece of iron weighing 500^g displace?

VI. DENSITY AND SPECIFIC GRAVITY.

§ **62. Density.** — We speak of a piece of cork as being heavier than a nail, at the same time that we speak of cork as light and iron as heavy. This seeming contradiction is accounted for by the different meanings which we attach to the terms *light* and *heavy*. In both cases, *light* and *heavy* are used as terms of comparison. In the former instance, we compare the weights of the two particular bodies, without reference to volume; in the latter, we call cork *light* and iron *heavy*, having no particular bodies in view, but because we know by experience that cork is not so *dense* as iron; *i.e.*, a given volume of cork contains less matter than an equal volume of iron. The term *weight* refers simply to the number of grams, kilograms, etc., that a particular body weighs without reference to the material or the volume. The *density* of a body can be stated only by expressing (or understanding) two quantities, *viz.*, mass and volume. For example, suppose that a block of wood measures $2 \times 10 \times 20^{cm}$ and has a mass of (*i.e.*, weighs) 300^g; its density is then $\frac{300}{2 \times 10 \times 20} = \frac{300}{400} = 0.75$ gram per cubic centimeter. When we speak of cork as lighter than iron, it is evident that we are comparing the densities of these two substances.

§ **63.** Specific gravity. — *The specific gravity of a substance is the ratio of the density of that substance to the density of another substance assumed as a standard;* in other words, it is the number which expresses how many times heavier a certain volume of a given substance is than an equal volume of another substance.

To facilitate comparison of densities, uniform standards are adopted. Distilled water at its maximum density, at 4° C., is the standard of specific gravity for all solids and liquids. Inasmuch as one cubic centimeter of water weighs one gram, when the weight of one cubic centimeter of any substance is given in grams, *i.e.*, when its density is given in its usual metric units, the same number also expresses its specific gravity. Thus one cubic centimeter of water weighs one gram; hence 1 is the specific gravity of water. The density of silver is 10.53^g per cubic centimeter; hence the specific gravity of silver is 10.53. The standard for gases is air at the average sea-level density, and at a temperature of 0° C. The weight of one cubic centimeter of air, under these conditions, is 0.0012932^g, or about $\frac{1}{773}$ of the weight of one cubic centimeter of water.

Let $G =$ the specific gravity of a substance; $D =$ its density

SPECIFIC GRAVITY. 81

in grams per cubic centimeter; V = the volume of a given body of it in cubic centimeters; W = the weight of the given body in grams; W' = the weight in grams of an equal volume of the standard. Then, as shown above, $D = \frac{W}{V}$, and, by definition, $G = \frac{W}{W'}$. G is *numerically* equal to D, and W' to V.

Since the loss of weight of a solid immersed in a liquid is just the weight of an equal volume of that liquid, it is evident that, *if we divide the weight of a solid in air by its loss in weight when immersed in water, the quotient will be its specific gravity.*

Experiment 1. Obtain small lumps of glass, iron, lead, marble, granite, etc., and weigh each in air. Partly fill with water a measuring-beaker graduated in cubic centimeters, and note the level of the water. Drop a lump into the water, and note the level again. The rise of water, as indicated by the graduated scale, gives the volume (V) of the specimen. With these data find the density (D), employing the formula $D = \frac{W}{V}$. Next weigh each of these lumps submerged in water, and find its loss in weight; and, from the data obtained, ascertain G from the formula $G = \frac{W}{W'}$. Prepare blanks, and tabulate your results thus: —

Name of Substance.	W g	V ccm	D or G	e	W g	W in water. g	W' g	G or D	e	Av.	e
Flint glass.	435	134	3.24	.09—	435	305	130	3.34	.01+	3.29	.04—

82   DYNAMICS.

When the result obtained differs from that given in the table of specific gravities (see Appendix, page 402), the difference is recorded in the column of errors (*e*). When the former is greater than the latter, it is indicated by a plus sign affixed to the number; when less, by the minus sign. The results recorded in the column of errors are not necessarily *real* errors; they may indicate the degree of impurity, or some peculiar physical condition, of the specimen tested.

Experiment 2. Obtain good specimens of cork, oak, elm, and poplar woods, all of which float on water. Tie to a specimen a piece of lead heavy enough to sink it; immerse the two, thus attached, in a measuring-glass, and find the number of cubic centimeters of water displaced by them. In the same way find the amount displaced by the lead alone. Subtract the amount displaced by the lead from the amount displaced by the two, and the remainder will be the amount displaced by the specimen. Then, regarding the number of centimeters of water displaced as so many grams, apply the formula $G = \dfrac{W}{W'}$.

Example. Find the specific gravity of a piece of elm wood. Attach to it a piece of lead weighing (say) 40g.

The combined solids displace	28.5g of water.
The lead displaces	3.5g "
The elm displaces	25.0g "
The elm weighs in air	20.0g "
The specific gravity of elm wood is	$20.0 \div 25 = .8$.

Experiment 3. Find the specific gravity of alcohol, a saturated solution of common salt, sea-water, naphtha, olive-oil, pure milk, and mercury in the following manner: ascertain the loss in weight of a sinker in each one of these liquids, also in water, and then apply the formula $G = \dfrac{W}{W'}$. Here W and W' represent the loss of weight of the sinker in the liquid and water respectively.

Example. Compute the specific gravity of alcohol from the following data: —

A piece of marble weighs in air	56.80g
The same weighs in water	36.80g
Loss in water	20.00g
	56.80g
The marble weighs in alcohol	40.96g
Loss in alcohol	15.84g

HYDROMETERS.   83

Since 20^s and 15.84^s are the weights respectively of equal volumes of water and alcohol, and since $G = \frac{W}{W'}$, then $\frac{15.84}{20} = .792$, the specific gravity of alcohol.

§ **64. Hydrometers.**—**Experiment.** Take a uniform rod of light wood about a foot long, and mark off on it a scale of equal parts. A convenient size is ½ inch square, and a suitable scale is inches and half inches. Coat the rod with paraffine to prevent its absorbing water and swelling. Bore into the end marked zero a hole about 2 inches deep, and drive in bullets till the rod will sink in water (Fig. 69) just to some inch-mark, and stop the end with paraffine. If it sinks too deep, cut off the upper end of the rod.

Fig. 69.

Suppose the rod sinks 8 inches in water; then, if it is ½ inch square, it displaces 2 cu. in. of water. The weight of the water displaced must just equal the weight of the rod (see page 78). Now immerse it in alcohol; it sinks deeper, say to the 10-inch mark; that is, $\frac{10}{4}$ cu. in. of alcohol weigh the same as $\frac{8}{4}$ cu. in. of water; therefore, $G = \frac{V}{V'} = \frac{8}{10} = .800$. If in brine it sinks only 6⅔ in., $G = \frac{8}{6\frac{2}{3}} = 1.20$.

Apparatus like that described is called a *hydrometer*. Instead of a rod of wood, a glass tube is generally used, terminating in a bulb containing shot or mercury. The tube contains a scale with numbers corresponding, which express the specific gravity, so that no computation is necessary. Make solutions of various substances, and test their specific gravity with your hydrometer, and test the accuracy of the results so obtained by other processes.

The most direct way of finding the specific gravity of liquids and gases is by employing vessels that hold definite weights of the two standards, water or air, and then weighing these vessels when filled with other liquids or gases; and, after deducting the weight of the vessel, applying the formula, $G = \frac{W}{W'}$.

The specific gravity of a solid that is dissolved by water may be found by weighing it in a liquid that will not dissolve it (*e.g.*, rock-salt in naphtha); and, having found its specific gravity as compared with the liquid used, multiply this result by the specific gravity of the liquid.

From the formula $D = \frac{W}{V}$, we have $V = \frac{W}{D}$; hence, *the volume of an irregular-shaped body may be found in cubic centimeters by dividing its weight in grams by its density.*

Again, from the formula $D = \frac{W}{V}$, we have $W = V \times D$. Hence, *when the volume and density of a body are known, its weight in grams may be found by multiplying its volume in cubic centimeters by its density.*

QUESTIONS AND PROBLEMS.[1]

1. How high can sulphuric acid be raised by a lifting-pump?
2. What is the weight of 50^g of water in water?
3. Find the specific gravity of wax from the following data: weight of a given mass of wax in air is 80^g; wax and sinker displace 102.88^{ccm} of water; sinker alone displaces 14^{ccm}.
4. Why does a light liquid (*e.g.*, oil), introduced under a heavier liquid (*e.g.*, water), rise?
5. Glass is about three times heavier than water; how, then, can a glass tumbler float in water?
6. How can iron vessels float in water?
7. A block of ice containing 500^{ccm} is floating on water; how many cubic centimeters are out of water?
8. Will ice float or sink in alcohol?
9. How much more matter is there in 500^{ccm} of sea-water than in the same volume of fresh water?
10. In 50^k of gold how many cubic centimeters?
11. What is the density of gold?
12. What is the density of cork?
13. What is the density of air at ordinary pressure, and at a temperature of 0° C?
14. An irregular piece of marble loses 53^g when weighed in water. How many cubic centimeters does it contain?
15. When will a body sink, and when float?
16. How many cubic centimeters of air at the sea-level does it take to weigh as much as 1^{ccm} of water?

[1] Consult the Tables of Specific Gravities, in the Appendix, Section C.

QUESTIONS AND PROBLEMS. 85

17. How much will 1^k of copper weigh in water?

18. What does a piece of lead $20 \times 10 \times 5^{cm}$ weigh?

19. What will it weigh in water?

20. What will it weigh in mercury?

21. What becomes of the weight that is lost?

22. If 15^g of salt be dissolved in 1^l of water, without increasing the volume of the liquid, what will be the specific gravity of the solution?

23. A mass of lead weighs 1^k in air. What will it weigh in a vacuum?

24. A mass whose weight in air is 30^g, weighs in water 26^g, and in another liquid 27^g. What is the specific gravity of the other liquid?

25. A silver spoon, weighing 150^g, is supported by a string in water. What part of the weight is sustained by the string, and what part is supported by the water?

26. A boat displaces 25^{cbm} of water. How much does it weigh?

27. If 50^k of stone were placed in the boat, how much water would it displace?

28. If the boat is capable of displacing 100^{cbm} of water, what weight must be placed in it to sink it?

29. An empty glass globe weighs 100^g; full of air it weighs 102.4^g; full of chlorine gas, it weighs 105.928^g. What is the specific gravity of chlorine gas?

30. What weight of alcohol can be put into a vessel whose capacity is 1^l.

31. You wish to measure out 50^g of sulphuric acid. To what number on a beaker graduated in cubic centimeters will that correspond?

32. State how you would measure out 80^g of nitric acid in a measuring-beaker.

33. A measuring-beaker contains 35^{ccm} of naphtha. What is the weight of the naphtha?

34. A lead pipe is carried 20^m below the surface of water in a reservoir. What bursting-force per square centimeter must it be capable of sustaining?

35. A cubical vessel, each of whose sides contains 2500^{qcm}, is filled with water. What pressure does its bottom sustain? One of its sides?

36. A solid floats at a certain depth in a liquid when the vessel which contains it is in the air; if the vessel is placed in a vacuum, will the solid sink, rise, or remain stationary?

VII. MOTION.

§ 65. Motion and rest relative terms.—To a person riding in a railway car, and confining his attention to objects in the car, everything appears to be at rest; but let him direct his attention to objects by the wayside, and at once he discovers that all in the car are in motion. Matter may be at rest with reference to certain objects, and in motion in regard to others. *Motion and rest are wholly relative terms*, and inapplicable to an object considered apart from all others. We cannot locate an object except with reference to another object, nor can we conceive of change or permanence of position of an object, except in relation to some other object. The aeronaut, moving at the rate of sixty miles an hour, knows not that he is moving at all, till he looks away from his balloon, and sees cities and towns passing in panorama beneath him.

§ 66. All matter is in motion.—*There is no such thing as absolute rest in the universe.* There is no use for the word *rest*, except to indicate, with reference to each other, the condition of objects that are moving in the same direction and with the same velocity. For example, a span of horses drawing a carriage, at the rate of ten miles an hour, are at rest with reference to each other and the carriage. The stars, that compose the heavenly constellations, maintain punctiliously their relative positions, while they sweep with prodigious velocities through space. The phrase "at rest" can only be used in an extremely limited sense, and in common language refers only to the condition of an object with reference to that on which it stands, as a car, deck of a ship, or surface of the earth. It is only by putting entirely out of mind the motions of the earth that we can speak of any terrestrial object as being at rest.

Not only is there motion of mass as a whole, or visible mechanical motion, but there is a motion of the molecules within the mass,—an invisible molecular motion called *heat*. We cannot see the movements of the molecules of steam, but we know that they exist by their great power, manifested in moving machinery.

§ 67. **Velocity. — Uniform and varied motion.** — All motion takes time; hence the term *velocity*, which refers to *the space traversed in a unit of time.* Motion may be uniform or varied: *uniform*, when an object traverses successively equal spaces in all equal intervals of time; *varied*, when unequal spaces are traversed in any equal intervals of time. Varied motion may be accelerated or retarded: *accelerated*, when the spaces traversed increase at each successive interval of time; *retarded*, when they diminish. The motion of a train of cars, in starting from a station is at first accelerated, afterwards tolerably uniform, and when the brakes are applied, it becomes retarded. Strictly speaking, all motions are varied; there is no illustration of absolutely uniform motion in Nature nor in art, though we may conceive of its possibility and have very closely approximated to it.

The velocity of a body having accelerated or retarded motion can be given only at some definite point by an estimate of the distance it *would* traverse in a unit of time, were it to continue in uniform motion at the speed it has at that point. For instance, a railway train passes us, and we estimate that its velocity is 30 miles an hour, although in a few minutes its speed may be reduced to 10 miles an hour, and a little later it may come to rest. When we assign a velocity of 30 miles an hour, we have no thought of whether it will run 30 miles during the next hour, or whether it will run an hour; we mean that, should it retain its present speed, it will be 30 miles away from us at the end of an hour.

VIII. FIRST LAW OF MOTION. — INERTIA.

Now, what is it that sets in motion that which was previously at rest? We may call it *force;* but what idea does this term convey? Let us question our own experience. We leave an apple lying upon a table; have we not entire confidence that it will continue to lie there, unless disturbed by some other body? If on returning we find it gone, are we not sure that it has been removed by the action of some body other than itself? An apple falls to the ground, and although the action is one of the most mysterious in all nature, yet do we not almost instinctively trace the cause to some action between the apple and the earth? The ball at rest is put in motion by a bat; but must not the bat first be put in motion? And when we find the cause of its motion, is it not an antecedent motion in some other object? We conclude, then (1), that *motion cannot originate in an object isolated from all others, but it always arises from* MUTUAL *action between at least two bodies.*

Again, the bat, having received motion, is capable of imparting motion to the ball; but, having set in motion one ball, is it equally capable of putting in motion another ball? Can a mass impart motion and retain all its motion? Is it not like a commercial transaction, a trade, to which there are two parties, one a buyer and the other a seller? that is, are not all transactions between the parties (*i.e.*, the mover and the moved) of the nature of a transfer, which should be entered on the debit side of one's account, and the credit side of the other's? We conclude (2) that *motion in one body is caused only by another body's parting with some of its power of producing motion.*

If a sled, on which a child is sitting, is suddenly put in motion, the child is left in the place from which the sled started. If the child and sled are both in motion, and the sled is suddenly stopped, the child lands some distance ahead. If the sled is started slowly, the child partakes of the motion of the sled, and is carried along with it; and if the sled gradually stops, the child's motion is gradually checked, and it retains its place on the sled. This shows (3) that *masses of matter receive motion gradually and surrender it gradually.*

Even very small bodies require time to start and to stop. The sand-blast, employed for engraving figures on glass, furnishes a fine illustration of this fact. A box of fine quartz-sand is placed in an elevated position. A long tube extends vertically down from the bottom of this box. The plate of glass to be engraved is covered with a thin layer of melted wax. When cool, the design is sketched with a sharp-pointed

FIRST LAW OF MOTION. 89

instrument, in the wax, leaving the glass exposed only where the lines are traced. The plate is then placed beneath the orifice of the tube, and exposed to a shower of sand. The velocity of the sand-grains is not at its maximum at the start, but is constantly accelerated till they reach the plate, where their velocity in turn is gradually given up. The wax, on account of its yielding nature, gradually brings them to rest; but the glass, notwithstanding its hardness, cannot stop them quite at its surface; and, therefore, it suffers a chipping action from the sand. Thus the soft wax affords a protection from the action of the falling sand of all parts except those intended to be cut. A still greater force is generally given to the sand by steam blown through the tube. For this reason the apparatus is called a *sand-blast*. Hard metals like steel are engraved in the same manner. Yet the hand may be held in the blast several seconds without injury. (What is the difference in the effects of catching a base-ball with hands held rigidly extended, and allowing the hands to yield somewhat to the motion of the ball?)

Roll a marble on a carpet, — it soon stops; roll it on a smooth marble floor, — it rolls much farther. On a perfectly smooth surface it might roll for hours. If we could provide such a surface, and dispense with the resistance of the air, how long would it roll? These conditions are impracticable? True. But have not the heavenly bodies rolled for millions of years through frictionless space, unchecked because unimpeded?

Motion unobstructed is perpetual. Motion undisturbed is in a straight line. Along which will a marble roll more nearly in a straight line, a smooth or a rough floor? What if the floor were perfectly smooth?

The relations between matter and force are admirably and concisely expressed in what are known as *Newton's Three Laws of Motion*.

§ 68. **First Law of Motion.**—*A body at rest remains at rest, and a body in motion moves with uniform velocity in a straight line, unless acted upon by some external force to change its condition.*

That part of the law which pertains to motion is briefly summarized in the familiar expression, "perpetual motion." "Is perpetual motion possible?" has been often asked. The answer is simple, — Yes, more than possible, *necessary*, *if no force interferes to prevent.* What has a person to do who would establish perpetual motion? Isolate a moving body from interference of all external forces, such as gravity, friction, and resistance of the air. *Can the condition be fulfilled?*

In consequence of its utter inability to put itself in motion or to stop itself, every body of matter tends to remain in the state that it is in with reference to motion or rest; this inability is called *inertia*. Evidently the term ought never to be employed to denote a hindrance to motion or rest. The First Law of Motion is often appropriately called the Law of Inertia.

IX. SECOND LAW OF MOTION, AND APPLICATIONS.

If a person wished to describe to you the motion of a ball struck by a bat, he would be obliged to tell you three things: (1) *where it started*, (2) *in what direction it moved*, and (3) *how far it went*. These three essential elements may be represented graphically by lines. Thus, suppose balls at A and D (Fig. 70) to be struck by bats, and that they move respectively to B and E in one second. Then the points A and D are their starting-points; the lines AB and DE represent the direction of their motions, and the lengths of the lines represent both the distances traversed and the relative intensities of the forces applied. In reading, the direction should be indicated by the order of the letters, as AB and DE.

Fig. 70.

Let a force whose intensity may be represented numerically by 8 (*e.g.*, 8^g), acting in the direction AB (Fig. 71), be applied continuously to a ball starting at A, and suppose this force capable of moving it to B in one second; now, at the end of the second let a force of the intensity 4, directed at right angles to the direction of the former force, act during a second, — it would move the ball to C. If, however, when the ball is at A, *both* of these forces should be applied at the same time, then at the end of a

COMPOSITION OF FORCES.

second the ball will be found at C. Its path will not be A B nor A D, but an intermediate one, A C. Still, each force produces in effect its own separate result, for neither alone would carry it to C, but both are required. Hence, the

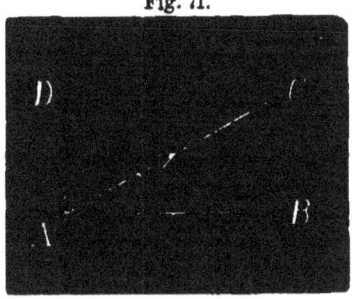

Fig. 71.

§ 69. **Second law of motion.** — *A given force has the same effect in producing motion, whether the body on which it acts is in motion or at rest; whether it is acted upon by that force alone, or by others at the same time.*

§ 70. **Composition of forces.** — It is evident that a single force, applied in the direction A C (Fig. 71), might produce the same result that is produced by the two forces A B and A D. Such a force is called a *resultant*. *A resultant is a single force, that may be substituted for two or more forces, and produce the same result that the combined forces produce.* The several forces that contribute to produce the resultant are called its *components*. When the components are given, and the resultant required, the problem is called *composition of forces*. *The resultant of two forces acting at an angle to each other is always a diagonal of a parallelogram, of which the components form two adjacent sides.* Thus, the lines A D and A B represent respectively the direction and relative intensity of each component, and A C represents the direction and intensity of the resultant.

The numerical value of the resultant may be found by comparing the length of the line A C with the length of either A B or A D, whose numerical values are known. Thus, A C is 2.23 times A D; hence, the numerical value of the resultant A C is $4 \times 2.23 = 8.92$.

When the components act at right angles to each other, as in Figure 71, the resultant divides the parallelogram into two equal right-angled triangles; and the intensity of the resultant may be

92   DYNAMICS.

found by calculating the hypothenuse, having two sides of either triangle given. Thus, $\sqrt{4^2 + 8^2} = 8.9+$ the numerical value of the resultant A C.

Copy upon paper and find the resultant of the components A B and A C, in each of the four diagrams in Figure 72. Also

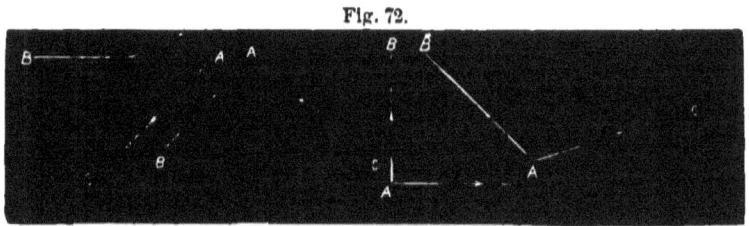

Fig. 72.

assign appropriate numerical values to each component, and find the corresponding numerical value of each resultant.

When more than two components are given, find the resultant of any two of them, then of this resultant and a third, and so on till every component has been used. Thus, in Figure 73, A C is the resultant of A B and A D, and A F is the resultant of A C and A E, *i.e.*, of the three forces A B, A D, and A E. (Invent several problems similar to this, in which three, four, or more forces are to be combined, and work out the results.)

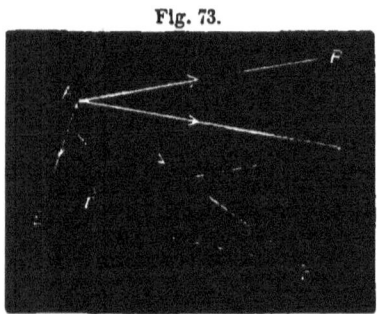

Fig. 73.

Generally speaking, a *motion may be the result of any number of forces.* When we see a body in motion, we cannot determine by its behavior how many forces have concurred to produce its motion.

§ 71. **Resolution of forces.** — Assume that a ball moves a certain distance in a certain direction, A C (Fig. 74), and that one of the forces that produces this motion is represented, in intensity and direction, by the line A B; what must be the

RESOLUTION OF FORCES. 93

intensity and direction of the other force? Since AC is the resultant of two forces acting at an angle to each other (§ 70), it is the diagonal of a parallelogram of which AB is one of the sides. From C, draw CD parallel and equal to BA, and complete the parallelogram by connecting the points B and C, and A and D.

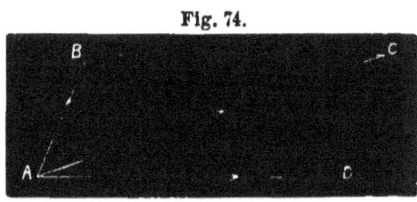

Fig. 74.

Then, according to the principle of composition of forces, AD represents the intensity and direction of the force which, combined with the force AB, would move the ball from A to C. The component AB being given, no other single force than AD will satisfy the question.

Had the question been, What forces can produce the motion AC? an infinite number of answers might be given. In a like manner, if the question were, What numbers added together will produce 50? the answer might be 20 + 30, 40 + 10, 20 + 20 + 10, and so on, *ad infinitum;* but if the question were, What number added to 30 will produce 50? only one answer could be given.

Experiment. Verify the preceding propositions in the following manner: From pegs A and B (Fig. 75), in the frame of a blackboard, suspend a known weight W, (say) 10 pounds, by means of two strings connected at C. In each of these strings insert dynamometers[1] x and y. Trace upon the blackboard short lines along the strings from the point C, to indicate the direction of the two component forces; also

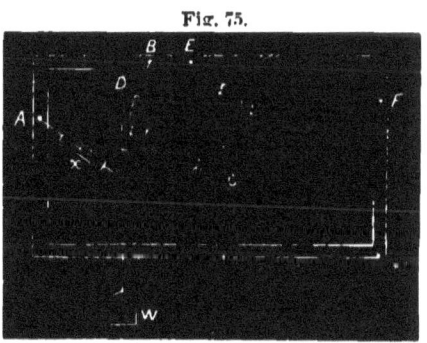

Fig. 75.

trace the line CD, in continuation of the line WC, to indicate the direction and intensity of the resultant. Remove the dynamometers,

[1] Dynamometer, *force-measurer.* The most common form is a spring balance.

extend the lines (as Ca and Cb), and on these construct a parallelogram, from the extremities of the line C D regarded as a diagonal. It will be found that 10 : number of pounds indicated by the dynamometer x :: C D : Ca; also that 10 : number of pounds indicated by the dynamometer y :: C D : Cb. Again, it is plain that a single force of 10 pounds must act in the direction C D to produce the same result that is produced by the two components. Hence, *when two sides of a parallelogram represent the intensity and direction of two component forces, the diagonal represents the resultant.* Vary the problem by suspending the strings from different points, as E and F, A and F, etc.

§ 72. Composition of parallel forces. — If the strings C A and C B (Fig. 75) are brought near to each other, as when suspended from B and E, so that the angle formed by them is diminished, the component forces, as indicated by the dynamometers, will decrease, till the two forces become parallel, when the sum of the components just equals the weight W. Hence, (1) *two or more forces applied to a body act to the greatest advantage when they are parallel, and in the same direction, in which case their resultant equals their sum.*

On the other hand, if the strings are separated from each other, so as to increase the angle formed by them, the forces necessary to support the weight increase until they become exactly opposite each other, when the two forces neutralize each other, and none is exerted in an upward direction to support the weight. If the two strings are attached to opposite sides of the weight (the weight being supported by a third string), and pulled with equal force, the weight does not move. But if one is pulled with a force of 15 pounds, and the other with a force of 10 pounds, the weight moves in the direction of the greater force ; and if a third dynamometer is attached to the weight, on the side of the weaker force, it is found that an additional force of 5 pounds must be applied to prevent motion. Hence, (2) *when two or more forces are applied to a body, they act to greater disadvantage the farther their directions are removed from one another; and the result of parallel forces acting in opposite directions is motion in the direction of the greater force, proportionate to their difference.*

COUPLE.

When parallel forces are not applied at the same point, the question arises, What will be the point of application of their resultant? To the opposite extremities of a bar AB apply two sets of weights, which shall be to each other as 3 : 1. The resultant is a single force, applied at some point between A and B. To find this point it is only necessary

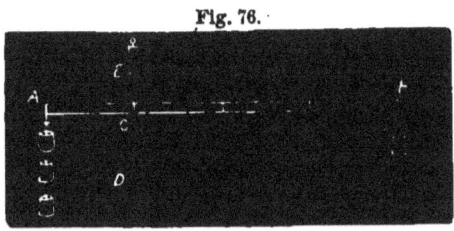

Fig. 76.

to find a point where a single force, applied in an opposite direction, will prevent motion resulting from the parallel forces; in other words, to find a point where a support may be applied so that the whole will be balanced. That point is found by trial to be at the point C, which divides the bar into two parts so that $AC : CB :: 1 : 3$. Hence, (3) *when two parallel forces act upon a body in the same direction, the distances of their points of application from the point of application of their resultant are inversely as their intensities.*

The dynamometer E indicates that a force equal to the sum of the two sets of weights is necessary to balance the two forces. A force whose effect is to balance the effects of several components is called an *equilibrant*. The resultant of the two components is a single force, equal to their sum, applied at C in the direction CD.

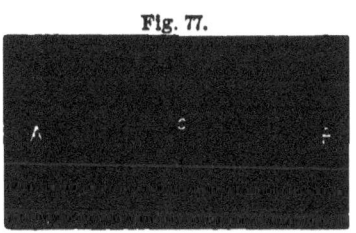

Fig. 77.

§ **73. Couple.**—If two equal, parallel, and opposite forces are applied to opposite extremities of a stick AB (Fig. 77), no single force can be applied so as to keep the stick from moving; there will be no motion of translation, but simply a *rotation* around its middle point C. Such a pair of forces, equal, parallel, and opposite, is called a *couple*.

DYNAMICS.

PROBLEMS, ETC.

1. A man and a boy, grasping opposite ends of a pole 3^m long, support thereon a weight of 50^k between them. Where should the weight be placed that the boy may support 20^k?

2. If the weight were placed 40^{cm} from the man, how much would each support?

3. Suppose that a boat is headed directly across a river half a mile wide, and is rowed with a velocity that would land it upon the opposite shore in half an hour, if there were no current; but the current carries the boat down the stream at the rate of one mile an hour. Where will the boat land?

4. How far will it travel?

5. How long will it be in crossing the river?

6. A ship is sailing due south-east at the rate of 10 miles per hour; what is its southerly velocity?

7. Find, both by construction and calculation, the intensity of two forces, acting at right angles to each other, that will support a weight of 15 pounds.

8. Verify the results with dynamometers.

X. OTHER APPLICATIONS OF THE SECOND LAW OF MOTION. — CENTER OF GRAVITY.

Let Figure 78 represent any body of matter; for instance, a stone. Every molecule of the body is acted upon by the force of gravity; the intensity of this force is measured by the *weight* of the molecule.

Fig. 78.

The forces of gravity of all the molecules form a set of parallel forces acting vertically downward, the resultant of which equals their sum, and has the same direction as its components. The resultant has a definite point of application in whatever position the body may be, and this point is called its *center of gravity*. The center of gravity (*c.g.*) *of a body is, therefore, the point of application of the resultant of all these forces;* and for many purposes *the whole weight of the body may be supposed to be concentrated at its center of gravity.* Hence mathematicians, by the *place* of a body, usually mean that point where the c. g. is situated.

Let G in the figure represent this point. For many practical purposes, then, we may consider that gravity acts only upon this point, and in the direction G F. If the stone falls freely, this point cannot, in obedience to the first law of motion, deviate from a vertical path, however much the body may rotate during its fall. Inasmuch, then, as the c.g. of a falling body always describes a definite path, a line G F that represents this path, or the path in which a body supported tends to move, is called the *line of direction*.

It is evident that if a force equal to its own weight and opposite in direction is applied to a body anywhere in the line of direction (or its continuation), this force will be the equilibrant of the forces of gravity; in other words, the body subjected to such a force is in equilibrium, and is said to be *supported*, and *the equilibrant* is called its *supporting force*. *To support any body, then, it is only necessary to provide a support for its center of gravity.* The supporting force must be applied somewhere in the line of direction, otherwise the body will fall.

Experiment. — Place a stick of wood, two meters long, horizontally across the tip end of a finger. When you succeed in getting the finger directly under its c.g., it will rest, but not till then. The difficulty of poising a book, or any other object, on the end of a finger, consists wholly in keeping the support under the center of gravity.

Figure 79 represents a toy called a "witch," consisting of a cylinder of pith terminating in a hemisphere of lead. The toy will not lie in the position shown in the figure on a horizontal surface *ab*, because the support is not applied immediately under its c.g. at G; but, when placed horizontally, it immediately assumes a vertical position. It appears to the observer to rise; but, regarded in a mechanical sense, it really falls, because its c.g., where all the weight is supposed to be concentrated, takes a lower position.

Fig. 79.

Whether a body will stand or fall depends upon whether or not its line of direction falls within its base. The base of a body is not necessarily limited to that part of the under surface of a body that touches its support. For example, place a string around the four legs of a table close to the floor: the rectangular figure bounded by the string is the base of the table. (What is the base of a man when standing on one foot? on two feet?)

§ 74. How to find the center of gravity of a body. — **Experiment.** Attach a string to a potato by means of a tack, as in Figure 80, and suspend from the hand. When the potato comes to rest there will be an equilibrium of forces, and the c.g. must be in the same line with the equilibrant of gravity; hence, if a knitting-needle is thrust vertically through the potato from a, so as to represent a continuation of the vertical line oa, the c.g. must lie somewhere in the path an made by the needle. Suspend the potato from some other point, as b, and a needle thrust vertically through the potato from b will also pass through the c.g. Since the c.g. lies in both the lines an and bs, it must be at c, their point of intersection. It will be found that, from whatever point the potato is supported, the point c will always be vertically under the point of support. On the same principle the c.g. of any body is found. But the c.g. of a body may not be coincident with any particle of the body; for example, the c.g. of a ring, a hollow sphere, etc.

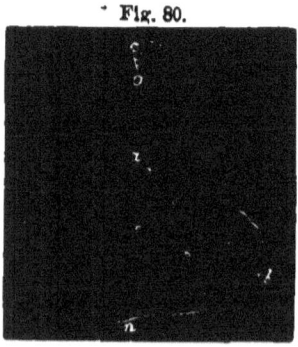
Fig. 80.

§ 75. Three states of equilibrium. — The weight of a body is a force tending downward; hence, *a body tends to assume a position such that its c.g. will be as low as possible.*

Experiment 1. Try to support a ring on the end of a stick, as at b (Fig. 81). If you can keep the support exactly under the c.g. of the ring, there will be an equilibrium of forces, and the ring will remain at rest. But if it is slightly disturbed, the equilibrium will be destroyed, and the ring will fall. Support it at a; in this position its c.g. is as low as possible, and any disturbance will raise its c.g.; but, in consequence of the tendency of the c.g. to get as low as possible, it will quickly fall back into its original position.

A body is said to be in *stable* equilibrium, if its position is such that a disturbance would raise its c.g., since in that event it would tend to return to its original position. On the other hand, a body is said to be in *unstable* equilibrium when a disturbance would lower its c.g., since it would not return to its original position.

Fig. 81.

A body is said to be in *neutral* or *indifferent* equilibrium when it rests equally well in any position in which it may be placed. A sphere of uniform density, resting on a horizontal plane, is in neutral equilibrium, because its c.g. is neither raised nor lowered by a change of base. Likewise, when the support is applied at the c.g., as when a wheel is supported by an axle, the body is in neutral equilibrium.

It is evident that, *if the c.g. is below the support*, as in the last experiment with the ring, *the equilibrium must be stable;* but, as in Figure 79, a body may be in stable equilibrium, though its c.g. is above the point of support. (When is this possible?)

It is difficult to balance a lead-pencil on the end of a finger; but by attaching two knives to it, as in Figure 82, the c.g. may be brought below the support, and it may then be rocked to and fro without falling.

Fig. 82.

§ 76. **Stability of bodies.** — The ease or difficulty with which bodies supported at their bases are overturned depends upon the hight to which their c.g. must be raised in overturning them. The letter c (Fig. 83) marks the position of the c.g. of each of the four bodies A, B, C, and D. To turn any one of these bodies over, its c.g. must pass through the arc ci, and be raised through the hight ai. By comparing A with B, and supposing them to be of equal weight, we learn that *of two bodies of equal hight and weight, the c.g. of that body which has the larger base must be raised higher, and is, therefore, overturned with greater difficulty.* A comparison of A and C, supposing them to be of equal weight, shows that *when two bodies have equal bases and weights, the higher body is more easily overturned.* D and C have equal bases and hights, but D is made heavy at the bottom, and this *lowers its c.g. and gives it greater stability.*

Fig. 83.

QUESTIONS.

1. Where is the c.g. of a box?
2. Why is a pyramid a very stable structure?
3. What is the object of ballast in a vessel?
4. State several ways of giving stability to an inkstand?
5. (*a*) In what position would you place a cone on a horizontal plane, that it may be in stable equilibrium? (*b*) That it may be in neutral equilibrium? (*c*) That it may be in unstable equilibrium?
6. In loading a wagon, where should the heavy luggage be placed? Why?
7. Why are bipeds slower in learning to walk than quadrupeds?
8. Why is mercury placed in the bulb of a hydrometer?
9. How will a man rising in a boat affect its stability?
10. Which is more liable to be overturned, a load of hay or a load of stone of equal weight?
11. (*a*) How would you place a book upon a table, that it may be in stable equilibrium? (*b*) That it may be in unstable equilibrium?

XI. OTHER APPLICATIONS OF THE SECOND LAW OF MOTION. — CURVILINEAR MOTION.

According to the first law of motion, every moving body proceeds in a straight line, unless compelled to depart from it by some external force. If the external force is continuous, *i.e.*, acts at every point, the direction is changed at every point, and the result is a *curvilinear motion;* and if the force is constant, and acts at right angles to the path, the curve becomes a *circle.*

Thus, suppose a ball at A (Fig. 84), suspended by a string from a point *d*, to be struck by a bat, in a manner that would cause it to move in the direction A*o*. At the same time it is restrained from taking that path by the tension of the string, which operates like a force drawing it toward *d*. It therefore takes, in obedience to the two forces, an intermediate course toward *c*. At *c* its motion is in the direction *cn*, in which path it would move, but for the string, in accordance with the first law of motion. Here, again, it is compelled to take an intermediate path toward *e*. Thus, at every point, the tendency of the moving body is to preserve the direction it has at that point, and consequently to move in a straight line. The only reason it does not so move, is that it is at every point forced from its natural path by the pull of the string. But if, when the ball reaches the point *i*, the string is cut, the ball, having no force operating to change its motion, continues in the direction in which it is moving at that point; *i.e.*, in the direction *ih*, which is a tangent to its former circular path.

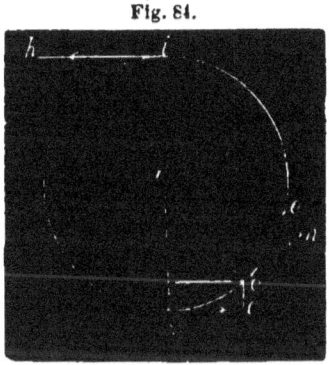

Fig. 84.

This tendency of a body moving in a curvilinear path to fly off in a straight line has been erroneously attributed to a supposed "centrifugal force," which is constantly urging it away from the center, its escape being prevented only by a force pulling it toward the center.

Centrifugal force has in reality no existence; the results that are commonly attributed to it are due entirely to the tendency of moving bodies to move in straight lines in consequence of their inertia. If a moving body is to describe a curvilinear path, a force called a *centripetal force* must be constantly applied to it at an angle to its otherwise straight path. [We shall make use of the expression *centrifugal force* for want of a better one, and because it has obtained universal currency.]

The greater the velocity of the moving body, the greater must be the force applied to produce a given departure from a straight line. This may be shown by suspending a weight to a dynamometer, and swinging them about the hand. If, when 30 revolutions are made in a minute, the force, as indicated by the dynamometer, is 4 pounds, then, on doubling its velocity, the force will be increased to 16 pounds. If the weight is doubled and the velocity remains the same, this force will be doubled. Hence, *to produce circular motion, the centripetal force must be increased as the square of the velocity increases, and as the mass increases.*

The farther a point is from the axis of motion,[1] the farther it has to move during a rotation, consequently the greater its velocity. Hence, bodies situated at the earth's equator have the greatest velocity, due to the earth's rotation, and consequently the greatest tendency to fly off from the surface, the effect of which is to neutralize, in some measure, the force of gravity. It is calculated that a body weighs about $\frac{1}{289}$ less at the equator than at either pole, in consequence of the greater centrifugal force at the former place. But 289 is the square of 17; hence, if the earth's velocity were increased seventeen-fold, objects at the equator would weigh nothing.

We have also learned (page 22) that a body weighs more at the poles in consequence of the oblateness of the earth. This is estimated to make a difference of about $\frac{1}{590}$. Hence, a body will weigh at the equator about $\frac{1}{590} + \frac{1}{289} = \frac{1}{194}$ less than at the poles.

---

[1] *Axis*, an imaginary straight line passing through a body about which it rotates.

QUESTIONS. 103

Experiment. Arrange some kind of rotating apparatus, *e.g.*, A, Figure 85. Suspend a skein of thread a by a string, and rotate; it assumes the shape of the oblate spheroid a'. This illustrates the probable method by which the earth, on the supposition that it was once in a fluid state, assumed its present spheroidal state. (Explain.) Suspend a glass fish aquarium e, about one-tenth full of colored water, and rotate. The liquid gradually leaves the bottom, rises, and forms an

Fig. 85.

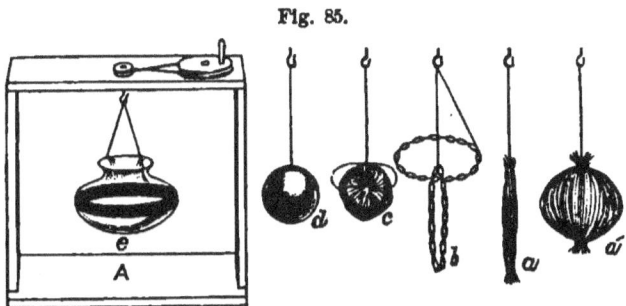

equatorial ring within the glass. Pass a string through the longest diameter of an onion c, and rotate; the onion gradually changes its position so as to rotate on its shortest axis. (Explain.) A chain b assumes on rotation a similar position.

QUESTIONS.

1. Why does not the sphere d (Fig. 85) change its position when rotated?

2. Why does the earth rotate on its shortest axis?

3. State the various facts illustrated in the act of slinging a stone.

4. (*a*) When will water and mud fly off from the surface of a revolving wheel? (*b*) Why do they fly off? (*c*) In what direction do they fly?

5. What is the force that keeps the earth and the other planets in their orbits?

6. How do you account for their curvilinear motion?

DYNAMICS.

XII. OTHER APPLICATIONS OF THE SECOND LAW OF MOTION. — ACCELERATED AND RETARDED MOTION.

§ 77. Accelerated motion or velocity. — So far the only case of motion under the action of a continuous force that we have studied is that of curvilinear motion, in which the force acts at an angle to the direction of the motion at every point, and so the direction of the force is constantly changing; but if the motion takes place in the same straight line as that in which the force acts, we shall have one of the cases of varied motion referred to on page 87.

Even if several men push against a heavy car we may be unable to recognize any motion for two or three seconds; but, if they continue to exert force upon the car, it will move with greater and greater velocity until the resisting force (which increases with the velocity) becomes equal to that applied by the men. This continually increasing velocity is termed *accelerated velocity*.

The most familiar illustration is that of falling bodies. We are sufficiently aware of the difference in the results that would follow a jump from a fifth-story window and a jump from a first-story window. Inasmuch as the velocity of falling bodies is so great that there is not time for accurate observation during their fall, we must resort to some method of checking their velocity, without otherwise changing the character of the fall.

Experiment. Take a smooth board (Fig. 86), about 4ᵐ long, and place it so that one end shall be about 4ᶜᵐ higher than the other. Suspend within easy view a string (about 1ᵐ long) and ball, as a pendulum. Set it in vibration, and, at the instant the ball reaches one extremity of its arc, let a marble begin to roll down the inclined plane. Let another person mark the point on the board that the ball reaches at the end of one swing of the pendulum. Repeat the operation several times, and mark the points that it reaches

Fig. 86.

ACCELERATED MOTION OR VELOCITY.

at the end of the second and third swings; also verify the preceding points by several trials; if there is a difference, take the mean distance between the points obtained at the end of a given swing for an approximate result. If the experiment is conducted with care, it will be found that during the first swing, which we call a unit of time (T), the marble moves through a certain space, which we represent by the expression $\frac{1}{2}k$; during the second unit of time it moves through $\frac{3}{2}k$, three times the space that it did in the first unit of time; and during the third unit of time it moves through $\frac{5}{2}k$.

Arrange the results of your observations in a tabulated form as follows: —

No. of units of time.	Total distance passed over.	Distance passed over in each unit; also *average velocity*.	Increase of velocity in each unit, i.e., *acceleration*.	Velocity at the end of each unit.
1	1 ($\frac{1}{2}k$)	1 ($\frac{1}{2}k$)	2 ($\frac{1}{2}k$)	2 ($\frac{1}{2}k$)
2	4 "	3 "	2 "	4 "
3	9 "	5 "	2 "	6 "
4	16 "	7 "	2 "	8 "
etc.	etc.	etc.	etc.	etc.

The marble, under the influence of gravity, starts from a state of rest, and moves through one space in a unit of time. Gravity, continuing to act, accomplishes no more nor less during any subsequent unit of time. But the marble moves through three spaces during the second unit; hence, two of the spaces must be due to the motion it had acquired during the first unit. In other words, if the action of gravity were suspended at the end of the first unit, the marble would still move on, and would pass through two spaces during the second unit. It therefore has at the end of the first unit a velocity (V) of two spaces (k). But it started from a state of rest; hence the constant action of gravity causes, during the first unit, an acceleration of velocity equal to two spaces (k); and it causes the same acceleration during every subsequent unit. The distance k is called the *acceleration* due to the constant force. *A body impelled by a single constant force, and encountering no resistances, always has a uniformly accelerated motion.*

§ **78. Formulas for uniformly accelerated motion.** — If we represent the distance traversed during a given unit of time by s, and the total distance the body has accomplished from the outset to the end of a given unit of time by S, we have the following formulas for solving problems of uniformly accelerated motion · —

(1) $V = (\tfrac{1}{2} k \times 2\,T) = k\,T$.
(2) $s = \tfrac{1}{2} k\,(2\,T - 1)$.
(3) $S = \tfrac{1}{2} k\,T^2$, or $S = \tfrac{1}{2} g\,T^2$.   (See § 79.)

§ **79. Velocity of a falling body independent of its mass and kind of matter.** — If we grasp a coin and a feather between the thumb and finger, and release both at the same instant, the coin will reach the floor first. It would seem as though a heavy body falls faster than a light body. Galileo was the first to show the falsity of this assumption. He let drop from an eminence iron balls of different weights: they all reached the ground at the same instant. Hence he concluded, that *the velocity of a falling body is independent of its mass.* (This celebrated experiment should be repeated by every student.)

He also dropped balls of wax with the iron balls. The iron balls reached the ground first. Are some kinds of matter affected more strongly by gravitation than others? If a coin and a feather are placed in a long glass tube (Fig. 87), and the air exhausted, and the tube turned end for end, it will be found that the coin and the feather will fall with equal velocities. Hence, *gravity attracts all matter alike;* but, inasmuch as a wax ball presents, according to the amount of matter in each, more surface for resistance of the air than an iron ball, it falls more slowly. We conclude, therefore, that *all bodies fall with equal velocities in a vacuum.*

Fig. 87.

When the body falls freely, and the unit of time is one second, we use the letter g instead of k to represent the acceleration. Experiments show that in the latitude of all the Northern States the value of g is 9.8^m, or about $32\tfrac{1}{8}$ ft.; that is, the velocity gained if the force of gravity acts one second is 9.8ᵐ per second, and the body would fall in the first second 4.9ᵐ or $16\frac{1}{12}$ ft.

§ 80. Retarded Motion. — If we reverse the order of the figures in Figure 86, the same diagram will represent the motion of a body rolling upward, or the motion of a body under the influence of a retarding force. The formulas given (§ 78) for finding velocities, etc., of bodies having uniformly accelerated motion, may be used for finding velocities, etc., of bodies having uniformly retarded motion; but the questions should be so framed as to be an exact converse of the questions to be solved. Thus, if we would find the velocity of a body at the end of the first second, or at the beginning of the second second, thrown upward by a force that would cause it to rise six seconds, we should calculate the velocity that a falling body has at the end of the fifth second, or at the beginning of the sixth second.

PROBLEMS.

(Solve these problems in both the metric and the English measures.)

1. Disregarding the resistance of the air, what distance will a body fall from a state of rest in five seconds?
2. What distance will it fall during the fifth second?
3. What is its velocity at the end of the fifth second?
4. A stone, dropped from a balloon, strikes the ground in seven seconds. How high is the balloon?
5. Under the influence of a constant force, a body moves 500ᵐ in a minute. How far will it go in an hour?
6. What will be its velocity at the end of the first half-hour?
7. How far will it move during the fifty-ninth minute?
8. A body falls four seconds; meantime it is acted on by a constant force which causes it to move in a horizontal direction 2ᵐ in the first second. Where will it strike the ground?
9. What is its horizontal velocity at the end of the fourth second?
10. What is its vertical velocity at the end of the fourth second?
11. With what vertical velocity must a body start that it may ascend three seconds?
12. How far does it rise during the first second?
13. At what point does a ball shot horizontally from a gun begin to fall?

108 DYNAMICS.

§ 81. Projectiles. — Experiment. Take a bottomless tin can A (Fig. 88), and connect a rubber tube C, 2^m long, with a glass tube passing through a stopper at B, and insert a short glass tube at D. Keep the can filled with water, bend the lower part of the rubber tube at D, so as to direct the stream at different angles of elevation, and observe the peculiarities of the curves formed by the streams, and the different vertical and horizontal distances reached by each.

In this experiment you have a miniature representation of the paths of all projectiles,[1] such as cannon-balls, stones thrown from the hand, etc. The horizontal distance that the projectile attains is called its *range* or *random*. Theoretically, the greatest range is obtained at an angle of 45°; but practically, on account of the resistance of the air, it is at a little less than 40°.

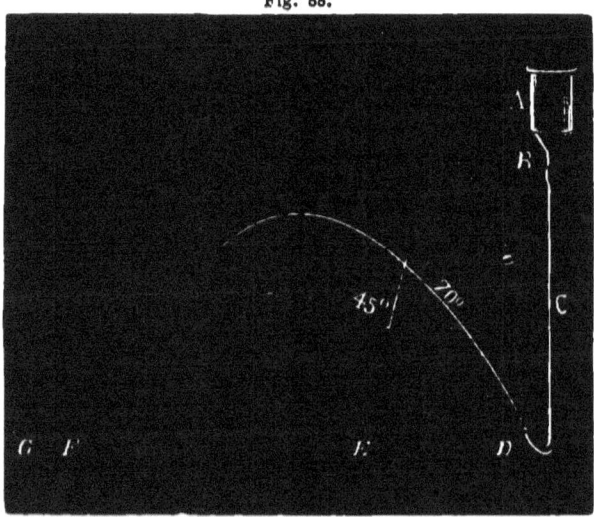

Fig. 88.

Every projectile is acted upon by two forces: (1) the force of gravity, and (2) the resistance of the air. It also has a certain velocity and direction at the instant of projection. If this velocity and direction are known, and the resistance of the air is disregarded, the path of a projectile can be determined. Thus, suppose that a projectile is thrown from A (Fig. 89) at an

[1] Projectile, *a body thrown.*

angle of 45°, that it is in the air six units of time, and that the vertical hights reached at the end of the first three units successively, are B, C, and D. Its horizontal motion, if unimpeded, is uniform, and the corresponding points reached in that direction at the same moments are (say) B', C', and D'. Combining these two motions, we obtain the points B'', C'', and D'', reached by the projectile successively, at the end of each of the first three units of time. The force of gravity constantly acting to change its direction, it must describe, during the first three units, the curved line AB''C''D''. Since the time of ascent and descent are equal, it must reach its greatest vertical hight at the end of the third unit, when it begins its descent. The path of descent D''E''F''G'' is found in a similar manner. The path thus described is known as a *parabolic curve;* but, inasmuch as this is practically modified by the resistance of the air, it in reality describes a peculiar path called a *ballistic curve.* The curve

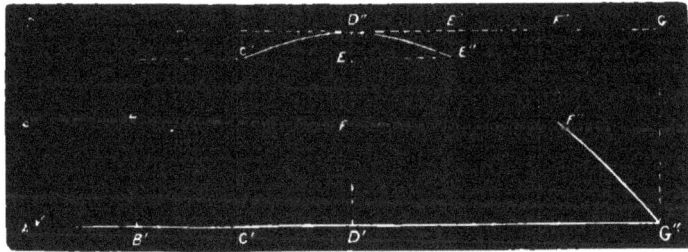

Fig. 89.

D''E''F''G'' represents also the path of a projectile thrown from D'', in the direction of the line D''G', with a horizontal velocity that it would cause it to reach G' at the end of the third unit of time.

An excellent verification of the second law of motion is found in the fact that a ball, projected horizontally, will reach the ground in precisely the same time that it would if dropped from a state of rest from the same hight. That is, any previous motion a body has in any direction does not affect the action of gravity upon the body.

110 DYNAMICS.

Experiment 2. Support two iron bars, a and b (Fig. 90), bent into the form of a curve, about 3^{cm} apart, and so situated that a ball n, rolling down them, will be discharged from them in a horizontal direction. So connect the wires of an electric battery c with these bars, that while the iron ball n rests upon them the circuit is closed, and the iron ball m is supported by the attraction of the electro-magnet e. Now allow n to roll down the curved path. When it leaves the bars, the circuit is broken, e instantly loses its power to hold m, and m drops. But both balls reach the floor at the same instant. If the horizontal velocity of n is varied, by allowing it to start at different points on the bars, so as to cause it to describe different paths, the two balls will, in every case acquire exactly equal vertical velocities.

Fig. 90.

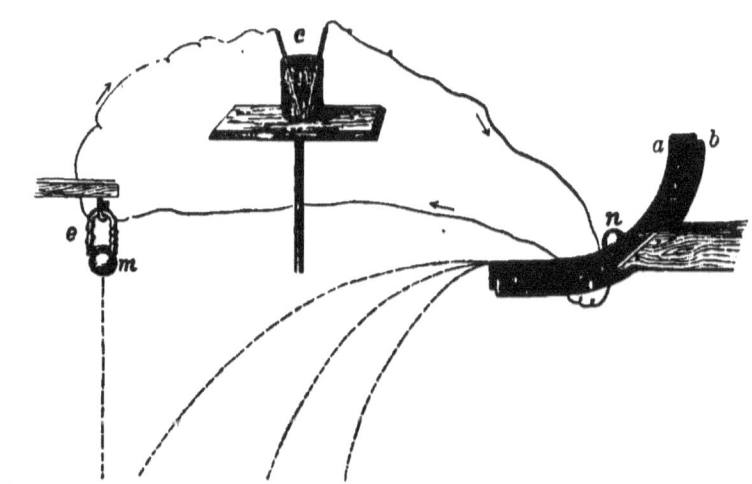

XIII. OTHER APPLICATIONS OF THE SECOND LAW OF MOTION. — THE PENDULUM.

Experiment 1. From a bracket suspend by strings leaden balls, as in Figure 91. Draw B and C one side, and to different hights, so that B may swing through a short arc, and let both drop at the same instant. C moves much faster than B, and completes a longer journey at each swing, but both complete their swing, or vibration, in the same time.

Hence, (1) *the time occupied by the vibration of a pendulum is independent of the length of the arc.* Of only very small arcs

CENTER OF OSCILLATION. 111

may this law be regarded as practically true. The pendulum requires a somewhat longer time for a long arc of vibration than for a short one, but the difference becomes perceptible only when the difference between the arcs is great, and then only after many vibrations.

Experiment 2. Set all the balls swinging; only B and C swing together; the shorter the pendulum, the faster it swings. Make B 1^m long, and F $\frac{1}{4}^m$ long. Watch in hand, count the vibrations made by B. It completes just 60 vibrations in a minute; in other words, it "beats seconds." A pendulum, therefore, to beat seconds must be 1^m long (more accurately, .993^m, or 39.09 in.). Count the vibrations of F; it makes 120 vibrations in the same time that B makes 60 vibrations. Make G one-ninth the length of B; the former makes three vibrations while the latter makes one, consequently the time of vibration of the former is one-third that of the latter.

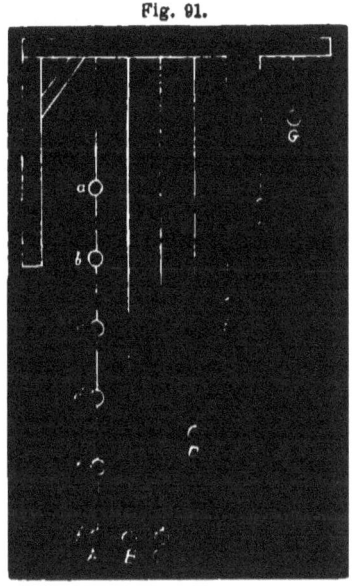

Fig. 91.

Hence, (2) *the time of one vibration of a pendulum varies as the square root of its length.*

QUESTIONS AND PROBLEMS.

1. What would be the effect if B were made twice as heavy as C? Why?

2. What is the length of a pendulum that beats half-seconds? Quarter-seconds? That makes one vibration in two seconds? That makes two vibrations per minute?

3. State the proportion that will give the number of vibrations per minute made by a pendulum 40^{cm} long.

§ 82. Center of oscillation. — **Experiment 1.** Connect six balls, at intervals of 15^{cm}, by passing a wire through them, after the manner of pendulum A. This forms a *compound pendulum* composed of six simple pendulums. Set A and B vibrating; A vibrates faster than B, although their lengths are the same. Why is this? If A were actuated only by the ball f, it would vibrate in unison with B. If the ball a were free, it would move much faster than f; but, as they are constrained to move together, the tendency of a is to quicken the motion of f, and the tendency of f is to check the motion of a. But e is quickened less than f, and d less than e; on the other hand, b is checked by f less than a, and c less than b. It is apparent that there must be some point between a and f, whose velocity is neither quickened nor checked by the combined action of the balls above and below it, and where, if a single ball were placed, it would make the same number of vibrations in a given time that the compound pendulum does. Shorten pendulum B, and find the required point. This point is called the *center of oscillation*.

Every compound pendulum is equivalent to a simple pendulum, whose length is equal to the distance between the center of oscillation and the point of suspension of the compound pendulum. Inasmuch as the distance between the point of suspension and the center of oscillation determines the rate of vibration, whenever the expression *length of pendulum* is used, it must be understood to mean this distance. Strictly speaking, *a simple pendulum is a heavy material point suspended by a weightless thread*. Of course such a pendulum cannot actually exist; but the leaden ball, suspended by a thread, is a near approximation to it.

Fig. 92.

Experiment 2. Suspend on the frame of Figure 91 a lath AB (Fig. 92), 1ᵐ long, and shorten the pendulum B till it swings in the same period as the lath; the ball of B marks the center of oscillation of the lath, which is found to be two-thirds the length of the lath below the point of suspension. Attach a pound-weight to the lower end of AB; its vibrations are now slower, and the simple pendulum B must be lengthened to vibrate in the same time as the lath and weight; hence the center of oscillation of the lath is lowered by the addition of the weight. Move the weight up the lath; the vibrations are quickened. (What is the office of a pendulum bob?)

Experiment 3. Remove the weight, bore a hole through the lath at

CENTER OF PERCUSSION. 113

its center of oscillation C, and, passing a knitting-needle through the hole, invert the lath and suspend it by the needle. The pendulum is now apparently shortened, and we naturally expect that its vibrations will be quicker than when suspended from A. But the part B C now vibrates in opposition to the part C A, rising as it sinks, and sinking as it rises. This tends to check the rapidity of the vibrations of C A, and it is found that the pendulum vibrates in the same time when suspended from C as when suspended from A.

The point of suspension and the center of oscillation are interchangeable; in other words, there are always two points in a compound pendulum about which it will oscillate in the same time.

This suggests a practical way of finding the center of oscillation, and the equivalent length of a compound pendulum. For we have only to find another point of suspension from which the pendulum makes the same number of vibrations, in a given time, as from its usual point of suspension: that point is its center of oscillation; and the distance between it and the *usual* point of suspension is, technically speaking, the length of the pendulum. It will be seen that these two points are unequally distant from the center of gravity.

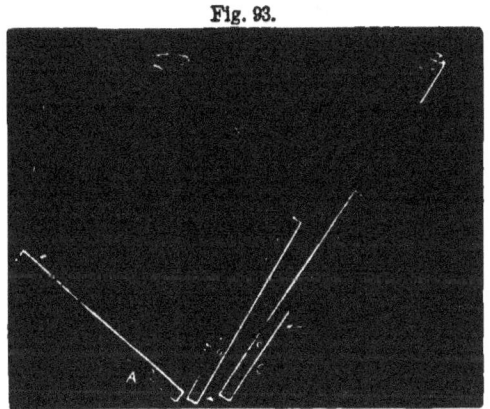

Fig. 93.

§ 83. **Center of percussion.** — Experiment. Suspend the lath by a string attached to one of its extremities, and with a club strike it horizontally near its upper extremity. This end of the lath moves in the direction of the stroke (A, Fig. 93), at the same time causing a sudden jerk on the string, which is felt by the hand. Strike the lath in the same direction, near its lower extremity; the upper end of the lath now moves in a direction opposite to the stroke (B), at the same time causing a similar jerk of the string. Next strike the lath successively at points higher and higher above its lower extremity; it is found that the jerk on the string becomes less till the center of oscillation is reached, when no pull on the string is felt, and neither end of the lath tends to precede the other, but both move on together (C). The full force of the blow is spent in moving the stick, and none is expended in pulling the string. This point is called the *center of percussion*.

The center of percussion is coincident with the center of oscillation. It is the point where a blow, given or received, is most effective, and produces the least strain upon the support or axis of motion. The base-ball player soon learns at what point on his bat he can deal the most effective blow to the ball, and at the same time feel the least tingle in his hands.

§ 84. Some useful applications of the pendulum. — The force that keeps a pendulum vibrating is gravity. Were it not for friction and resistance of the air, a pendulum, once set in motion, would never cease vibrating. Since the force of gravity keeps the pendulum in motion, it follows that the rate of vibration of a given pendulum must be determined by the intensity of this force. Hence it is apparent, that if the rate of vibration is known, the intensity of the force of gravity may be calculated. It is found by experiment that *the time of vibration varies inversely as the square root of the force of gravity.*

So the pendulum becomes a most serviceable instrument for measuring the intensity of gravity at various altitudes and at different latitudes on the earth's surface. (Compare § 21). It is also the most accurate instrument for measuring time that has been invented. Its value, as a time-measurer, depends upon the absolute uniformity of the rate of vibration as long as its length is constant, and the length of its arc very small. But as heat is ever modifying the dimensions of all visible bodies, various devices have been called into existence by which heat may be made to correct automatically its own mischief. Clocks that do not have self-regulating pendulums are fast in winter and slow in summer. (How would you regulate them?)

MOMENTUM. 115

QUESTIONS.

1. Where is the center of percussion in a hammer or axe ? Why ?
2. At what point (disregarding the length and weight of the arm that swings it) should a blow be dealt with a bat of uniform dimensions when held in the hand at one extremity ?
3. What change in the location of the center of percussion is produced by making one end of a bat heavier than the other ?
4. Which end of a bat, the heavier or lighter, should be held in the hands ? Why ?

XIV. MOMENTUM. — THIRD LAW OF MOTION.

§ 85. Momentum. — A small stone dropped upon a cake of ice produces little effect; a large stone dropped upon the ice crushes it. An empty car in motion is much more easily stopped than a loaded car. We dread the approach of large masses because we instinctively associate with them a large amount of motion or force. It is evident that if two bodies move with the same speed, there is a greater quantity of motion in that which contains the greater quantity of matter, just as there is more heat in a gallon of water than in a pint of water, when both have the same temperature.

Again, we have a similar dread of masses moving with great velocities. A ball tossed is a different affair from a ball thrown. Our experience, then, teaches us that *the quantity of motion*, or, in a word, the *momentum a body may have, depends upon its mass and velocity.* For example, a large mass, moving slowly, has great momentum, but the same mass will have twice the momentum if its velocity is doubled; again, a small mass, moving swiftly, has great momentum, but its momentum is increased in proportion as its mass is increased.

If the motion of a mass weighing 1^k, having a velocity of 1^m per second, is taken as a unit of momentum, then a mass weighing 5^k, moving with the same velocity, would have a momentum of 5; and if the latter mass should have a velocity of 10^m per second, its momentum would be $5 \times 10 = 50$. Hence, *the numerical value of momentum is found by multiplying units of mass by units of velocity.*] There is no name for the unit of momentum. We return to this subject on page 123.

QUESTIONS AND PROBLEMS.

1. Compare the momenta of a car weighing 50 tons, moving 10 ft. per minute, and a lump of ice weighing 5 cwt., at the end of the third second of its fall.

2. Why are pile-drivers made heavy? Why raised to great hights?

3. A boy weighing 25^k must move with what velocity to have the same momentum that a man has weighing 80^k running at the rate of 10^{km} per hour?

4. A body has a certain momentum after falling through a certain space. How many times this space must it fall to double its momentum?

§ 86. Third law of motion. — It has been shown (page 88) that motion cannot originate in a single body, but arises from mutual action between two bodies. For example, a man can lift himself by pulling on a rope attached to some other object, but not by his boot-straps, or a rope attached to his feet. Whenever one body receives motion, another body always parts with motion, or is set in motion in an opposite direction; that is, [*in every change in regard to motion there are always at least two bodies oppositely affected.*]

Experiment. Float two blocks of wood of unequal masses on water, connecting them by a stretched rubber band. Let go the blocks, and the band will set both in motion, but the smaller block will have the greater velocity.

A man in a boat weighing one ton pulls at one end of a rope, the other end of which is held by another man, who weighs twice as much as the first man, in a boat weighing two tons: both boats will move towards each other, but in opposite directions; the lighter boat will move twice as fast as the heavier, but with the same momentum.

If the boats are near each other, and the men push each other's boats with oars, the boats will move in opposite directions, though with different velocities, yet with equal momenta.

The opposite impulses received by the bodies concerned are usually distinguished by the terms *action* and *reaction*. We measure these by their momenta. As every force is either a push or a pull (§ 12), and produces equal momenta in two bodies in opposite directions, hence, the

THIRD LAW OF MOTION: *To every action there is an equal and opposite reaction.*

The application of this law is not always obvious. Thus, the apple falls to the ground in consequence of the mutual attraction between the apple and the earth. The earth does not appear to fall toward the apple. But, allowing that their momenta are equal, we are not surprised that the motion of the earth is imperceptible, when we reflect that the velocity of the earth must be as many times less than that of the apple as the mass of the apple is less than that of the earth. (Compare § 20.)

QUESTIONS.

1. The velocity of the rebound or "kick" of a gun is slight when compared with the velocity of the ball. Why?

2. In rowing a boat, what are the opposite results of the stress between the oar and the water?

3. Point out the results of the action and reaction that occur when a person leaps from the ground.

Fig. 94.

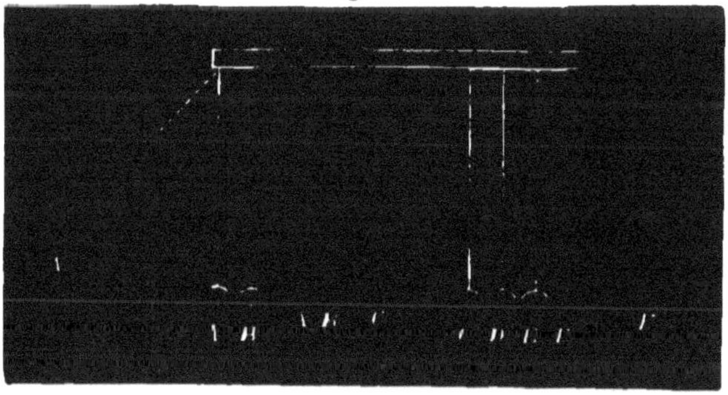

4. If there were no ground or other object beneath him, and he were motionless in space, could he put himself in motion? Why?

5. A boy, running, strikes his head against another boy's head. Which is hurt? Why?

6. Suspend two balls of soft putty of equal weight, A and B (Fig. 94). Draw A one side, and let it fall so as to strike B. Both balls

will then move on together; with what momentum compared with A's momentum when it strikes B?

7. What will be the momentum of each ball after A strikes B, compared with A's momentum when it strikes B?

8. How will their velocity compare with A's velocity when it strikes B?

9. Raise A and B equal distances in opposite directions, and let fall so as to collide. Both balls will instantly come to rest after collision. Show that this result is consistent with the third law of motion.

10. Substitute, for the inelastic putty balls, ivory billiard balls, which are highly elastic. Let A strike B. Then B goes on with A's original velocity, while A is brought to rest. Show that this result is consistent with the third law of motion.

11. Suspend four ivory balls, C, D, E, and F. Let C strike D. D eventually receives all of C's momentum, and instantly communicates it to E, E to F, and F, having nothing to which to communicate it, moves with C's original velocity. Trace the actions and reactions throughout.

12. What would happen if the four balls were inelastic?

§ 87. **Law of reflection.** — **Experiment 1.** Hold D (Fig. 94) firmly in its place, and allow C to strike it. D being immovable, C's entire momentum is spent in compressing the balls, and, on recovering their shape, C is thrown back to its starting-point at C'. But in this case the hand exerts as much force to prevent the motion of D as would be necessary to project C to C'. *When an elastic body strikes another fixed elastic body, it rebounds with its original force.*

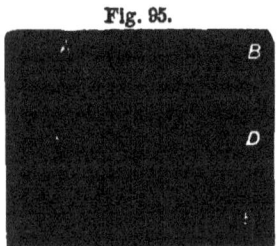

Fig. 95.

Experiment 2. Lay a marble slab A (Fig. 95) upon a table, and roll an ivory ball in the line DC, perpendicular to the surface of the slab; the ball rebounds in the same line to D. Roll the ball in the line BC; it rebounds in the line CE. The angle BCD, which its forward path makes with DC, a perpendicular to the surface struck, is called the *angle of incidence*. The angle ECD, which its retreating path makes with the same perpendicular, is called the *angle of reflection*.

It is found by measurement that these angles are equal when the two bodies are perfectly elastic. This equality is expressed by the LAW OF REFLECTION: *When the striking body and the body struck are perfectly elastic, the angle of reflection is equal to the angle of incidence.*

XV. WORK AND ENERGY.

§ 88. Work. — We have learned (page 44) that a force may produce either motion or pressure (or tension), or it may produce both effects at the same time and in the same body. But *a force does work*, in the sense in which this term is used in science, *only when it produces motion*. A person may support a weight for a time and become weary from the continuous application of force to prevent the weight from falling, or, in other words, to prevent the force of gravity from doing work, but he accomplishes no work, because he effects no change, *i.e.*, causes no motion. *The body that is moved is said to have work done upon it; and the body that moves another body is said to do work upon the latter.* When the heavy weight of a pile-driver is raised, work is done upon it; when it descends and drives the pile into the earth, work is done upon the pile, and the pile in turn does work upon the matter in its path.

Whenever a force causes motion, it does work. A force may act for an indefinite time without doing any work; but *whenever a force acts through space, work is done.* Force and space (or distance) are essential elements of work, and are naturally the quantities employed in estimating work. A given force acting through a space of one meter will do a certain amount of work; it is evident that the same force acting through a space of two meters will do twice as much work. Hence the general formula,

$$W = FS, \qquad (1)$$

in which W represents the work done, F the force employed, and S the space through which the force acts.

In case a force encounters resistance, the magnitude of the force necessary to produce motion depends upon the amount of resistance. Indeed, in cases in which the body having been moved through a given space comes to rest in consequence of resistance, the entire work done upon the body is often more conveniently determined by *multiplying the resistance by the space through which it is overcome*, and our formula becomes by substitution of resistance, R, for the force which overcomes it,

$$W = RS. \qquad (2)$$

For example, a ball is shot vertically upward from a rifle in a vacuum; the work done upon the ball may be estimated by multiplying the average force (difficult to ascertain) exerted upon it by the space through which the force acts (a little greater than the length of the barrel), or by multiplying the resistance offered by gravity, *i.e.*, its weight (easily ascertained) by the distance the ball ascends. Also, in case the motion produced is uniform, the resistance and the force are equal, and it is immaterial which formula is used. When there is no resistance and the only effect is acceleration, as when a body falls freely in a vacuum, we must estimate the work done (in this case by gravity) by the first formula. When it is required to estimate only that part of the work done in producing acceleration, the formulas given on page 124 will be found convenient, work being substituted for energy, inasmuch as both are measured by the same units.

§ 89. Unit of work.—We shall first consider the unit employed when resistance is taken as one of the elements of work. (In § 96 will be defined the unit usually employed when the force is employed as a factor of work.) The unit of work adopted by the French is the work done in raising 1^k through a vertical hight of 1^m. It is called a *kilogrammeter* (abbreviated ^{kgm}). The English unit of work is that done in raising one pound one foot, and is called a *foot-pound*. The kilogrammeter is about $7\frac{1}{4}$ (more accurately, 7.233) times greater than the foot-pound. Now, since the work done in raising 1^k 1^m high is 1^{kgm}, the work of raising it 10^m high is 10^{kgm}, which is the same as the work done in raising 10^k 1^m high; and the same, again, as raising 2^k 5^m high.

There are many other kinds of work besides that of raising weights. But since, with the same resistance, the work of producing motion in any other direction is just the same as in a vertical direction, it is easy, in all cases in which the two elements of work (viz., resistance and space) are known, to find the equivalent in work done in raising a weight vertically. By thus securing a common standard for measurement of work, we are able to compare any species of work with any other. For instance, let us compare the work done by a man in sawing through a stick of wood, whose saw must move 10^{m} against an average resistance of 12^k, with that done by a bullet in penetrating a plank to a depth of 2^{cm} against an average resistance of 200^k. Moving a saw 10^m against 12^k resistance is equivalent to raising 12^k 10^m high, or doing 120^{kgm} of work; a bullet moving 2^{cm} against 200^k resistance does as much work as is required to raise 200^k 2^{cm} high, or $200 \times .02 = 4^{kgm}$ of work. $120 \div 4 = 30$ times as much work done by the sawyer as by the bullet.

§ 90. Rate of doing work.—In estimating the total amount of work done, the time consumed is not taken into consideration. The work done by a hod-carrier, in carrying 1,000 bricks to the top of a building, is the same whether he does it in

POTENTIAL AND KINETIC ENERGY. 121

a day or a week. But in estimating the power of any agent to do work, as of a man, a horse, or a steam-engine, in other words, *the rate* at which it is capable of doing work, it is evident that time is an important element. The work done by a horse, in raising a barrel of flour 20 feet high, is about 4000 ft.-lbs.; but even a mouse could do the same amount of work in time. The unit in which rate of doing work is usually expressed is a *horse-power*. Early tests showed that a very strong horse may perform 33,000 ft.-lbs. of work in one minute. So 1 horse-power = 33,000 ft.-lbs. per minute = 550 ft.-lbs. per second = about 4570^{kgm} per minute = about 76^{kgm} per second.

§ 91. Energy. — The *energy* of a body is its capacity of doing work, and is measured by the work it can do. *Doing work usually consists in a transfer of motion, or energy, from the body doing work to the body on which work is done.* Wherever we find matter in motion, whether in the solid, liquid, or gaseous state, we have a certain amount of energy which may often be made to do useful work.

§ 92. Potential and kinetic energy. — Place a stone, weighing (say) 10^k, on the floor before you; it is devoid of energy, powerless to do work. Now raise it, and place it on a shelf (say) 2^m high; in so doing you perform 20^{kgm} of work on it. As you look at it, lying motionless on the shelf, it appears as devoid of energy as when lying on the floor. Attach one end of a cord 3^m long to it, and, passing it over a pulley, wind 2^m of the string around the shaft connected with a sewing-machine, coffee-mill, lathe, or other convenient machine. Suddenly withdraw the shelf from beneath the stone. It moves, it sets in motion the machine, and you may sew, grind coffee, turn wood, etc., with the power given to the machine by the stone.

Surely, the work done on the stone in raising it was not lost; the stone pays it back while descending. There is a very important difference between the stone lying on the floor, and the stone lying on the shelf: the former is powerless to do work; the latter can do work. Both are alike motionless, and you can see no difference, except an *advantage* that the latter has over the former *in position*. What gave it this advantage? Work. *A body, then, may possess energy due merely to* ADVANTAGE OF POSITION, *derived always from work bestowed upon it*. So a body at rest is not necessarily devoid of energy. In the stone lying passively on the shelf there exists a power to do work as real as that possessed by the stone which, falling freely, has acquired great velocity.

We see, then, that energy may exist in either of two widely different states, and yet be as real in one case as in the other. It may exist as *actual motion*, either visible, as in mechanical motion, or invisible, as in the molecular motions called heat; or it may exist in a *stored-up condition*, as in the stone lying on the shelf. In the former case it is called *kinetic* (moving) or *actual* energy; in the latter, it is called *potential* energy, or *energy of position*.

We are as much accustomed to store up energy for future use as provisions for the winter's consumption. We store it when we wind up the spring or weight of a clock, to be doled out gradually in the movements of the machinery. We store it when we bend the bow, raise the hammer, condense air, and raise any body above the earth's surface.

How, then, is energy stored in a body? *Only at the expense of work done upon it*. The force of gravitation is employed to do work, as when mills are driven by the power of falling water; but the water is first deposited on the hillside by the energy of the sun's heat. Elasticity of springs is employed as a motive power; but elasticity is due to an advantage of position which the molecules of springs have acquired in consequence of force applied to them.

We conclude, then, that *a body possesses potential energy when, in virtue of work done upon it, it occupies a position of advantage, or its molecules occupy positions of advantage, so that the energy expended can be at any time recovered by the return of the body to its original position, or by the return of its molecules to their original positions.*

§ 93. Energy contrasted with momentum. — Problem. A bullet weighing 30^g is shot with a velocity of 98^m per second from a gun weighing 4^k; required the momentum and the energy of both the bullet and the gun, and the velocity of the gun. *Solution:* Using the kilogram, the meter, and the second as units, the momentum of the ball is $.03 \times 98 = 2.94$ units. If the ball were shot vertically upward, its velocity would diminish 9.8^m per second; so it would rise $\frac{98}{9.8} = 10$ seconds, and, therefore, before its energy is expended, to a hight of (§ 78) $4.9^m \times 10^2 = 490^m$. Hence, its energy at the outset is $.03 \times 490 = 14.7^{kgm}$. Similarly for the gun, by the third law of motion its momentum must be just the same as that of the ball, 2.94 units; its velocity is therefore $2.94 \div 4 = .735^m$ per second. Then $T = \frac{.735}{9.8} = .075$ second; the hight (supposing the gun to be raised vertically by the impulse received) $= 4.9 \times .075^2 = .02766^m$; and its energy $= 4 \times .02766 = .1102^{kgm}$.

While, therefore, the momenta generated in the two bodies by the burning of the powder are equal, the energy of the bullet is $\frac{14.7}{.1102} = 133\frac{1}{3}$ times that of the gun. (Why are the effects produced by the bullet more disastrous than those produced by the recoil of the gun?)

§ 94. Formula for energy. — We can find, as in the above example, to what vertical hight a body having a given velocity would rise, and thus in all cases determine its energy; but a formula may be obtained which will give the same result with less trouble: thus, substituting g for k in Formula 1 (§ 78), $V = g T$; hence,
$$T = \frac{V}{g}, \text{ or } T^2 = \frac{V^2}{g^2}.$$
Again, $S = \frac{1}{2} g T^2$; substituting the value of T^2 in this equation, we have
$$S = \frac{1}{2} g \times \frac{V^2}{g^2} = \frac{V^2}{2g}.$$

But energy $= WS$ (weight into hight); substituting for S in this equation its value, we have,

(1) $\text{Energy} = \dfrac{WV^2}{2g}$.

Farther on, we shall see that $W = Mg$; substituting for W in the last equation its value, we have, also,

(2) $\text{Energy} = \dfrac{MV^2}{2}$.

It is evident that, *when the weight* (W) *or mass* (M) *of a body remains the same, its energy is proportional to the square of its velocity, while its momentum*, as we have learned, *is proportional to its velocity*. In other words, the effect of increasing the velocity of a moving body would seem to be to increase its working power much more rapidly than its momentum. Is this *practically* true?

Experiment. Fill an ordinary water-pail with moist clay. Let a leaden bullet drop upon the clay from a hight of .5^m. Then drop the same bullet from a hight of 2^m, or four times the former hight, in order that it may acquire *twice* the velocity. In the latter case it penetrates to four times the depth that it did in the former.

So it appears that *the energy of a moving body varies*, not as its velocity, but *as the square of its velocity*. Doubling the velocity multiplies the energy fourfold; trebling the velocity multiplies it ninefold, and so on; but the corresponding momentum is multiplied only twofold, threefold, etc. A bullet moving with a velocity of 400 feet per second, will penetrate, not twice, but four times, as far into a plank as one having a velocity of 200 feet per second. A railway train, having a velocity of 20 miles an hour, will, if the steam is shut off, continue to run four times as far as it would if its velocity were 10 miles an hour. The reason is now apparent why light substances, even so light as air, exhibit great energy when their velocity is great.

§ 95. Measure of a force. — Commonly we measure forces by a spring balance, and say that the force, for instance, with

MEASURE OF A FORCE. 125

which a horse draws a wagon is 50^k; that is, a spring interposed between the horse and the wagon is stretched just as much as it would be by the force of gravity acting on a mass of 50^k hung from the spring. But often it is impossible to measure the force except by the motion it produces. Experience has shown that *a useful and accurate measure of a force is the momentum it produces or destroys in a second;* if the body is already in motion, we must say the *change of momentum produced in a second.*

For example, gravity we know will impart in three seconds, to a body having a mass of (say) 5^g, and free to fall, a velocity of 3×980^{cm} per second; that is, the momentum generated is $5 \times 3 \times 980$. Then, by definition above, the measure of the force of gravity on the body is $\frac{5 \times 3 \times 980}{3} = 5 \times 980$. When the centimeter, gram, and second are taken as the units of length, mass, and time respectively, the system of units of measurement based on them is called the C.G.S. system, and in it the unit of force is called a *dyne*.

A dyne is that force which, acting for a second, will give to a gram of matter a velocity of one centimeter per second. In the example above we have a force of $5 \times 980 = 4900$ dynes.

We can almost as easily graduate a spring balance to indicate forces in dynes as in pounds; and then we have a unit which is constant wherever we go on the earth or above it. (Compare § 21.)

The *gravity unit of force* is the weight of any unit of mass, *e.g.*, a gram, kilogram, pound, or ton. In distinction from gravity units, the C.G.S. units are called *absolute units*. Gravity units are easily changed to absolute units; thus in the Northern States the force of gravity acting upon 1^g of matter free to fall will give it an acceleration of velocity of 980^{cm} per second; hence in these latitudes the gravity unit is equal to 980 absolute units.

Returning to our example, represent 5^g, the mass of the body moved by M; by g, 980^{cm} per second, the acceleration produced by gravity; and by W, the weight, or F, the force: then

$$W = F = Mg.$$

The equation is a general one; that is, whenever any two of the three quantities specified are known, the third may be computed

If the force acts, not against gravity, but against resistances considered as constant, such as the forces shown in cohesion, elasticity, etc., the equation will still be true, only g should be replaced by some other letter, as a.

Now let us learn what is the

§ 96. Measure of the effect of a force. — One measure we know already, — the product of the force into the distance through which it acts; that is, *the work done, or the energy imparted* to the body moved, *is a measure of the effect of a force.* If the force is measured in dynes, and the distance is centimeters, the work done will be expressed in a C.G.S. unit called an *erg*. *An erg is the work done or energy imparted by a force of one dyne working through a distance of one centimeter.* Besides the erg we have the common gravitation units, the kilogrammeter, and foot-pound; that is, we have another measure just as we ma have various kinds of measures for common things; just as, for instance, we may express lengths in inches, meters, or miles; masses, in grains or pounds, etc.

Experiment 1. Suspend by a long cord a heavy body, — 10^k or more, — and with a string attached to the body draw it to one side, pulling for two, four, and six seconds, and let go. The longer you pull the greater is the velocity given to the body, provided it is not moved far from its place of rest.

Experiment 2. Suspend by a string 1^m long a stone whose mass is (say) 5^k. Attach to the stone a No. 36 cotton thread; this will support about 1^k. Pull the ball slowly to one side; when it has been drawn about 20^{cm} from its place of rest, the thread will break. and the ball will swing back to the other side like a pendulum, and so when it passes through its lowest point it has a definite momentum.

Attach new pieces of thread, and pull more and more quickly, breaking the thread each time; the motion produced is less and less. As the string is straightened the pull on it increases from zero to 1^k; so the average force each time is about the same; in gravitation units, nearly or exactly $\tfrac{1}{2}^k$. Here, as before, *with the same force, the momentum produced varies as the time during which the force acts.*

MEASURE OF THE EFFECT OF A FORCE. 127

But if we use stronger and stronger threads, we may pull more and more quickly than at first, and yet give to the ball just the same momentum as at first; that is, *the effect of a greater force acting for a shorter time is to produce the same momentum.*

So far then as our experiments go, they teach that the product of a force into the time it acts, or *the momentum produced, is a measure of the effect of a force.* We may draw the same conclusion from our last equation, $F = Mg$: multiply both sides by T, the time during which the force acts, and we have $FT = MgT = MV =$ Momentum (§ 85). If T equals one second, we see that the momentum of a moving body is the measure of the force that would in one second give it this motion. It is evident that if motion is to be produced by a force acting for a very short time, the force must be enormous.

We have, then, two measures of the effect of a force, — *momentum* and *energy*. The first is found by multiplying the force by the *time* it acts; the second, by multiplying the force by the *space* through which it acts. The latter can also be found by multiplying the momentum by one-half the velocity. One is MV; the other is $\frac{1}{2}MV^2$. Which is the correct measure? Both are correct; so the question now is, Which is the more useful? Experience shows that momentum is a useful measure only in cases where the force acts all the time in the line of motion, as in falling bodies, or where it acts for so short a time that the body does not sensibly change its position during the action, as in the cases of a blow, a jerk, collision between balls, etc. Experience further shows that energy in *all* cases gives a useful measure.

§ 97. Summary of mechanical units, and formulas for their determination.[1] — The following tables show the quantities measured, the unit of each in the C.G.S. system, and the formulas for the determination of the derived quantities: —

[1] It is not expected that pupils of the ordinary high school will master this section; yet they may frequently find it convenient for reference, while the more advanced student cannot fail to be greatly profited by its careful study.

DYNAMICS.

FUNDAMENTAL QUANTITIES AND UNITS.

Length (L or S) .... 1^{cm}.
Mass (M) .... 1^g.
Time (T) .... 1 sec.

DERIVED QUANTITIES, UNITS, AND FORMULAS.

Velocity (V) = rate of motion; unit, 1^{cm} per sec.; in uniform motion,
$$V = \frac{S}{T}. \quad (1)$$

Acceleration (A) = rate of change in velocity; unit, an increase of velocity in 1 sec. of 1^{cm} per sec.; body starting from rest under constant force, $A = \frac{V}{T}$. (2)

Force (F); unit, 1 dyne = a force that in 1 sec. imparts to 1^g a velocity of 1^{cm} per sec.; $\therefore F = MA$. (3)

Work or Energy (E); unit, 1 erg = the work done by 1 dyne working through 1^{cm}; $\therefore E = MAS = FS$. (4)

Rate of doing work, or Work Power (P); unit, 1 erg per sec.;
$$P = \frac{MAS}{T}. \quad (5)$$

Momentum; unit, 1^g moving with a velocity of 1^{cm} per sec., or that produced by 1 dyne in 1 sec.; Momentum = MV.

From (2) and (3) we have the very useful equations, $F = \frac{MV}{T}$ and
$$V = \frac{FT}{M}. \quad (6) \text{ and } (7)$$

A body, mass M, acted upon by the force F, starting from rest will acquire in time T a velocity $V = \frac{FT}{M}$. The acceleration, which from (3) is $= \frac{F}{M}$, is a constant quantity, and the whole space passed over will be equal to the time T multiplied by the *mean* velocity. The latter is one-half the final velocity; hence, mean $V = \frac{FT}{2M}$, and $S = \frac{FT^2}{2M}$ (an equation of great importance). (8)

To find an expression for the energy of a moving body combine (4) and (8): $W = \frac{F^2 T^2}{2M}$; but $FT = MV$, $\therefore E = \frac{MV^2}{2}$. (9)

Anywhere in the Northern States, the weight of 1^g = 980 dynes.
1^{kgm} = 98,000,000 ergs; 1 foot-pound = 13,550,000 ergs.
1 horse-power = 447,000,000,000 ergs per min.

§ **98. Transformation of energy.** — In the operation of raising the stone (§ 92), kinetic energy is transformed into potential energy. During its descent it is re-transformed into kinetic energy. If, instead of being attached to machinery, and thereby made to do work, the stone is allowed to fall freely, it acquires great velocity. On striking the ground, its motion as a body suddenly ceases, but its molecules have their quivering motions accelerated. Mechanical motion is, thereby, transformed into heat. We shall often have occasion to examine the transformations of energy, as into electric energy, heat, etc., but never of momentum. We shall study Joule's equivalent (page 174), expressing the relation between the unit of energy, or work, and the unit of heat; but it is certain that there is no relation between the latter and the unit of momentum.

§ 99. **Physics defined.** — All physical phenomena consist either alone in transferences of energy from one portion of matter to another, or in both transferences and transformations of energy. Transformations may be from one condition of energy to another, as from kinetic to potential; or from one phase of kinetic energy to another, as from mechanical motion to heat; or both may occur, as when the falling stone does work, a part of its energy being expended in producing mechanical motion, and a part being transformed into heat, occasioned by friction of the moving parts.

Physics is that branch of natural science which treats of transferences and transformations of energy. It does not, however, in its usual limitation, include a group of phenomena which occur outside the earth, and also a group whose essential characteristic is an alteration in the nature of the material considered. The study of the former group is the object of *Astronomy;* of the latter, that of *Chemistry.*

QUESTIONS AND PROBLEMS.

1. Does the energy expended in raising the stones to their places in the Egyptian pyramids still survive?
2. What kind of energy is that contained in gunpowder?
3. What transformation of energy takes place in burning coal?
4. When steam works by expansion, its temperature is reduced. Why?

5. How much work is done per hour if 80^k are raised 4^m per minute?

6. (*a*) What energy must be imparted to a body weighing 50^g that it may rise 4 seconds? (*b*) How many times as much energy must be imparted to the same body that it may ascend 5 seconds? (*c*) Why?

7. Compare the momenta, in the two cases given in the last question, at the instants the body is thrown.

8. How much energy is stored in a body which weighs 50^k, at a hight of 80^m above the earth's surface?

9. How much energy would the same body have if it had a velocity of 100^m per second?

10. Suppose it to fall in a vacuum, how much kinetic energy would it have at the end of the fourth second?

11. If it should fall through the air, what would become of a part of the energy?

12. A projectile weighing 25^k is thrown vertically upward with an initial velocity of 29.4^m per second. How much energy has it?

13. What becomes of its energy during its ascent?

14. (*a*) Compare the momentum of a body weighing 50^k, and having a velocity of 2^m per second, with the momentum of a body weighing 50^g, having a velocity of 100^m per second. (*b*) Compare their energies.

15. Which, momentum or energy, will enable one to determine the amount of resistance that a moving body may overcome?

16. Explain how a child who cannot lift 30^k can draw a carriage weighing 150^k.

17. A car weighing 6000^k is drawn by a horse with a speed of 100^m per minute. The index of the dynamometer to which the horse is attached stands at 40^k. (*a*) At what rate is the horse working? (*b*) Express the rate in horse-powers. (See § 90.)

18. A dynamometer shows that a span of horses pull a plow with a constant force of 70^k. What power is required to work the plow if they travel at the rate of 3^{km} per hour?

19. What horse-power in an engine will raise $1,350,000^k$ 5^m in an hour?

20. How long will it take a 3 horse-power engine to raise 10 tons 50 feet?

21. How far will a 2 horse-power engine raise 1000^k in 10 seconds?

22. How much work can a 5 horse-power engine do in an hour?

23. How long would it take a man to do the same work, the amount of work a man can do in a day being about $90,000^{kgm}$?

24. If you would increase the energy of a moving body fourfold, how much must you increase its velocity?

XVI. MACHINES.

§ 100. Uses of machines. — **Experiment 1.** Obtain from a hardware store two or three pulleys, and arrange apparatus as in Figure 96. The dynamometers a and b read 4 lbs. each, showing that the power (P) employed to support each weight (W) of 8 lbs. is just one-half of the weight.[1] If the power applied in each instance is slightly increased, the weights will rise. Raise each of the weights and measure the distances traversed respectively by W and P in each instance. It will be found that the distance that W moves is just one-half the distance that P moves; *i.e.*, if W rises 2 ft., P must move 4 ft. Now, 8 (lbs.) × 2 (ft.) = 16 foot-pounds of work done on W. Again, 4 (lbs.) × 4 (ft.) = 16 foot-pounds of work performed by P. It thus seems that the work applied by the power is just equal to the work done upon the weight. What advantage is derived from the use of the apparatus? It has been proved that no advantage is gained, so far as the amount of work is concerned. But suppose that W is 400 lbs., and that the utmost power (P) that one man can exert is 200 lbs. Then, without this apparatus, the services of two men would be required; whereas one man could raise the weight with the apparatus. The advantage gained in this case would seem to be one of *convenience*.

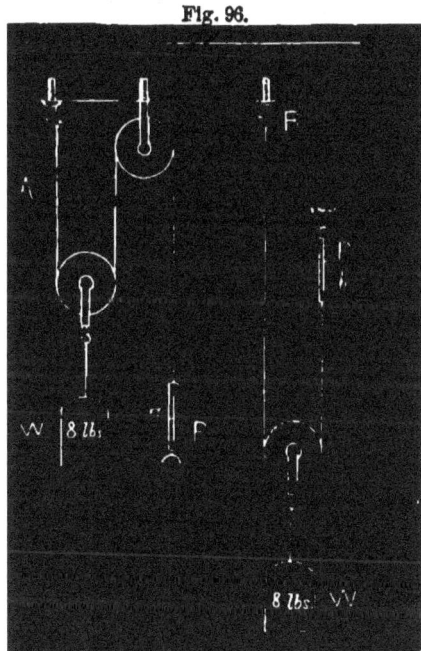

Fig. 96.

Experiment 2. Let P and W of A exchange places. The index of the dynamometer a now reaches 16 lbs. There seems to be in this case a loss of power, for a power of 16 lbs. is only able to sustain a weight of 8 lbs. But so far no work has been done. (Why?) Raise W, and measure the distance traversed respectively by P and W. P moves only 2 ft. for every 4 ft. that W moves. Now, 2 (ft.) × 16 (lbs.) = 32 foot-pounds of work

---

[1] A small allowance must be made for the weight of the movable pulleys.

done by P. And 4 (ft.) × 8 (lbs.) = 32 foot-pounds of work done upon W. We thus learn that, when the power is employed in doing work, there is really no loss of power in this method of applying the apparatus. Is there any advantage gained in this case by the use of apparatus? We found that W moved twice as far, and consequently with twice the velocity, that P moved.

It thus appears that, if it should be desirable to move a weight with greater velocity than it is possible or convenient for the power to move, it may be accomplished through the mediation of a machine, by applying to it a power proportionately greater than the weight. This apparatus is one of many *contrivances called machines, through the mediation of which power can be applied to resistance more advantageously than when it is applied directly to the resistance.* Some of the many advantages derived from the use of machines are:—

(1) *They may enable us to overcome a large resistance with a comparatively small power by causing the power to move through a proportionately greater distance, (i.e. with greater velocity); or, conversely, they may enable us to secure great velocity (i.e. to do work with great speed) by employing a power proportionately greater than the resistance.*

(2) *They may enable us to employ a force in a direction that is more convenient than the direction in which the resistance is to be moved.*

Fig. 97.

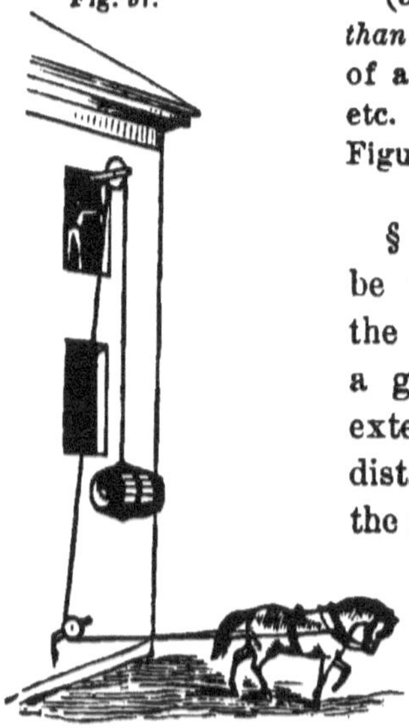

(3) *They may enable us to employ other forces than our own in doing work;* e.g., the strength of animals, the forces of wind, water, steam, etc. (How are the last two uses illustrated in Figure 97?)

§ **101. Law of machines.** — Let P be the power applied to a machine, p the distance through which it moves in a given time, W the weight moved or external resistance overcome, and w the distance through which it is moved in the same time; then the mechanical work applied to the machine is Pp (e.g., in kilogrammeters or foot-pounds), and the mechanical work done by the machine is Ww. Now we have learned from the above experiments that (1) Pp = Ww.

Hence we have for all machines, without exception, the following general law: *The work applied to a machine is equal to the work done by the machine.*

No machine, therefore, creates or increases energy. No machine gives back more energy than is spent upon it. P can be made as small as we please by taking p great enough: in this case we see that *in proportion as power is gained, time, distance, or velocity is lost.* On the other hand, W remaining the same, w (the distance traversed by W in a given time, *i.e.*, its velocity) may be increased indefinitely by taking P large enough: in this case, *as velocity, time. or space is gained, power is lost.* A machine, then, is much like a bank: it pays out no more than it receives. A bank will give you in exchange for a fifty-dollar note fifty one-dollar notes; or, for fifty one-dollar notes, deposited successively, it will return to you a fifty-dollar note. In a similar manner, if you apply to a machine a power sufficient to move 50 lbs. 1 ft., you may get from it the ability to move 1 lb. 50 ft.; or, if you apply to a machine a force of 1 lb. successively through 50 ft. of space, you may get from it the ability to move 50 lbs. through 1 ft. of space.

In our discussion hitherto we have ignored the internal resistances, chiefly due to friction, which exist in every machine. The whole work done by a machine is practically divided into two parts, — the *useful* part and the *wasted* part; the former, expressed as a fraction of the whole, is usually called the *efficiency* or *modulus* of the machine. But energy is indestructible. That portion of the visible energy that is apparently destroyed by friction is transformed into heat, which is wasted, so far as the work to be done by the machine is concerned. Let I represent *internal work* performed *in* the machine, *i.e.*, the *wasted work*, and W w the external work; then our general formula for machines, as modified in its practical applications, becomes.

(2) $P p = W w + I$;

that is, *the work applied to a machine is equal to the effective work, plus the internal work done by the machine.* So that, so far from any machine being a *source of power*, as is sometimes erroneously supposed, no machine practically returns as much power as is applied to it.

By division, Formula (1) $Pp = Ww$ becomes

(3) $$\frac{W}{P} = \frac{p}{w};$$

i.e., *weight : power : : the distance through which the power moves : the distance through which the weight is moved in the same time.*

Fig. 98.

Problems pertaining to machines may generally be solved by Formula (3), and afterwards suitable allowances may be made for the internal work done. Thus, suppose that P (Fig. 99) is 10 lbs., and it is required to find what weight (W) it will raise. By experiment we find that P travels 8 ft. while W travels 4 ft. Then, x (W) : 10 (P) : : 8(p) : 4(w) ; whence $x = 20$ lbs. The 20 lbs. in W is just sufficient to balance the 10 lbs. in P; *anything less* than 20 lbs. will be raised.

It is to be observed that, as we saw on page 119, work is not always, or even usually, expended in raising a weight, but in overcoming resistance of any kind ; so we may interpret Formula (3) thus ; *resistance : power : : the distance through which the power moves : the distance through which the resistance is overcome.*

QUESTIONS AND PROBLEMS.

1. If the power applied to any machine is 2^k, and it moves with a velocity of 10^m per second, with what velocity can it move a resistance of 10^k? To how great a load could it give a velocity of 50^m per second?

2. A power of 50^k, moving through a space of 100^m, is capable of moving how many kilograms through a space of 2^m? What advantage would be gained by the use of the machine?

3. Watch the movements of the foot in working the treadle of a sewing-machine, also the movements of the needle in sewing, and determine what mechanical advantage is gained by the machine.

4. Arrange three levers, as in Figure 98; and, calling the distance (ab) of the power from the prop the *power-arm* of the lever, and the distance (bc) of the weight from the prop the *weight-arm*, verify by experiment the following special formula for levers: —

$$\frac{W}{P} = \frac{p}{w} = \frac{\text{power-arm}}{\text{weight-arm}}.$$

N.B. — Equilibrium must first be established between the two arms of the first lever, by placing weights on the short arm.

5. Ascertain the advantage that may be gained by each lever.

6. A lever is 75^{cm} long; where must the prop be placed in order that a power of 2^k at one end may move 4^k at the other end? What will be the pressure on the prop?

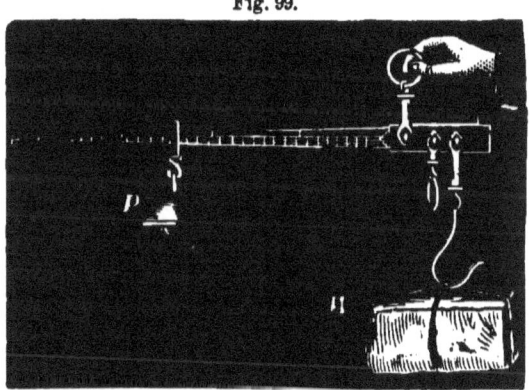

Fig. 99.

7. Show that the results obtained in the last problem are consistent with the third law of parallel forces (page 95).

8. What advantage is gained by a lever, when its power-arm is longer than its weight-arm? What, when its weight-arm is longer?

9. Two weights, of 5^k and 20^k, are suspended from the ends of a lever 70^{cm} long. Where must the prop be placed that they may balance?

10. What mechanical advantage is gained by a lemon-squeezer?

11. If P (Fig. 99), weighing 1 lb., is suspended 15 spaces from the fulcrum of the steelyard, what weight (W) suspended 3 similar spaces the other side of the fulcrum will balance it?

12. How would you weigh out 6 lbs. of tea with the same steelyard?

13. If the circumference of the axle, Figure 100, is 60^{cm}, and the power applied to the crank travels 240^{cm} during each revolution, what power will be necessary to raise the bucket of coal weighing (say) 40^k?

14. How many meters must the power travel (Fig. 100) to raise the bucket from a cavity 10^m deep?

15. (a) In the train of wheels (Fig. 101), if the circumference of the wheel a is 36 in., and that of the pinion b is 4 in., a power of 1 lb. at P will exert what force on the circumference of the wheel d? (b) If the circumference of the wheel d be 30 in., and that of the pinion c 6 in., the power of 1 lb. at P will exert what force on the circumference of the wheel f? (c) If the circumference of the wheel f be 40 in., and that of the axle e 8 in., how many pounds in W will be necessary to prevent motion of the train of wheels, when P weighs 1 lb.? (d) If W has a velocity of 5 ft. per second, what will be P's velocity?

Fig. 100.

Fig. 101.

16. Prepare a special formula for the solution of problems pertaining to the wheel and axle.

17. The weight W (Fig. 102), in traversing the inclined plane AB, only rises through the vertical hight CB, while P must move through a distance equal to AB. Let L represent the length of an inclined plane, and H its hight, and prepare a special formula for the solution of problems pertaining to the inclined plane.

18. A skid 12 ft. long rests one end on a cart 3 ft. high, and the other end on the ground. What force must a boy exert while rolling a barrel of flour weighing 200 lbs. over the skid into the cart?

19. During one revolution a screw advances a distance equal to the

distance between two turns of the thread, measured in the direction of the axis of the screw. Suppose the screw in the letter-press, Figure 103, to advance ¼ in. at each revolution, and a power of 25 lbs. to be applied to the circumference of the wheel b, whose diameter is 14 in. What pressure would be exerted on articles placed beneath the screw. [The circumference of a circle is 3.1416 times its diameter.]

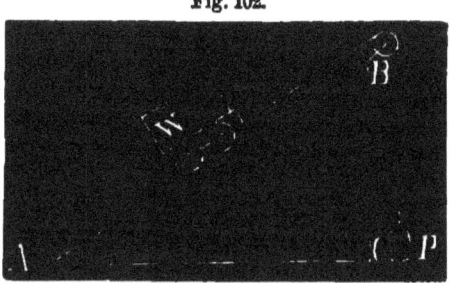

Fig. 102.

20. The toggle-joint (Fig. 104) is a machine employed where great pressure has to be exerted through a small space, as in punching and shearing iron, and in printing-presses, in pressing the types forcibly against the paper. An illustration may be found in the joints used to raise carriage-tops. Force applied to the joint c will cause the two links ac and bc to be straightened, or carried forward to d, while the guides move through a distance equal to $(ac + bc) - ab$. If $dc = 10^{cm}$, $ab = 98^{cm}$, and $ac + bc = 100^{cm}$, then a force of 80^g applied at c would exert what average pressure on obstacles in the path of the guides?

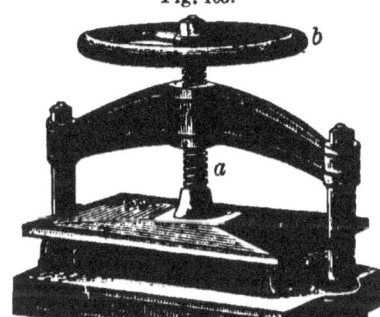

Fig. 103.

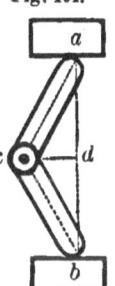

Fig. 104.

21. Show that the hydrostatic press, page 65, conforms in its operation to the general law of machines.

CHAPTER III.

MOLECULAR ENERGY. — HEAT.

XVII. WHAT HEAT IS. — SOME SOURCES OF HEAT.

In the preceding pages the theory of heat has been several times anticipated; we are now better qualified to judge of its truth or falsity.

§ 102. Mechanical motion convertible into heat. — Experiments. Hold some small steel tool upon a rapidly revolving dry grindstone; a shower of sparks flies from the stone. Place a tenpenny nail upon a stone and hammer it briskly; it soon becomes too hot to be handled with comfort, and we may conceive that if the blows were rapid and heavy enough, it might soon become red hot. Rub a desk with your fist, and your coat-sleeve with a metallic button; both the rubbers and the things rubbed become heated.

You observe that in every case heat is generated at the expense of work or mechanical motion, i.e., *mechanical motion checked becomes heat.* When the brakes are applied to the wheels of a rapidly moving railroad train, its motion is all converted into heat, much of which may be found in the wheels, brake-blocks, and rails. The meteorites, or "shooting-stars," which are seen at night passing through the upper air, sometimes strike the earth, and are found to be stones heated to a light-giving state. They become heated when they reach our atmosphere, in consequence of their motion being checked by the resistance of the air.

§ 103. Heat convertible into mechanical motion. — Experiment. Take a thin glass flask A, Figure 105, and half fill it with water; fit a cork air-tight[1] in its neck. Perforate the cork,

[1] A good way to make a cork air-tight is to soak it in melted paraffine.

HEAT DEFINED.

insert a glass tube bent as indicated in the figure, and extend it into the water. Apply heat to the flask; soon the liquid rises in the tube, and flows from its upper end.

Here heat produces mechanical motion, and does work in raising a weight in opposition to gravity. Every steam engine is a *heat engine*. All the power of steam consists in its heat. The steam which leaves the cylinder of an engine (see page 176), after it has set the piston in motion, is cooler than when it entered, and *cooler in proportion to the work done*. Furthermore, it will be shown (page 174) that *heat and work are so related to each other that a definite quantity of the one is always equal to a definite quantity of the other*.

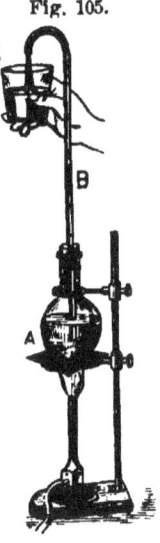

Fig. 105.

Now, when the appearance of one thing is so connected with the disappearance of another, that the quantity of the thing produced can be calculated from the quantity of that which disappears, we conclude that the one has been formed at the expense of the other, and that they are only diferent forms of the same thing. We have, therefore, reason to believe that *heat is of the same nature as mechanical energy*, i.e., *it is only another form of kinetic energy*.

§ **104. Heat defined.** — A body loses motion in communicating it (page 88). The hammer descends and strikes the anvil; its motion ceases, but the anvil is not sensibly moved; the only observable effect produced is heat. Instead of the progressive motion of the hammer as a *whole*, there is now, according to the modern view, an increased vibratory motion of the *molecules* that compose the hammer, — *a mere change of motion in kind and locality*. Of course, this latter motion is invisible. The conclusion is that *heat is molecular motion*. A body is heated by having the motion of its molecules quickened, and cooled by parting with some of its molecular motion. *One body is hotter than another when the average energy of each molecule in it is greater than in the other.*

§ **105. Heat generated by chemical action.** — Experiment. Take a glass test-tube half full of cold water, and pour into it one-fourth its volume of sulphuric acid. The liquid almost instantly becomes so hot that the tube cannot be held in the hand.

When water is poured upon quicklime heat is rapidly developed. The invisible oxygen of the air combines with the various fuels, such as wood, coal, oils, and illuminating gas, and gives rise to what we call *burning* or *combustion*, by which a large amount of heat is generated. In all such cases the heat is generated by the combination or clashing together of molecules of substances that have an affinity (*i.e.*, an attraction) for each other. Before their union they are in the condition of a weight drawn up; while approaching each other, they are like the falling weight; and when they collide, their motion, like that of the weight when it strikes the earth, is converted into heat. The chemical potential energy of the molecules is converted, in the act of combination, into kinetic energy, — into molecular motion.

§ **106. Origin of animal heat and muscular motion.** — The plant finds its food in the air (principally the carbonic acid in the air) and in the earth in the condition of a fallen weight; but, by the agency of the sun's radiation, work is performed upon this matter during the growth of the plant; potential energy is stored in the plant, — the weight is drawn up. The animal now finds its food in the plant, appropriates the energy stored in the plant, and converts it into energy of motion in the form of heat and muscular motion. The plant, then, may be regarded as a machine for converting energy of motion received from the sun into potential energy; the animal, as a machine for transforming it again into the energy of motion.

§ **107. The sun as a source of energy.** — Not only is the sun the source of the energy exhibited in the growth of plants, as well as of the muscular and heat energy of the animal, but it is the source, directly or indirectly, of very nearly all the energy employed by man in doing work. Our coal-beds, the results of the deposit of vegetable matter, are vast storehouses of the sun's energy, rendered potential during the growth of the plants many ages ago. Every drop of water that falls to the earth, and rolls its way to the sea, contributing its mite to the unbounded water-power of the earth, and every wind that blows, derives its power directly from the sun.

XVIII. TEMPERATURE.

§ **108. Temperature defined.** — If body A is brought in contact with body B, and A loses and B gains in heat, then A is said to have had originally a higher *temperature* than B. If neither body gains or loses, then both had the same temperature. *Temperature is the state of a body with reference to its power of communicating heat to or receiving heat from other bodies.* The direction of the flow of heat determines which of two bodies has the higher temperature.

§ **109. Temperature distinguished from quantity of heat.** — The term *temperature* has no reference to *quantity* of heat. If we mix together two equal quantities of a substance at the same temperature, the temperature of the mixture is not the sum of the temperatures, — it is not greater or less than either before they were mixed; but evidently the mixture contains twice as much heat as either alone. If we dip from a gallon of boiling water a cupful, the cup of water is just as hot, *i.e.*, has the same temperature, as the larger quantity, although of course there is a great difference in the quantities of heat the two bodies of water contain. *Temperature depends upon the average kinetic energy of the individual molecule, while quantity of heat depends upon the average kinetic energy of the individual molecule multiplied by the number of molecules.*

XIX. DIFFUSION OF HEAT.

There is always a tendency to *equalization of temperature;* that is, heat has a tendency to pass from a warmer body to a colder, or from a warmer to a colder part of the same body, until there is an equilibrium of temperature.

§ 110. Conduction. — **Experiment 1.** Place one end of a wire about 15^{cm} long, in a lamp-flame, and hold the other end in the hand. Heat gradually travels from the end in the flame toward the hand. Apply your fingers successively at different points nearer and nearer the flame; you find that the nearer you approach the flame the hotter the wire is.

The flow of heat through an unequally-heated body, from places of higher to places of lower temperature, is called *conduction;* the body through which it travels is called a *conductor.* The molecules of the wire in the flame have their motion quickened; they strike their neighbors and quicken their motion; the latter in turn quicken the motion of the next; and so on, until some of the motion may be finally communicated to the hand, and creates in it the sensation of heat.

Experiment 2. Hold wires of different metals of the same length, also a glass tube, a pipe-stem, etc., in the flame, and notice the difference in time that elapses before the sensation of heat is felt in the different bodies.

Experiment 3. Go into a cold room, and place the bulb of a thermometer in contact with various substances in the room; you will probably find that they have the same, or very nearly the same, temperature. Place your hand on the same substances; they appear to have very different temperatures. This is due to the fact that some substances conduct heat away from the hand faster than others. Those substances that appear coldest are the best conductors. If you go into a room warmer than your body, all this is reversed; those substances which feel warmest are the best conductors, because they conduct their own heat to your hand fastest.

Experiment 4. Twist together at one end similar wires or strips of iron, copper, brass, etc., 10 or 15^{cm} long, and introduce them into a small flame. After a few minutes you can tell approximately the order of their conducting powers, by moving a match along each wire, and seeing how far from the flame it will light.

You learn that some substances conduct heat much more rapidly than others. The former are called *good conductors*, the latter *poor conductors*. Metals are the best conductors, though they differ widely among themselves.

Experiment 5. Fill a test-tube nearly full of water, and hold it somewhat inclined (Fig. 106), so that a flame may heat the part of the tube near the surface of the water. The water may be made to boil near its surface for several minutes before any change of the temperature at the bottom will be perceived.

Fig. 106.

Liquids, as a class, are poorer conductors than solids. Gases are much poorer conductors than liquids. It is difficult to discover that pure, dry air possesses any conducting power. The poor conducting power of our clothing is due to the poor conducting power of the fibres of the cloth in part, but chiefly to the air which is confined by it. (Why is loose clothing warmer than that closely fitting?)

Bodies are surrounded with bad conductors, to *retain* heat when their temperature is above that of surrounding objects, and to *exclude* it when their temperature is below that of surrounding objects.

§ 111. **Convection.** — When a hot brick, or a bottle of hot water, is placed at one's feet, heat is also conveyed to the feet. When heat is transferred from one place to another by the bodily moving of heated substances, the operation is called *convection;* but this term is rarely applied to solids. Solids require some external force to effect the conveyance; fluids do not necessarily, as may be seen by the following experiments: —

Experiment 1. Arrange apparatus as in Fig. 107. Fill the large beaker nearly full of water, and elevate it so that the tip of a Bunsen flame may just touch the middle of the bottom. Fill a glass tube B

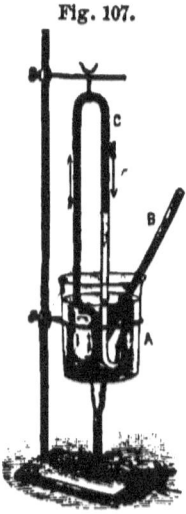

Fig. 107.

with a deeply-colored aniline solution, stop one end with a finger, and thrust the other end into the water to the bottom of the beaker; remove the finger, and allow the solution to flow out and color the water at the bottom for a little depth. Soon the colored liquid immediately over the flame becomes heated, expands, and thereby becomes less dense than the liquid above; consequently it rises and forms an upward current through the colorless liquid. At the same time the cooler liquid on the sides descends to take the place of that which rises, and soon the descending currents become visible by the coloration of the water. By this means heat is conveyed to all parts of the liquid, which would otherwise become much hotter at the bottom than at the top in consequence of the poor conducting power of water.

If a glass tube C, bent as shown in the figure, is filled with water, and introduced into the beaker so that the orifice of the short arm shall be just beneath the surface of the colored water, the colored liquid will be seen slowly to ascend the short arm, while the colder water will descend the longer arm.

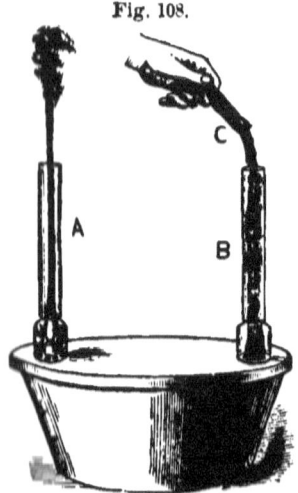

Fig. 108.

Experiment 2. Provide a tightly-covered tin vessel (Fig. 108) and two lamp-chimneys A and B. Near one side of the top of the cover cut a hole a little smaller than the large aperture of chimney B. Near the opposite side of the cover cut a series of holes of about 7^{mm} diameter, arranged in a circle, the circle being large enough to admit a candle without covering the holes. Light the candle, and cover it with chimney A, which should be outside the circle of holes. Fasten both chimneys to the cover with wax. Hold smoking touch-paper C (see page 278) near the top of chimney B. The smoke, instead of rising, as it usually does, rapidly descends the chimney, and in a few seconds will be found ascending chimney A. The air around the flame becomes heated, expands, and rises, while air from the outside rushes down the other chimney to supply the deficiency in the rarefied space. Thus heat from the flame is conveyed away to distant places. Cover the orifice of chimney B with the hand, and the flame will quickly go out.

The last experiment furnishes an explanation of many familiar phenomena. It explains the cause of chimney drafts, and shows the necessity of providing a means of ingress as well as egress of air to and from a confined fire. It explains the method by which air is put in motion in winds. It illustrates a method often adopted to ventilate mines. Let the interior of the tin vessel represent a mine deep in the earth, and the chimneys two shafts sunk to opposite extremities of the mine. A fire kept burning at the bottom of one shaft will cause a current of air to sweep down the other shaft, and through the mine, and thus keep up a circulation of pure air through the mine.

Liquids and gases are heated by convection. (Why not solids?) The heat must be applied at the bottom of the body of liquid or gas. (Why not at the top?) There is a still more important method by which heat is diffused, called *radiation*, which will be treated of in its proper place, under the head of *radiant energy*.

§ **112**. Ventilation. — Intimately connected with the topic *Convection*, is the subject (of vital importance) *Ventilation*, inasmuch as our chief means of securing the latter is through the agency of the former. The chief constituents of our atmosphere are nitrogen and oxygen, with varying quantities of water vapor, carbonic acid gas, ammonia gas, nitric acid vapor, and other gases. The atmosphere also contains in a state of suspension varying quantities of small particles of free carbon in the form of smoke, microscopic organisms, and dust of innumerable substances. All of these constituents except the first three are called *impurities*. Carbonic acid is the impurity that is usually the most abundant and most easily detected; so it has come to be taken as the measure of the purity of the atmosphere, though not itself the most deleterious constituent. Pure out-door air contains about 4 parts of it by volume in 10,000. If the quantity rises to 10 parts, the air becomes unwholesome.

Experiment 1. Place a teaspoonful of unslacked lime in a tumbler of water; a part of it will be dissolved. Filter the solution through unsized paper, and into the clear liquid blow breath *from the lungs* through a glass tube. The liquid turns milky-white in appearance, because the carbonic acid in the breath unites with the lime dissolved in the water, and forms the insoluble carbonate of lime, which remains suspended for a time in the liquid, but finally settles as a white powder at the bottom.

Experiment 2. Take a fresh quantity of lime water in each of two glasses, and in any poorly-ventilated room which has been occupied by several persons for a short time (unfortunately almost any school-room will answer the purpose), place one glass near the floor, and with a bellows blow into the liquid a few puffs of the lower stratum of air. Then place the other glass near the top of the room, and blow with the bellows some of the upper stratum of air into the lime water. In both cases carbonic acid will be found to be present, but it will be much more abundant in the upper stratum, as shown by the greater rapidity with which the cloudiness is produced in the upper stratum.

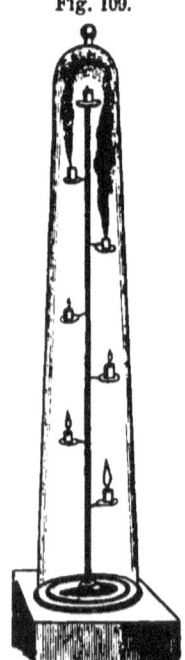

Fig. 109.

Experiment 3. In the center of a small circular plank (Fig. 109) insert an iron wire 60^{cm} long and 7^{mm} in diameter. At intervals of 9^{cm} solder to the wire short pieces of small wire, so as to project horizontally from the large wire; and to the free extremities of these short wires solder small circular pieces of tin 3^{cm} in diameter. Arrange these little platforms spirally around the vertical wire. Fix stumps of candles upon these platforms by means of a little melted tallow. Light the candles, and carefully cover the whole with a tall glass jar. Heated air, from which the life-sustaining oxygen has been largely extracted and replaced by carbonic acid, rises from each flame and accumulates at the top of the jar. This air will neither support life nor combustion, consequently the highest candle flame is quickly extinguished. The colder and purer air descends and feeds the lower flames, while flame after flame, from the top downward, is successively extinguished, the lowest flame being the last to go out.

Carbonic acid gas is about one and one-half times heavier than air at the same temperature; consequently, when both have the same temperature, and the former is very abundant, it tends to settle to the bottom, as in the vicinity of lime-kilns, in which large quantities of this gas are generated.

The knowledge of this fact has led many to suppose that a means for the escape of impure air need only be provided near the floor of a room. But it should be remembered (1) that the tendency of carbonic acid gas, unless present in excessive quantities, is to diffuse itself equally through a body of air; but (2) when it is heated to a temperature above that of the surrounding air, as when generated by flames, or when it escapes in the warm breath of animals, it is lighter than the air, and consequently rises. If this impure air could escape at the ceiling while fresh air entered at the floor, the ventilation would be good. But usually this fresh air must be warmed; and in passing over a stove, furnace, or steam radiator, its temperature will generally become higher than that of the impure air, so that it will rise above the latter, and pass out at a ventilator in the ceiling, leaving the floor cold; hence, the most impure air is often found in high school-rooms half-way up.

Experience shows that, with the ordinary means of heating, it is usually best, in cold weather, to provide for the escape of the foul air at the floor into a flue, in which a draft is maintained by a neighboring hot chimney-flue, or a gas-burner, while the warm, fresh air is introduced at the floor, on the opposite side of the room, or sometimes at the ceiling.

The quantity of fresh air introduced must be great enough to dilute the impurities till they are harmless. An adult makes about 18 respirations per minute, expelling from his lungs at each inspiration about 500^{ccm} of air, over 4 per cent of which is carbonic acid. At this rate, about 9,000^{ccm} of air per minute become unfit for respiration; and to dilute this sufficiently, good authorities say that about 100 times as much fresh air is needed; or, for proper ventilation, about *a cubic meter of fresh air per minute is needed for each person*, or, in English measures, 2,000 cubic feet per hour.

If the heating could be so arranged as to keep the floor properly warmed, the vitiated air might pass out at the ceiling, and the quantity of fresh air entering at the floor might be much less than that just stated. In mild weather, when the fresh air does not require warming, the inlet may be at the floor and the outlet at the ceiling.

QUESTIONS AND PROBLEMS.

1. How would you ventilate the tall jar in Experiment 3?
2. At evening assemblies in lighted halls, what two fruitful sources of carbonic acid are ever present?
3. Why are gas burners frequently placed under the orifices of ventilators?
4. A bed room is 3^m square and 2.5^m high; how long would the enclosed air supply two persons on the supposition that none was to be re-breathed?
5. A hall contains a thousand persons, and its dimensions are $35 \times 18 \times 7^m$. How often should a complete change of air be effected that it may not become vitiated?

XX. EFFECTS OF HEAT. — EXPANSION.

Having learned something of the nature of heat, and how it passes from point to point, let us examine the *effects* it produces on bodies: these are *expansion* and *change of state*. The first gives a means of measuring *temperature*, and leads to a fuller study of gases than we have yet made. Under the second effect of heat we study *liquefaction* and *vaporization*. A third effect that is very obvious, the *change of temperature*, will be found to depend in part on what is called *specific heat*, to be studied on page 170.

§ **113**. **Expansion of solids, liquids, and gases.** — **Experiment 1.** Obtain two short brass tubes, — one of a size that will permit it just to enter the bore of the other. Heat the smaller tube; it will no longer enter the larger.

Experiment 2. Fit stoppers tightly in the necks of two similar thin glass flasks (or test-tubes), and through each stopper pass a glass tube about 60^{cm} long. The flasks must be as nearly alike as possible. Fill one flask with alcohol and the other with water, and crowd in the stoppers so as to force the liquids in the tubes a little way above the corks. Set the two flasks into a basin of hot water, and note that, at the instant the flasks enter the hot water, the liquids sink a little in the tubes, but quickly begin to rise, until perhaps they reach the top of the tubes and run over.

When the flasks first enter the hot water they expand, and thereby their capacities are increased; meantime the heat has not reached the liquids to cause them to expand, consequently the liquids sink momentarily to accommodate themselves to the enlarged vessel. Soon the heat reaches the liquids, and they begin to expand, as shown by their rise in the tubes. The alcohol rises faster than the water. Different substances, both in the solid and liquid states, expand unequally on experiencing equal changes of temperature.

Fig. 110.

Experiment 3. Take one of the flasks used in the last experiment, dry it well inside and outside, invert the flask, insert the end of the tube in a bottle of colored water (Fig. 110), and apply heat to the flask; the enclosed air expands and comes out through the colored liquid in bubbles. After a few minutes, withdraw the heat, keeping the end of the tube in the liquid; as the air left in the flask cools, it contracts, and the water is forced by atmospheric pressure up the tube into the flask, and partially fills it.

§ 114. Coefficients of expansion.—There being generally greater cohesive force between the molecules of solids than between the molecules of liquids, the former expand less than the latter on receiving the same amount of heat, and for the same reason liquids expand less than gases. (See page 18.) *All gases expand alike for equal differences of temperature, and the expansion is uniform at all temperatures.* Under uniform pressure the volume of any body of gas is increased by

$\frac{1}{273}$ its volume at the freezing point of water for every degree centigrade, or $\frac{1}{491}$ for every degree Fahrenheit, its temperature is raised. These fractions are called the *coefficients of expansion*. Not only do the coefficients of expansion of liquids and solids vary with the substance, but the coefficient for the same substance varies at different temperatures, being greater at high than at low temperatures.

In the expansion of fluids we have only to do with increase of volume, called *cubical expansion*. In the expansion of solids, we have frequent occasion to speak of expansion in one direction only, and this is called *linear expansion*.

§ **115. Power of expansion and contraction.** — The force which may be exerted by bodies in expanding or contracting may be very great, as shown by the following rough calculation: If an iron bar, 1 sq. in. in section, is raised from 0° C. (freezing point of water) to 500° C. (a dull, red heat), its length, if allowed to expand freely, will be increased from 1 to 1.006, its coefficient of expansion being about .000012. Now, a force capable of stretching a bar of iron of 1 sq. in. section this amount, is about 90 tons, which represents very nearly the force that would be necessary to prevent the expansion caused by heat. It would require an equal force to prevent the same amount of contraction (caused by what?) if the bar is cooled from 500° to 0° C.

Boiler plates are riveted with red-hot rivets, which, on cooling, draw the plates together so as to form very tight joints. Tires are fitted on carriage-wheels when red hot, and, on cooling, grip them with very great force.

§ **116. Abnormal expansion and contraction of water.** — Water presents a partial exception to the general rule that matter expands on receiving heat and contracts on losing it. If a quantity of water at 0° C., or 32° F., is heated, it contracts as its temperature rises, until it reaches 4° C., or about 39° F., when its volume is least, and therefore it has its *maximum density*. If heated beyond this temperature it expands, and at about 8° C. its volume is the same as at 0°. On cooling, water reaches its maximum density at 4° C., and expands as the temperature falls below that point. It is probable tnat crystallization, and consequently expansion (see page 26), begins at 4° C. (What is the temperature at the bottom of a pond when water begins to freeze at the surface?)

XXI. THERMOMETRY.

§ 117. Temperature measured by expansion. — The effects of expansion by heat are well illustrated in the common thermometer. As its temperature rises, both the glass and the mercury expand; but, as liquids are more expansible than solids, the mercury expands much more rapidly than the glass, and *the apparent expansion of the mercury, shown by its rise in the tube, is the difference between the actual increase of volume of the mercury and that of the part of the glass vessel containing it.* The thermometer, then, primarily indicates changes in volume; but as changes of volume in this case are caused by changes of temperature, it is commonly used for the more important purpose of measuring *temperature.* (Will a thermometer measure *quantity* of heat?)

§ 118. Construction of a thermometer. — A thermometer generally consists of a glass tube of capillary bore, terminating at one end in a bulb. The bulb and part of the tube are filled with mercury, and the space in the tube above the mercury is usually, but not necessarily, a vacuum. On the tube, or on a plate of metal behind the tube, is a scale, to show the hight of the mercurial column.

§ 119. Standard temperatures. — That a thermometer may indicate any definite temperature, it is necessary that its scale should relate to some definite and unchangeable points of temperature. Fortunately Nature furnishes us with two convenient standards. It is found that under ordinary atmospheric pressure ice always melts at the same temperature, called the *melting point*, or, more commonly, *the freezing point* (inasmuch as water freezes and ice melts at the same temperature). Again, the temperature of steam rising from boiling water under the same pressure is always the same.

§ **120. Graduation of thermometers.** — The bulb of a thermometer is first placed in melting ice, and allowed to stand until the surface of the mercury becomes stationary, and a mark is made upon the stem at that point, and indicates the *freezing point*. Then the instrument is suspended in steam rising from boiling water, so that all but the very top of the column is in the steam. The mercury rises in the stem until its temperature becomes the same as that of the steam, when it again becomes stationary, and another mark is placed upon the stem to indicate the *boiling point*. Then the space between the two points found is divided into a convenient number of equal parts called *degrees*, and the scale is extended above and below these points as far as desirable.

Fig. 111.

	F.	C.	Abs. temp.
Water boils	212°	100°	373°
Blood heat	98°	37°	310°
Max. den. of water	39.2°	4°	277°
Water freezes	32°	0°	273°
Mercury freezes	—37.8°	—38.8°	234.2°
No heat	—460°	—273°	0°

In centigrade degrees.

Two methods of division are adopted in this country: by one, the space is divided into 180 equal parts, and the result is called the *Fahrenheit* scale, from the name of its author; by the other, the space is divided into 100 equal parts, and the resulting scale is called *centigrade*, which means *one hundred steps*. In the Fahrenheit scale, which is generally employed for ordinary household purposes, the freezing and boiling points are marked respectively 32° and 212°. The 0 of this scale (32° below freezing point),

CONVERSION FROM ONE SCALE TO THE OTHER. 153

which is about the lowest temperature that can be obtained by a mixture of snow and salt, was incorrectly supposed to be the lowest temperature attainable. The centigrade scale, which is generally employed by scientists, has its freezing and boiling points more conveniently marked, respectively 0° and 100°. A temperature below 0° in either scale is indicated by a minus sign before the number. Thus, $-12°$ F. indicates 12° below 0° (or 44° below freezing point), according to the Fahrenheit scale. Under F. and C., Figure 111, the two scales are placed side by side, so as to exhibit at intervals a comparative view.

§ **121. Conversion from one scale to the other.** — Since $100°$ C. $= 180°$ F., $5°$ C. $= 9°$ F., or $1°$ C. $= \frac{9}{5}$ of $1°$ F. Hence, to convert centigrade degrees into Fahrenheit degrees, we multiply the number by $\frac{9}{5}$; and to convert Fahrenheit degrees into centigrade degrees we multiply by $\frac{5}{9}$. In finding the temperature on one scale that corresponds to a given temperature on the other scale, it must be remembered that the number that expresses the temperature on a Fahrenheit scale does not, as it does on a centigrade scale, express the number of degrees above freezing point. For example, 52° on a Fahrenheit scale is not 52° above freezing point, but $52° - 32° = 20°$ above it.

Hence, to reduce a Fahrenheit reading to a centigrade reading, *first subtract* 32 *from the given number, and then multiply by* $\frac{5}{9}$. Thus,

$$\tfrac{5}{9}(F - 32) = C.$$

To change a centigrade reading to a Fahrenheit reading, *first multiply the given number by* $\frac{9}{5}$, *and then add* 32. Thus,

$$\tfrac{9}{5} C + 32 = F.$$

PROBLEMS.

1. The difference between two temperatures is 80 centigrade degrees. What is the difference in Fahrenheit degrees?

2. When the temperature of a room falls 30 Fahrenheit degrees, how many centigrade degrees is its temperature lowered?

3. Suppose the temperature of the above room before the fall was 68° F., (*a*) what was its temperature after the fall? (*b*) What were the

temperatures of the room before and after the fall, according to a centigrade thermometer?

4. Express the following temperatures of the centigrade scale in the Fahrenheit scale: 100°; 40°; 56°; 60°; 0°; − 20°; − 40°; 80°; 150°.

NOTE. — In adding or subtracting 32°, it should be done *algebraically*. Thus, to change − 14° C. to its equivalent on the Fahrenheit scale: $\frac{9}{5}$ × (− 14) = − 25.2; − 25.2° + 32° = 6.8°, the required temperature on the Fahrenheit scale. Again, to find the equivalent of 24° F. in the centigrade scale: 24 − 32 = − 8; − 8 × $\frac{5}{9}$ = − 4$\frac{4}{9}$; hence, 24° F. is equivalent to − 4.4° + C.

5. Express the following temperatures of the Fahrenheit scale in the centigrade scale: 212°; 32°; 90°; 77°; 20°; 10°; − 10°; − 20°; − 40°; 40°; 59°; 329°.

§ **122. Air thermometer.** — Prepare apparatus as shown in Figure 112. A is a glass flask of about one-fourth liter capacity, tightly stopped. Through the stopper extends a glass tube about 60^{cm} long, which also passes through the stopper of a bottle B, partly filled with colored water. The latter stopper is pierced by a hole a to allow air to pass in and out freely. A strip of paper C, containing a scale of equal parts, is attached to the tube by means of slits cut in the paper.

Fig. 112.

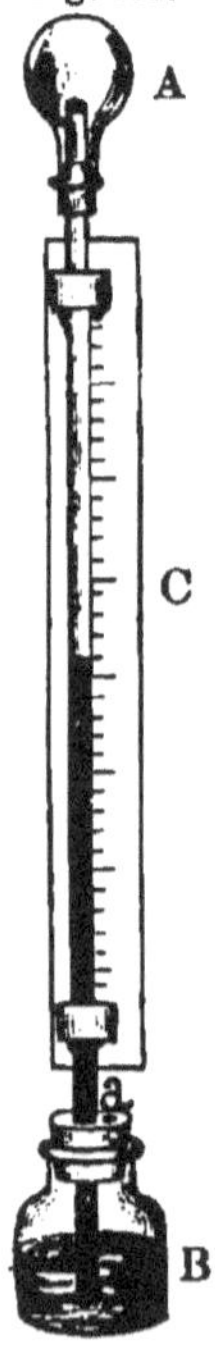

Grasp the flask with the palms of both hands, and thereby heat the air in the flask and cause it to expand and escape through the liquid in bubbles. When several bubbles have escaped, remove the hands, and the air, on cooling, will contract, and the liquid will rise and partly fill the tube.

The apparatus described is usually called an *air thermometer;* but it is, more correctly speaking, a *thermoscope*. It renders slight changes of temperature much more perceptible than a mercury thermometer, and therefore is said to be more *sensitive*. For instance, if an air thermometer and a mercury thermometer, whose bulbs are of the same size, are carried from a cold room into a warm room, or vice versa, the changes in the hight of the liquid column in the air thermometer will be much greater and more rapid than in the mercury thermometer. In the former, the temperature is measured by the expansion of air; in the latter, by the expansion of mercury. (Why is the former more sensitive than the latter?) This simple air thermometer cannot have a fixed scale showing the temperature in Fahrenheit or centigrade degrees as a mercury thermometer does, inasmuch as the hight of the liquid column is affected by atmospheric pressure as well as by temperature, so that when the temperature remains the same, variations occur corresponding to the changes of the barometric column. But in many scientific investigations a good air thermometer is better than one containing mercury. The thermopile and galvanometer (see page 236) constitute a still more sensitive apparatus for showing changes in temperature.

§ **123. Measurement of extreme temperatures.** — Mercury boils at 350° C. (662° F.) and freezes at about −39° C., and therefore cannot be used for indicating temperatures above or below these points. Extremely high temperatures are measured by the expansion of solids, usually a rod of platinum, and the instrument used for this purpose is called a *pyrometer*. Alcohol is used in thermometers employed to measure extremely low temperatures. The air thermometer may be used at any temperature that will not soften the bulb and tube.

§ **124. Absolute temperature.** — If a body of air at 0° C. is heated, its volume is increased $\frac{1}{273}$ of the original volume for every degree its temperature is raised. At 273° C. its volume is consequently doubled. If a body of air is cooled below 0° C., its volume is diminished for every degree its temperature is lowered $\frac{1}{273}$ of its volume at 0°; and so, if its volume were to continue to decrease at that rate until it should reach − 273° C., mathematically speaking its volume would become nothing; but, practically, the air would cease to be a gas, and would become a compact, motionless mass; that is, all molecular motion would cease at that point, and so the point of no heat would be reached. This point is called the *absolute zero*, and temperature reckoned from this point is called *absolute temperature*. On this scale all temperatures would be positive.

NOTE. — Air and all other gases we know (see page 20) are converted into liquids and solids long before they reach the temperature of $-273°$ C.; so, of course, they cease to obey the law of Mariotte (page 59). Though a body has never been cooled to the absolute zero, there are reasons, far more conclusive than the one given, which justify us in believing that all molecular motion would cease at a point very near $-273°$ C. In the further study of heat, the use of the scale of absolute temperature is a great convenience.

The absolute temperature (based on the above theory) may be found by adding 273 to its reading on a centigrade thermometer, or 459 to its reading on a Fahrenheit thermometer. (See Figure 111.)

§ 125. **Laws of gaseous bodies.** — It follows, from the above discussion, that *the volume of a given mass of gas at constant pressure is proportional to its absolute temperature.* This is called the *Law of Charles.*

If, however, a body of gas at 0° C. is enclosed in a vessel of rigid sides, its volume must remain constant at all temperatures. In this case the pressure on the sides is increased by $\frac{1}{273}$ of the pressure at 0° for every degree its temperature rises, and is diminished $\frac{1}{273}$ for every degree its temperature falls; and if it were to continue to decrease at this rate, at $-273°$ C., it would become nothing. Hence, *the pressure of a given body of gas, whose volume is kept constant, is proportional to its absolute temperature.*

Mariotte's law states that *at a constant temperature the volume of a given body of gas is inversely proportional to the pressure to which it is subjected;* i.e., *the product of the pressure and the volume is constant.* Now, when both the pressure and the volume vary at the same time, it is evident that *the product of the pressure and the volume of a given body of gas is proportional to its absolute temperature.*

PROBLEMS.

1. Find in both centigrade and Fahrenheit degrees the absolute temperatures at which mercury boils and freezes.
2. At 0° C. the volume of a certain body of gas is 500^{ccm} under a constant pressure; (*a*) what will be its volume if its temperature is raised

to 75° C.? (b) What will be its volume if its temperature becomes −20° C.?

3. If the volume of a body of gas at 20° C. is 200^{ccm}, what will be its volume at 30° C.? · *Solution:* 20° C. is equivalent to (20 + 273) 293 abs. temp.; then 293 : 303 : : 200 : 206.8^{ccm}. *Ans.*

4. To what volume will a liter of gas contract if cooled from 30° C. to −15° C.?

5. One liter of gas under a pressure of one atmosphere will have what volume if, at a constant temperature, the pressure is reduced to 900^g per square centimeter?

6. The volume of a certain body of air at a temperature of 17° C., and under a pressure of 800^g per square centimeter, is 500^{ccm}; what will be its volume at a temperature of 27° C. under a pressure of 1200^g per square centimeter? *Solution:* 17° C. is equivalent to 290° abs. temp.; 27° C. is equivalent to 300° abs. temp. Then $290 : 300 : : 500 \times 800 : x \times 1200$. Whence $x = 344.8^{cc}$. *Ans.*

7. If the volume of a body of gas under a pressure of 1^k per square centimeter, and at a temperature of 0° C., is 1 liter, at what temperature will its volume be reduced to 1^{ccm} under a pressure of 200^k per square centimeter? *Ans.:* 54.6° abs. temp., or −218.4° C.

8. Find the temperatures on the absolute scale at which bodies named on page 161 melt or boil.

9. If a cubic foot of coal-gas at 32° F., when the barometer is at 30 in., weighs $\frac{1}{25}$ lb., how much will an equal volume weigh at 68° F. when the barometer is at 29 in.?

§ 126. **Kinetic theory of gases.** — This theory claims that in gases the molecules are so far separated from each other that their motions are not generally influenced by molecular attractions. Hence, in accordance with the first law of motion, the molecules of gases move in straight lines and with uniform velocity, until they collide with each other or strike against the walls of the containing vessel, when, in consequence of their elasticity, they at once rebound and start on a new path. We may picture to ourselves what is going on in a body of calm air, for instance, by observing a swarm of bees, when every individual bee is flying with great velocity, first in one direction and then in another, while the swarm either remains at rest or sails slowly through the air.

158 MOLECULAR ENERGY. — HEAT.

§ 127. Pressure of a gas due to the kinetic energy of its molecules. — Consider, then, what a molecular storm must be raging about us, and how it must beat against us and against every exposed surface. According to the kinetic theory, the pressure of a gas (or its expansive power as it is sometimes called), is entirely due to the striking of the molecules against the surfaces on which the gas is said to press, the impulses following each other in such rapid succession that the effect produced cannot be distinguished from constant pressure. Upon the kinetic energy of these blows, and upon the number of blows per second, must depend the amount of pressure. But we saw on page 141, that on the energy of the individual molecules depends that condition of a gas called its *temperature;* so, it is apparent, as stated above, that *the pressure of a given quantity of gas varies as its temperature.* Again, as at the same temperature the number of blows per second must depend upon the number of molecules in the unit of space, it is apparent that *the pressure varies as the density.*

The following estimates[1] made for hydrogen molecules at 0° C., and under a pressure of one atmosphere, may prove interesting : —

Mean velocity, 6100 feet per second.
Mean path without collision, 38 ten-millionths of an inch.
Collisions, 17,750 millions per second.
Mass, 216,000 million million million weigh 1 gram.
Number, 19 million million million fill 1 cubic centimeter.

§ 128. Diffusion of gases and liquids. — The kinetic theory of gases explains why gases penetrate into any spaces open to them, and likewise the phenomenon known as the *diffusion of gases* (see page 41). The presence of a gas in a given space only delays the spread of another gas in the same space by collision between the molecules of the inter-diffusing gases. The diffusion between liquids, though not so well understood, is undoubtedly due in part to similar molecular motions.

[1] Maxwell.

XXII. EFFECTS OF HEAT CONTINUED. — LIQUEFACTION AND VAPORIZATION.

Experiment 1. Melt separately tallow, lard, and beeswax. When partially melted, stir well with a thermometer, and ascertain the melting points of each of these substances.

Experiment 2. Place a test tube (Fig. 113), half filled with ether, in a beaker containing water at a temperature of 60° C. Although the temperature of the water is 40° below its boiling point, it very quickly raises the temperature of the ether sufficiently to cause it to boil violently. Introduce a chemical thermometer[1] into the test tube, and ascertain the boiling point of ether.

Fig. 113.

Experiment 3. Half fill a glass beaker of a liter capacity with fragments of ice or snow, and set the beaker into a basin of boiling-hot water. Stir the contents of the beaker with a thermometer until the ice is all melted, observing from time to time the temperature of the contents. The temperature remains constant at 0° C. until the ice is all melted.

Experiment 4. As soon as the last piece of ice disappears, remove the flask from the warm water, wipe the outside, and place it over a Bunsen burner and heat. Observe that the temperature rises constantly until the water begins to boil; but after it begins to boil, the temperature remains constant as long as it boils. Place more burners under the beaker; the water boils more violently, but the temperature is not raised.

Experiment 5. Place in contact the smooth, dry surfaces of two pieces of ice; press them together for a few seconds; remove the pressure, and they will be found firmly frozen together. The ice at the surfaces of contact melts under the pressure, but when the pressure is removed the liquid instantly freezes and cements the pieces together. It is in this manner that snow-balls are formed.

NOTE. — If a thermometer is placed in a mixture of ice and water, and the mixture is subjected to great pressure, some of the ice will melt and the temperature will fall; but when the pressure is removed, a portion of the water freezes and the temperature rises. From this we learn that *the melting (or freezing) point of water is very slightly lowered by pressure.* The depression is about $\frac{1}{135}$ of 1° C for each atmosphere. On the other hand, it is found that *substances which, unlike ice, expand in melting, have their melting points raised by pressure.*

---

[1] A chemical thermometer has its scale on the glass stem, instead of a metal plate, and is otherwise adapted to experimental use.

Experiment 6. Half fill a thin glass flask with water. Boil the water over a Bunsen burner; the steam will drive the air from the flask. Withdraw the burner, quickly cork the flask very tightly, and plunge the flask into cold water, or invert the flask and pour cold water upon the part containing steam, as in Figure 114; the water in the flask, though cooled several degrees below the usual boiling point, boils again violently. The application of cold water to the flask condenses some of the steam, and diminishes the tension of the rest, so that the pressure upon the water is diminished, and the water boils at a reduced temperature.

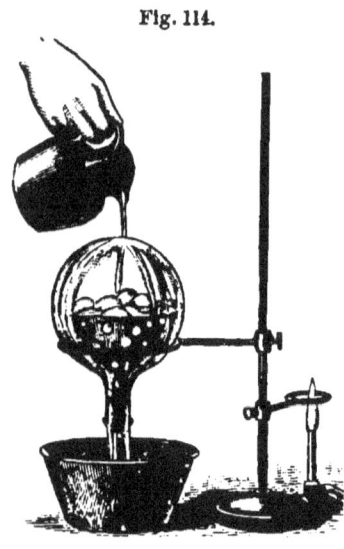

Fig. 114.

If hot water is poured upon the flask, the water ceases to boil. (Why?) Under the receiver of an air-pump, water may be made to boil at any temperature between 0° and 100° C.; indeed, if exhaustion is carried far enough, boiling and freezing may be going on at the same time. When high temperature is objectionable, apparatus is contrived for boiling and evaporating in a vacuum; as, for instance, in the vacuum pans used in sugar refineries. As water boils more easily under diminished pressure, so it boils with more difficulty when the pressure is increased; and the temperature to which water may be raised under the pressure of its own steam is only limited by the strength of the vessel containing it. Vessels of this kind are often employed to effect a complete penetration of water into solid and hard substances. By this means gelatine is extracted from the interior of bones. In the boiler of a locomotive, where the pressure is sometimes 150 lbs. above the atmosphere, the boiling point rises to about 180° C. (360° F.).

Experiment 7. Dissolve table-salt in water, and you may raise its boiling point till it reaches 108° C. With saltpetre it may reach 115° C.

LAWS OF FUSION AND BOILING. 161

On the other hand, it is well known that sea-water, which contains saline matter in solution, freezes at a lower temperature than 0° C. From the above experiments, and others of a similar nature, we derive the following

LAWS OF FUSION AND BOILING.

1. *The temperature at which solids melt differs for different substances, but is invariable for the same substance, if the pressure is constant. Substances solidify usually at the same temperatures as those at which they melt.*

2. *After a solid begins to melt, the temperature remains constant until the whole is melted.*

3. *Pressure lowers the melting (or solidifying) point of substances that expand on solidifying, and raises the melting point of those that contract.*

4. *The freezing point of water is lowered by the presence of salts in solution.*

1. *The temperature at which liquids boil differs for different substances, but is invariable for the same substance if the pressure is constant. Vapors liquefy at the same temperatures as those at which they boil.*

2. *After a liquid begins to boil the temperature remains constant until the whole is vaporized.*

3. *Pressure raises the boiling point of all substances.*

4. *The boiling point of water is raised by the presence of salts in solution.*

REFERENCE TABLES.

Melting Points.

Alcohol............Never frozen.
Mercury ....— 38.8° C.
Sulphuric acid ....— 34.4°
Ice.................... 0°
Phosphorus............ 44°
Sulphur .... 115°
Tin ....about 233°
Lead............. " 334°

Zinc............about.... 425° C.
Silver .... " ....1000°
Gold ....,..... " ....1200°
Cast-iron .... " 1050–1250°
Wrought-iron .... " 1500–1600°
Iridium (the most infusible metal)
about................1950°

Boiling Points under a Pressure of one Atmosphere.

Carbonic acid ....— 78° C.
Ammonia................— 40°
Sulphurous acid..........— 10°
Ether .... 35°

Carbon bisulphide........ 48° C.
Alcohol.................. 78°
Water .... 100°
Mercury .... 350°

MOLECULAR ENERGY. — HEAT.

Boiling Points of Water at Different Pressures.

Barometer.		Atmospheres.	
184° F	16.68 inches.	212° F	1
190°	18.99 "	249.5°	2
200°	23.45 "	273.3°	3
210°	28.74 "	306°	5
212°	29.92 "	356.6°	10

The temperature of the boiling point of water varies with the altitude of places, in consequence of the different atmospheric pressure. A difference of altitude of 533 ft. causes a variation of 1° F. in the boiling point.

Boiling Points of Water at Different Altitudes.

	Above the sea-level.	Mean hight of Barometer.	Temperature.
Quito	+ 9,500 ft.	21.53 in.	195.8° F.
Mont Blanc	15,650 "	16.90 "	186°
Mt. Washington	6,290 "	22.90 "	200°
Boston	0 "	30. "	212°
Dead Sea (below)	− 1,316 "	31.50 "	214°

§ **129. Distillation.** — Apparatus like that represented in Figure 115 may be easily constructed. The following experiment will be found interesting and instructive.

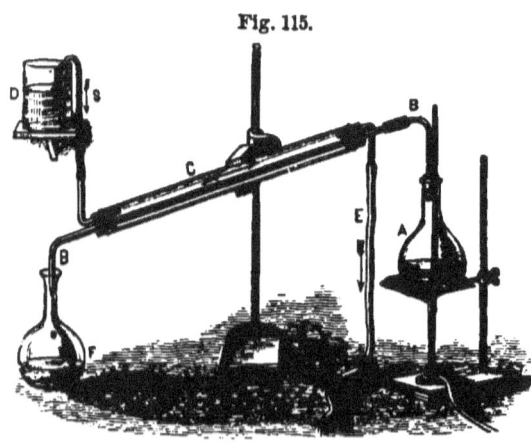

Fig. 115.

Experiment. Half fill the flask A with water colored with a few drops of ink. Boil the water, and the steam arising will escape through the glass delivery tube BB. This tube is surrounded in part by a larger tube C, called a *condenser*, which is kept filled with cold water flowing from a vessel D through a siphon S, the water finally escaping through the tube E.

EVAPORATION. 163

The steam is condensed in its passage through the delivery tube, and the resulting liquid is caught in the vessel F. The liquid caught is colorless. A complete separation of the watery portion of the colored liquid from the other ingredients of the ink is effected, the latter being left in the flask A.

The separation is accomplished on the principle that the temperature of the boiling points of different substances differ. The water is raised to its vaporizing point, but the other substances are not. The apparatus is called a *still*, and the operation *distillation*.

If a volatile liquid, such as alcohol, is to be separated from water, the mixture is heated to the temperature at which the volatile liquid boils, but not to the boiling point of water, when the alcohol will pass into the vessel F, and the water, for the most part, will remain in the flask.

§ **130. Evaporation.** — In boiling, the heat, usually applied at the bottom, rapidly converts the liquid into vapor, which, rising in bubbles and breaking at or near the surface, produces a violent agitation in the liquid, sometimes called *ebullition*. *Evaporation* is that form of vaporization which takes place quietly and slowly at the surface. The phenomena and laws of vaporization of all liquids are similar, but we will study only the important case of water. Although hastened by heat, the evaporation of water occurs at any temperature, however low; even ice and snow evaporate.

The rapidity of evaporation varies directly with the temperature, amount of surface exposed, and dryness of the atmosphere, and inversely with the pressure upon the liquid. This vapor of water mixes freely with the air, and diffuses readily through it, acting like another gas (compare pages 42 and 158). The air does not take up water like a sponge, as is commonly imagined; for, if the air could be removed from a room, where there is a large vessel of water, every cubic foot of the space in the room would be found to contain just as much water-vapor as it does when the air is present, — probably a very little more. In either case, only a definite quantity would be found in each cubic foot, a quantity depending on the temperature of the space. Thus, at 0° C., each cubic foot can contain 0.14^g; at 10°, 0.26^g; at 20°, 0.49^g; and at 30°, 0.85^g. Evidently the capacity is nearly doubled by a rise of 10° in temperature.

§ 131. **Dew point.** — When a space contains such an amount of water-vapor, whether it contains other gases or not, that its temperature cannot be lowered without some of the water being precipitated in the form of a liquid, the space is said to be *saturated*, and the temperature is called the *dew point*. The form in which the condensed vapor appears is, according to its location, *dew*, *fog*, or *cloud*. The atmosphere is said to be *dry* or *humid*, according as the difference between the dew point and the temperature of the atmosphere is great or little.

QUESTIONS.

1. Why does our breath produce a cloud in winter and not in summer?

2. (*a*) If air at 0° is warmed to 20° C., how will its dryness be affected? (*b*) What effect would such warmed air have on wet clothes?

3. If saturated air at 20° is blown into a cellar where the temperature is 10°, what will happen?

4. What is the cause of the general complaint of dryness of air in rooms heated by stoves or furnaces?

5. Does a given mass of air in such a room contain less water-vapor than an equal mass of cold out-door air at the same time?

XXIII. HEAT CONVERTIBLE INTO POTENTIAL ENERGY, AND VICE VERSA.

§ 132. Heat units. — It is frequently necessary to measure quantity of heat, and for this purpose a standard unit of measurement is required. The heat unit generally adopted is *the amount of heat required to raise the temperature of one kilogram of water from $0°$ to $1°$ C.* This unit is called a *calorie*.

Let it be required to find the amount of heat that disappears (Exp. 3, p. 159) during the melting of one kilogram of ice.

Experiment 1. Place 1^k of ice at $0°$ C. in a beaker, and the beaker in a large basin of boiling water (Fig. 116), and at the same instant place in the hot water another beaker containing 1^k of water at $0°$ C. Place in each beaker a thermometer, and at the instant that the ice disappears note the temperature of the water in each; it will be found that while the temperature of the former has not changed, the latter has risen to about $80°$ C.

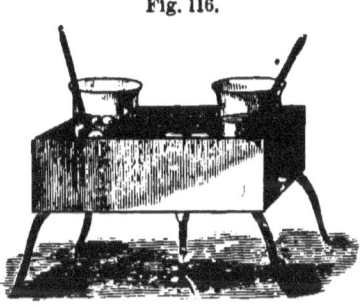

Fig. 116.

It is evident that the contents of both beakers must have received the same amount of heat; hence, the amount of heat received by the water being 80 calories, *the amount of heat that disappears or is lost during the melting of one kilogram of ice is 80 calories.*

Next, let it be required to find the amount of heat that disappears (Exp. 4, p. 159) during the conversion of 1^k of water into steam.

Experiment 2. Place 1^k of water at $0°$ C. in a beaker, and heat the same with a Bunsen burner. Note the time that it takes to raise the water from $0°$ C. to $100°$ C., also the time during which the temperature of the water remains stationary while the water is boiling away. The latter time will be found to be about five times the former.

Now, as the water receives 100 calories during the time it is rising from the freezing to the boiling point, it must receive about 500 calories during the time it is converted into steam : but the temperature of the water is not changed during the latter operation. More accurate methods have the number 537 ; so it follows that *537 calories disappear, or are lost during the conversion of 1 kilogram of water into steam.*

§ **133. Two questions answered.** — Inasmuch as none of the heat applied during the melting of ice and the conversion of water into steam raises the temperature of the body to which it is applied, the question arises, *What does the heat do?* Again, *Why is not ice instantly converted into water on reaching the melting point, and water instantly converted into steam on reaching the boiling point?*

The answer to the first question is, All of the heat applied in melting ice is consumed in doing *interior work*, as it is called. The molecules that were firmly held in their places by molecular forces are now moved from their places, and so work is done against these forces, just as work is done against gravity when a weight is lifted. In the conversion of water into steam, a similar action goes on ; the heat is expended in separating the molecules so far that the molecular attractive forces are no longer sensible, all except the small fraction used in overcoming atmospheric pressure.[1] Heat, the energy of motion, in both instances does important work, and is thereby converted into the energy of position, or potential energy, — energy of the same kind as that of a raised weight.

The answer to the second question is, The amount of work done in both instances is great, as shown by the amount of heat consumed in doing the work ; 80 calories per kilogram of ice being required in the first instance, and 537 calories per kilogram of water in the second ; hence it requires a long time to acquire the requisite amount of heat. It is fortunate that it takes a large quantity of heat to melt ice ; otherwise, on a single

---

[1] This fraction is about $\frac{1}{15}$.

warm day in winter, all the ice and snow would melt, creating most destructive freshets. The heat which disappears in melting and boiling is generally, but with our present knowledge of the subject, rather objectionably, called *latent* (hidden) *heat*. The error consists in calling that heat which has ceased to be heat, *i.e.*, has ceased to be molecular motion.

§ 134. **Methods of producing artificial cold.** — The fact that heat must be consumed because work is done, in the conversion of solids into liquids and liquids into vapors, and in the simple expansion of gases, is turned to practical use in many ways for the purpose of producing *artificial cold.* They are embraced under three heads; viz., *Cold produced by solution, by evaporation, and by expansion of gases.* The following experiments will illustrate.

§ 135. Cold by solution. — **Freezing mixtures.** — Experiment. Prepare a mixture of 2 parts by weight of pulverized ammonium nitrate and 1 part of ammonium chloride, and dissolve in 3 parts of water (not warmer than 10° C.), stirring the same while dissolving with a test-tube containing a little water. The water in the test tube will be quickly frozen. A finger placed in the solution will feel a painful sensation of cold, and a thermometer will indicate a temperature of about —10° C.

Fig. 117.

One of the most common freezing mixtures consists of 3 parts of snow or broken ice and 1 part of common salt. The affinity of salt for water causes a liquefaction of the ice, and the resulting liquid dissolves the salt, both operations requiring heat.

§ 136. Cold by evaporation. — Experiment 1. Fill the palm of the hand with ether; the ether quickly evaporates and produces a painful sensation of cold.

Experiment 2. Place water at about 10° C. in a thin porous cup, such as is used in the Grove's battery (see page 190), and introduce the bulb of a thermometer; although the experiment be conducted in a warm room, the large surface exposed by means of the porous vessel will so hasten evaporation that in the course of fifteen minutes there will be a very sensible fall in temperature.

Experiment 3. Cover closely the bulb of an air thermometer (Fig. 117) with thin muslin, and partly fill the stem with water. Let one person slowly drop ether on the bulb while another briskly blows the air charged with vapor away from the bulb with a bellows. (Why?) The water in the stem will quickly freeze even in a warm room.

QUESTIONS.

1. Why do we bathe the fevered forehead with alcohol and water?
2. How does perspiration contribute to our comfort?
3. Why do we fan ourselves?
4. Why does a windy day seem colder to us than a still day, although the temperature is the same on both days?
5. Why do we blow our hot tea, and why pour it into a saucer?
6. How does sprinkling a floor cool the air of a room?

§ **137. Cold by expansion of gases.** — When a beer bottle is opened, a fog is suddenly produced in the neck of the bottle due to the chill of an expanding gas.

The work done in the expansion of a gas consists only in forcing back the surrounding air. If confined air is allowed to expand into a vacuum, no work is done, and the temperature is not changed. By allowing condensed air containing, as it usually does, watery vapor, to escape suddenly from the vessel in which it is confined, icicles have been formed around the orifice whence it escapes.

§ **138. Potential energy converted into heat by the solidification of liquids and the liquefaction of vapors.**—
Experiment 1. Boil about ½ liter of water in a glass flask, and add, slowly, pulverized sodium sulphate until the boiling water refuses to dissolve more (hot water will dissolve about twice its weight of this substance). Then set the hot solution in a place where it will not be disturbed, and let it stand for about 24 hours, that it may acquire the temperature of the room. Thrust the bulb of a thermometer into the solution,[1] and at the same time drop in a lump of sodium sulphate;

---
[1] The solution is now said to be *supersaturated*.

solidification instantly sets in, and in a few seconds the liquid mass will be almost wholly replaced by crystals. At the same time the temperature, as indicated by the thermometer, rapidly rises.

The heat which is consumed in dissolving a solid, and in giving the molecules an advantage of position, is restored when the molecules are allowed to resume their original positions, as a falling weight restores the kinetic energy consumed in raising it.

Experiment 2. Place water at about 10° C. in a bottle, and introduce a thermometer. Surround the bottle with a snow and salt freezing mixture; the temperature of the water rapidly falls until it reaches 0° C.

The heat which the water loses is consumed in melting the ice and dissolving the salt. At 0° C. the water begins to freeze, and the temperature remains stationary until all the water is frozen, when its temperature again falls. The temperature of the freezing mixture is much lower than that of the water while freezing; the latter, then, must give heat to the former. That the mixture receives heat is shown by the continuation of the melting and dissolving. But as the temperature of the water while freezing does not fall, it must be that the heat which it surrenders during solidification arises from the conversion into heat of the potential energy possessed by the molecules of the liquid.

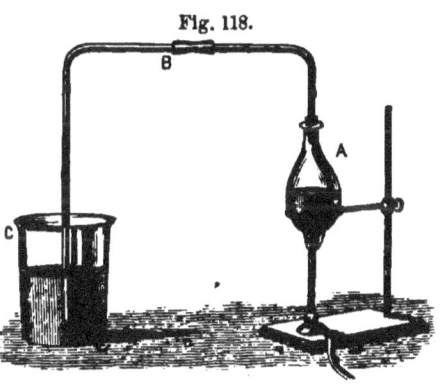

Fig. 118.

Experiment 3. Arrange apparatus as in Figure 118. When water in the flask A begins to boil, introduce the end of the delivery tube B into a vessel C of water at 0° C. The steam that passes through the tube is condensed on entering the cold water and heats the water. When a considerable portion of the water has boiled away, weigh the water remaining in A, and ascertain the quantity that has been converted into steam; also ascertain the temperature of the water in C, and the number of calories which it has received.

For every kilogram of water that is converted into steam, 5.37^k of water (practically, considerably less than this quantity, in consequence of loss of heat by radiation and evaporation from C) will be raised from 0° to 100°. As 1^k requires 100 units of heat to raise it to 100°, the 5.37^k must require 537 units of heat. But the steam raises the water to its own temperature without having its own temperature lowered. (Whence come the 537 units of heat that raise the temperature of the water?)

Heat that is consumed in liquefying solids, and in vaporizing liquids, is always restored when the reverse change takes place. Farmers well understand that water, in freezing, gives out a great deal of heat, — at a low temperature, it is true, but still high enough to protect vegetables which freeze only when considerably colder than melting ice. The fact that steam, in condensing, generates a large amount of heat, is turned to practical use in heating buildings by steam.

XXIV. SPECIFIC HEAT.

§ **139. Temperatures of different substances raised unequally by equal quantities of heat.** — Will equal quantities of heat applied to equal weights of different substances raise their temperatures equally?

Experiment 1. Mix 1^k of water at 0° with 1^k at 20°; the temperature of the mixture becomes 10°. The heat that leaves 1^k of water when it falls from 20° to 10° is just capable of raising 1^k of water from 0° to 10°.

Experiment 2. Take (say) 300^g of sheet lead, and make a loose roll of it, and suspend it by a thread in boiling water for about five minutes, that it may acquire the same temperature (100° C.) as the water. Remove the roll from the hot water, and immerse it as quickly as possible in 300^g of water at 0°, and introduce the bulb of a thermometer. Note the temperature of the water when it ceases to rise, which will be found to be about 3° (accurately 3.3°+). The lead cools

SPECIFIC HEAT DEFINED. 171

very much more than the water warms. Lead falls about 33° for every degree an equal weight of water is warmed.

From the first experiment we infer that a body, in cooling a certain number of degrees, gives to surrounding bodies as much heat as it takes to raise its temperature the same number of degrees. From the second experiment we learn that the quantity of heat that raises 1^k of lead from 3.3°+ to 100°, when transferred to water, can raise 1^k of water only from 0° to 3.3°. Hence we conclude that *equal quantities of heat, applied to equal weights of different substances, raise their temperatures unequally.*

§ 140. Capacity for heat. — If equal weights of mercury, alcohol, and water are exposed to the same heat, the mercury will rise 30°, and the alcohol nearly 2°, while the water is rising 1°. From this we infer that to raise a kilogram of each of these substances from 0° to 1° requires 30 times as much heat for the water as for the mercury, and twice as much as for the alcohol. Since heat affects the temperature of water less than mercury and alcohol, the first is said to have a *greater capacity for heat*. *The number of units of heat required to raise the temperature of a body 1°C., is called its capacity for heat.*

§ 141. Specific heat defined. — It is a great convenience to be able to compare the capacities of different substances for heat. The standard employed is water, and the ratio which expresses the comparison is called *specific heat*.

The specific heat of a body is the ratio of its capacity for heat to that of an equal weight of water.

From the data obtained in the last experiment we may calculate the specific heat of lead as follows: The same quantity of heat that raises the water 3.3° (from 0° to 3.3°) raises the lead 96.70° (from 3.3° to 100°); hence, to raise the lead 1° requires $\frac{3.3}{96.7} = .034+$ as much heat as to raise the water 1°.

The specific heat of all solids and liquids, and most gases, increases slightly with the temperature. Thus water at 0° C. has a specific heat of 1; at 40°, 1.0013; at 80°, 1.0035. Substances in the liquid state usually have a higher specific heat than in the solid or gaseous state. Thus water has nearly double the specific heat of ice, and a little more than double the specific heat of steam.

REFERENCE TABLES.

Table of mean specific heat between 0° C. and 100° C.

Hydrogen	3.4090	Iron	.1138
Air	.2375	Copper	.0952
Sulphur	.2026	Mercury	.0333
Glass	.1770	Lead	.0314

Specific heat of the same substance in different states.

	Solid.	Liquid.	Gaseous.
Water	.5040	1.0000	.4805
Bromine	.0833	.1060	.0555
Lead	.0314	.0402	
Alcohol		.55–.77.	.45

§ 142. One cause of difference in capacity for heat. — Of the whole quantity of heat applied to a solid or liquid body, only a part goes to increase the heat of the body, and thereby to raise its temperature; the remainder performs *interior work*, in overcoming cohesion between the molecules of the body, and in forcing them to take up new positions. (Since, then, some of the heat is converted into potential energy, we may properly introduce the subject of specific heat at this place.) The greater the portion of heat consumed in interior work upon a body, the less there is left to raise its temperature, and consequently the greater its capacity for heat. Thus, when equal quantities of heat are applied to equal masses of water and lead, more is consumed (*i.e.*, converted in potential energy) in interior work upon the water than upon the lead; consequently the temperature of the former is not raised as much as that of the latter. The limits of this work forbid the discussion of the causes of the difference of capacity for heat of different gases.

§ 143. **Great capacity of water for heat.** — Water requires more heat to warm it, and gives out more in cooling through a given range of temperature, than any substance except hydrogen. The quantity of heat that raises a kilogram of water from 0° to 100° C. would raise a kilogram of iron from 0° to 800° or 900° C., or above a red heat: Conversely, a kilogram of water in cooling from 100° to 0° C. gives out as much heat as a kilogram of iron in cooling from about 900° to 0° C.

QUESTIONS AND PROBLEMS.

1. How much heat is required to change 100^k of ice at 0° into steam at 100° C.?

2. (a) 1000^k of steam at 100° C. is conveyed by pipes through a building, and the water resulting from its condensation returns to the boiler at a temperature of 80°; how much heat is given out in the building. (b) The same quantity of heat would raise how many kilograms of water from 0° to 100°?

3. 50^k of water at 100° will melt how many pounds of ice at 0° C.?

4. How much heat is required to raise 1^k of ice from $-10°$ to 10° C.?

5. (a) Apply the same quantity of heat to equal weights of ice and water, each at a temperature of 0° C.; when the latter reaches the boiling point what will be the temperature of the former? (b) Why will not both have the same temperature?

6. What effect on the temperature of the air has the freezing of the water of lakes and other bodies of water?

7. If 1^k of iron at 100° is immersed in 1^k of water at 0° C., what will be the resulting temperature?

8. What is the specific heat of a substance, 1^k of which at 100°, when put into 1^k of water, at 0° raises its temperature to 5° C.?

9. 50^k of mercury at 80° will melt what weight of ice at 0° C.?

10. Why is hot *water* in bottles often used to warm beds in preference to other substances?

11. If there were no water on the earth, why would the difference in temperature between day and night, and between summer and winter, far exceed what it is now?

12. Why are places in vicinity of water less subject to extremes of heat and cold than places inland?

XXV. THERMO-DYNAMICS.

§ 144. Thermo-Dynamics defined. — *Thermo-dynamics is that branch of science that treats of the relation between heat and mechanical work.* One of the most important discoveries in science is that of the *equivalence of heat and work;* that is, that *a definite quantity of mechanical work can always produce a definite quantity of heat; and conversely, this heat, if the conversion were complete, can perform the original quantity of work.*

§ 145. Correlation and conservation of energy. — The proof of the facts just stated was one of the most important steps in the establishment of the grand twin conceptions of modern science. (1) That *all kinds of energy are so related to one another that energy of any kind can be changed into energy of any other kind,* — known as the doctrine of CORRELATION OF ENERGY; (2) That *when one form of energy disappears, an exact equivalent of another form always takes its place, so that the sum total of energy is unchanged,* — known as the doctrine of CONSERVATION OF ENERGY. These two principles constitute the corner-stone of physical science.

§ 146. Joule's experiment. — The experiment to ascertain the "mechanical value of heat," as performed by Dr. Joule of England, was conducted about as follows. He caused a paddle-wheel to revolve in water by means of a falling weight attached to a cord wound around the axle of a wheel. The resistance offered by the water to the motion of the paddles was the means by which the mechanical motion of the weight was converted into heat, which raised the temperature of the water. Taking a body of a known weight, *e.g.*, 80^k, he raised it a measured distance, *e.g.*, 53^m high; by so doing 4240^{kgm} of work were performed upon it, and consequently an equivalent amount of energy was stored up in it ready to be converted, first, into mechanical motion, then into heat. He took a definite weight of water to be agitated, *e.g.*, 2^k, at a temperature of $0°$ C. After the descent of the weight, the water was found to have a temperature of $5°$ C. ; consequently the 2^k of water must have received 10 units of heat (careful allowance being made for all losses of heat), which is the amount of heat-energy that is equivalent to 4240^{kgm} of work, *or 1 unit of heat is equivalent to 424^{kgm} of work* (more accurately 423.985^{kgm}).

§ 147. Mechanical equivalent of heat. — As a converse of the above it may be demonstrated by actual experiment that the quantity of heat required to raise 1^k of water from $0°$ to $1°$ C. will, if converted into work, raise a 424^k weight 1^m high, or 1^k weight 424^m high. According to the English system, the same fact is stated as follows: The quantity of heat that will raise 1 lb. of water $1°$ F. will raise 772 lbs. 1 ft. high. The quantity, 424^{kgm}, or 772 ft. lbs., is called the *mechanical equivalent of heat, or Joule's equivalent* (abbreviated, simply J.).

XXVI. STEAM ENGINE.

§ 148. Description of a steam engine. — A steam engine is a machine in which the elastic force of steam is the motive power. Inasmuch as the elastic force of steam is entirely due to heat, *the steam engine is properly one form of a heat engine;* that is, it is a machine by means of which heat is continuously transformed into work or mechanical motion.

The modern steam engine consists essentially of an arrangement by which steam from a boiler is conducted to both sides of a piston alternately ; and then, having done its work in driving the piston to or fro, is discharged from both sides alternately, either into the air or into a condenser. The diagram in Figure 119 will serve to illustrate the general features and the operation of a steam engine. The details of the various mechanical contrivances are purposely omitted, so as to present the engine as nearly as possible in its simplicity.

In the diagram, B represents the *boiler*, F the *furnace*, S the *steam pipe* through which steam passes from the boiler to a small chamber VC, called the *valve chest*. In this chamber is a *slide valve* V, which, as it is moved to and fro, opens and closes alternately the passages M and N leading from the valve chest to the *cylinder* C, and thus admits the steam alternately each side of the *piston* P. When one of these passages is open the other is always closed. Though the passage between the valve chest and the space in the cylinder on one side of

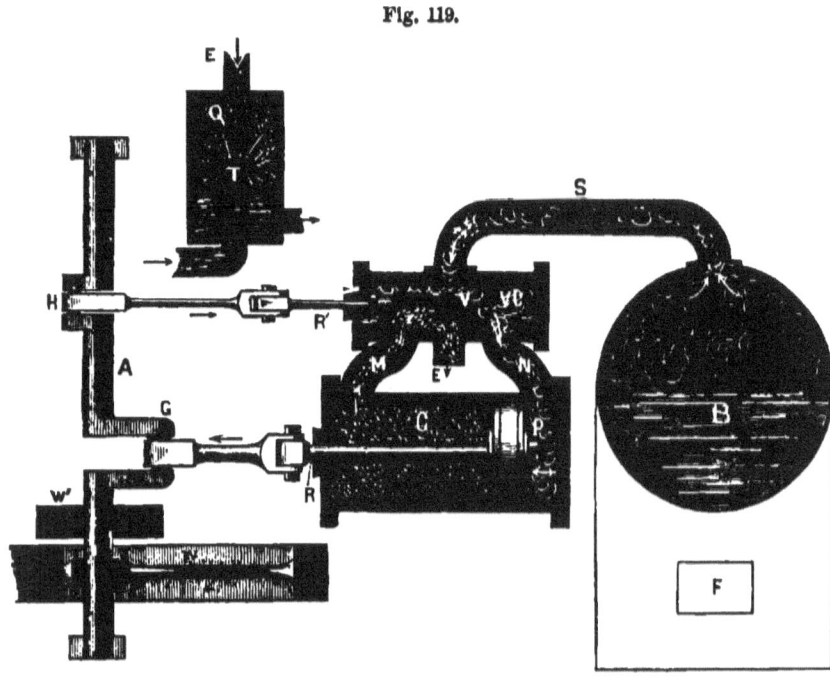

Fig. 119.

the piston is closed, thereby preventing the entrance of steam into this space, the passage leading from the same space is open through the interior of the valve so that steam can escape from this space through the *exhaust pipe* E. Thus, in the position of the valve represented in the diagram, the passage N is open, and steam entering the cylinder at the top drives the piston in the direction indicated by the arrow. At the same time the steam on the other side of the piston escapes through the passage M and the exhaust pipe E. While the piston moves to the left, the valve moves to the right, and eventually closes the passage

N leading from the valve chest, and opens the passage M into the same, and thus the order of things is reversed.

Motion is communicated by the piston through the *piston rod* R to the *crank* G, and by this means the *shaft* A is rotated. Connected with the shaft by means of the crank H, is a rod R' which connects with the valve V, so that as the shaft rotates, the valve is made to slide to and fro, and always in the opposite direction to that of the motion of the piston.

The shaft carries a *fly-wheel* W. This is a large, heavy wheel, having the larger portion of its weight located near its circumference; it serves as a reservoir of energy which is needed to carry the shaft past two points (called the *dead points*) in each revolution of the shaft, where the power communicated directly by the steam is ineffectual in moving the shaft. It also assists to make the rotation of the shaft and all other machinery connected with it uniform, so that sudden changes of velocity resulting from sudden changes of the driving power or resistances are avoided. (Why should the wheel be heavy? Why should it be large? Why should the rim be heavy? See p. 102.) By means of a belt passing over the wheel W' motion may be communicated from the shaft to any machinery desirable.

§ 149. Condensing and non-condensing engines.[1] — Sometimes steam, after it has done its work in the cylinder, is conducted through the exhaust pipe to a chamber Q called a *condenser*, where, by means of a spray of cold water introduced through a pipe T, it is suddenly condensed. This water and the condensed steam must be pumped out of the condenser by a special pump called technically the *air-pump;* thus a partial vacuum is maintained. Such an engine is called a *condensing engine*. The advantage of such an engine is obvious, for, if the exhaust pipe, instead of opening into a condenser, communicates with the outside air as in the *non-condensing engine*, the steam is obliged to move the piston constantly against a resistance arising from atmospheric pressure of 15 pounds for every square inch of the surface of the piston. But in the condensing engine no resistance arises from atmospheric pressure, and so with a given

---

[1] The terms, *low pressure* and *high pressure* engines, are not distinctive as applied to engines of the present day.

steam pressure in the boiler the effective pressure on the piston is considerably increased; hence, condensing engines are usually more economical in their working.

§ **150. The locomotive.**—The distinctive feature of the locomotive engine is its great steam-generating capacity, considering its size and weight, which are necessarily limited. To do the work ordinarily required of it, from three to six tons of water must be converted into steam per hour. This is accomplished in two ways: viz., first, by a rapid combustion of fuel (from a quarter of a ton to a ton of coal per hour); second, by bringing the water in contact with a large extent (about 800 sq. ft.) of heated surface. The fire in the "fire-box" A (see cut on the opposite page) is made to burn briskly by means of a powerful draft which is created in the following manner: The exhaust steam, after it has done its work in the cylinders B, is conducted by the exhaust pipe C to the smoke box D, just beneath the smoke stack E. The steam as it escapes from the blast pipe F pushes the air above it, and drags by friction the air around it, and thus produces a partial vacuum in the smoke box. The external pressure of the atmosphere then forces the air through the furnace grate and hot-air tubes G, and thus causes a constant draft. The large extent of heated surface is secured as follows: The water of the boiler is brought not only in contact with the heated surface of the fire box, but it surrounds the pipes G (a boiler usually contains about 150). These pipes are kept hot by the heated gases and smoke, all of which must pass through them to the smoke box and smoke stack.

Study the cut carefully, trace the course of the steam from the boiler H through the throttle valve I (under the control of the engineer), steam pipe J, etc., to its exit from the smoke stack. Ask some engineer to explain from the object the offices of such parts as you do not understand.

The steam engine, with all its merits and with all the improvements which modern mechanical art has devised, is to-day an exceedingly wasteful machine. The best engine that has been constructed utilizes only twenty per cent of the heat-power used.

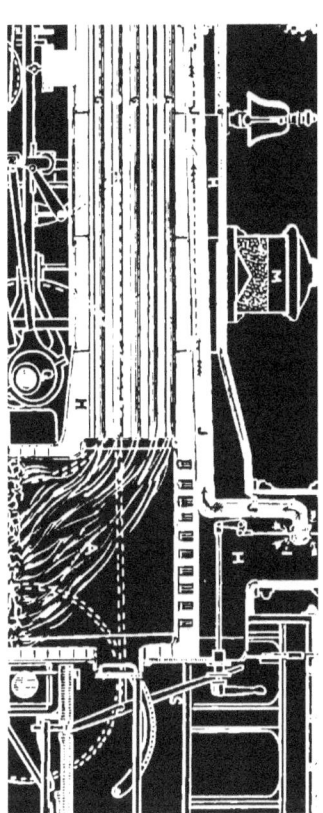

CHAPTER IV.

ELECTRICITY AND MAGNETISM.

THERE is a large and important class of phenomena depending on new principles that we have now to study; among these are lightning, the actions of telegraph instruments, the electric light, magnetic attraction and repulsion, etc. We shall inquire whether energy is involved in these actions as in all those we have so far studied; and, if so, where it comes from; and under what laws it acts, and what finally becomes of it.

If we chose to begin with those experiments easiest to perform, we should take those with magnets, and some of those to be studied under the head of Frictional Electricity; but we should find it difficult to see clearly how the subject of energy was to be introduced. So let us take first some experiments that will lead us more easily to this great central idea.

XXVII. CURRENT ELECTRICITY.

§ **151. Introductory experiments.** — **Experiment 1.** Take a strip of sheet copper and a strip of sheet zinc, each about 10^{cm} long and 4^{cm} wide. Take also a tumbler two-thirds full of water, and to it add about two tablespoonfuls of sulphuric acid. Place the zinc strip in the liquid; instantly bubbles of gas collect on the surface of the zinc, break away from it, rise to the surface of the liquid, and are rapidly replaced by others. These are bubbles of hydrogen gas, and may be collected and burned. It is soon found that the zinc wastes away, or is dissolved in the liquid.

Fig. 120.

Experiment 2. Place the copper strip in the liquid a little way from the zinc, but nowhere touching it; no bubbles are formed. Now bring the extremities of the two strips that project from the liquid into contact, as in Figure 120; quickly a change takes place;

torrents of bubbles now rise from the copper, and only a very few from the zinc; still it is found, after a lapse of time, that the copper has undergone no change, while the zinc has wasted away.

Experiment 3. Withdraw the zinc from the liquid, and while it is yet wet rub a little mercury over its surface, so that it may become completely wet with the liquid metal. Now repeat the above experiments in order. First, it is found that the zinc, when alone in the liquid, is not affected by it, and no bubbles of gas are formed. But when the two metals are immersed in the liquid, and are brought into contact, bubbles of gas quickly appear on the copper as before, but none appear on the zinc, although the zinc is still the metal that wastes away, while the copper remains unchanged.

Experiment 4. Instead of placing the metals in contact, connect them by means of a wire of any metal, the points of contact being clean; the bubbles are given off at the copper as before. Cut the connecting wire at any point, or separate it from the zinc or copper; all evolution of bubbles ceases, but begins again the instant the contact is made.

Experiment 5. Interpose between the connecting wire and the plates, or between the cut ends of the wire, a piece of paper, wood, or rubber, or use some one of these, instead of a wire, to connect the two plates; no action appears in the cell.

Thus it appears that there must be a connection, and that too of a particular kind, between the two metals, in order that action may occur. The connecting wire, then, is an important factor in the changes that occur, and it seems altogether probable that some influence is exerted by the metals upon one another through the wire; in other words, that something unusual is going on in the wire when so used.

Does the connecting wire possess any unusual properties during this use?

Experiment 6. Take an ordinary compass, or poise a magnetic needle at its center, either by a pivot, as in Figure 121, or by a fine, untwisted silk thread, and arrange the connecting wires as in the figure. The needle, when at rest, points north and south. The connecting wire being over the needle, and parallel with it, bring the two extremities of the wire into contact; instantly the needle turns on its axis, tending to place itself at right angles to the wire, and, after a few

INTRODUCTORY EXPERIMENTS. 181

vibrations, takes up a permanent position, forming an angle with the wire. This deviation from its normal position is called a *deflection of the needle.* Separate the two extremities of the wires; the needle swings back to its normal position.

Experiment 7. Bring the ends of the wires together as before, interposing a piece of paper between them; the needle is not moved. This is another illustration of the necessity of employing a suitable substance for a connector in order that any action may take place.

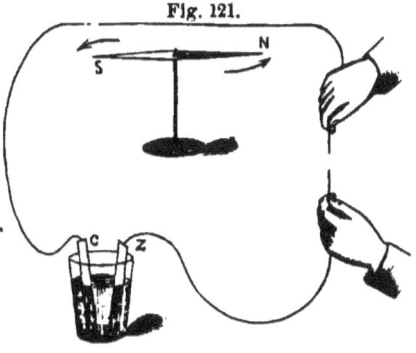

Fig. 121.

Experiment 8. Take a large iron nail, and plunge one end of it into iron filings, and then remove it; no filings cling to the nail. Next, wrap a piece of paper around the nail, leaving the ends exposed, and wind around it 20 or more turns of copper wire, taking pains that the coils do not touch each other. Now connect the wire with the zinc and copper just used, so that there will be a continuous connection from one strip to the other through the coil, and dip one end of the nail again into the filings; raise the nail, and a considerable quantity of filings cling to the nail.

From these experiments, and others which will be performed later, it appears that when the zinc and copper are thus placed in acid and connected by a wire, the wire exhibits unusual properties. The cause of these and many other allied phenomena is called *electricity*, and these properties in the wire are attributed to the passage of an *electric current* through it.

Almost from the dawn of the science of electricity there have been many who have believed in the existence of an "electric fluid"; but it is not yet claimed that there is any *positive* proof of its existence, and therefore we cannot *affirm* that a current passes through the wire. Yet the theory upon which these terms are based is at least a convenient one by which to explain the various phenomena, and the terms are therefore universally used.

§ 152. Some definitions. — Experiments (not easily performed by the pupil) show that the current traverses the liquid between the metallic plates in the battery at the same time that it traverses the connecting wire, so that the current makes a complete circuit. The term *circuit* is applied to the entire path along which electricity is supposed to flow, and the wire along which it flows is called the *conductor*. Bringing the two extremities of the wires in contact, and separating them, is called, technically, *making and breaking, or closing and opening, the circuit*.

Our arrangement of acidulated water and two metals is called a *voltaic*[1] *cell*, *element*, or *pair*. A series of cells, properly connected, is called a *battery*, though this term is sometimes applied to a single cell.

§ 153. Direction of the current. — It is evidently necessary, in defining a current, to know its direction; but as no

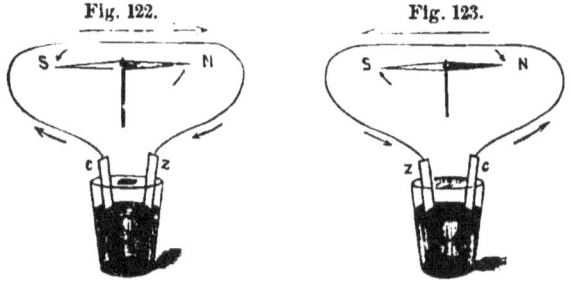

Fig. 122. Fig. 123.

phenomena known serve to indicate the direction, electricians have universally agreed to assume that in such a cell as described the electricity flows *from the copper to the zinc in the wire*.

Experiment. Place the conducting wire over and parallel with a magnetic needle, in the manner represented in Figure 122; the north end of the needle is deflected toward the west. Turn the cell half-way around so as to have the position in Figure 123; a deflection of the needle toward the east shows that the current is reversed.

[1] Voltaic, *from Volta*, an Italian, who devised the voltaic pile, which is the *parent* of all batteries.

§ **154. Poles or electrodes.** — The copper strip is frequently called the *negative plate*, and the zinc strip the *positive plate*, and the end of any conductor connected with the copper or negative plate is called the *positive pole*, or *electrode*, while the end connected with the zinc or positive plate is called the *negative pole*, or *electrode*. Then, by our assumption, if we bring together the + and − electrodes, the current passes from the former to the latter, across the junction; and generally that plate and that electrode is + *from* which the current goes, and that plate and that electrode is − *to* which the current goes.

§ **155. Potential.** — If a current of water is to flow from one vessel A to another B through a pipe, we know that there must be a greater pressure at the end of the pipe next A than at the other end; *i.e.*, in ordinary language, the *head* of water in A is higher than in B. So in the study of electricity we find two bodies in different conditions such that a current of electricity flows from one (A) to the other (B), and we say that A has a higher *potential* than B. In the experiments already tried the +electrode, or the wire connected with the copper, has a higher potential (according to our assumption for the direction of the current) than the −electrode or the wire connected with the zinc.

It is not necessary that we know the hight from the center of the earth, or above the level of the sea, of a reservoir, and the tank it is to fill; what we want to know is the *difference* in hight between the two. Just so *it is difference of potential that determines the direction of the flow, and the quantity of electricity that is to flow through a given conductor in a given time*. Sometimes the potential of a body is expressed as so many units above or below that of the earth, assumed as zero.

§ **156. Ampère's rule for determining deflection, etc.** — If the magnetic needle is placed over the current, its deflection is the reverse of that produced when placed beneath it. This tends to confuse; but an artifice, proposed by Ampère, will readily enable us to determine the deflection, when the direction of the current is known, and to determine the direction of the current when that of the deflection is known. He suggests that *we imagine ourselves to be swimming in the current, and with the current, and facing the needle; in which case the north end of the needle will always be deflected towards our left.* (The pupil should test this rule experimentally in various ways and many times, till he is familiar with its application.)

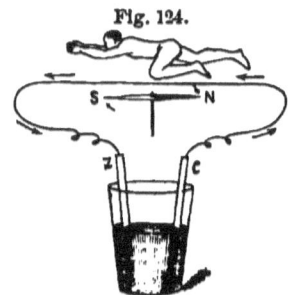

Fig. 124.

§ **157. Galvanoscope.** — The magnetic needle serves the double purpose of determining both the presence and direction of a current in a wire. A needle used for these purposes is called a *galvanoscope*.[1] Electricity set in motion by a voltaic battery is called *galvanic* or *voltaic* and sometimes *current electricity*.

EXERCISES.

1. Let the current be above the needle, and go from N to S; what will be its deflection?
2. Let the current be below the needle, and go from S to N; what deflection will it cause?
3. Let the needle be above the current; what must be the direction of the current when the north end is deflected to the east?
4. Let the needle be below the current, and the deflection toward the east; what is the direction of the current?
5. What is the effect when the current is at the side of the needle?

§ **158. How electric energy originates.** — If you take the liquid from a battery after considerable zinc has disappeared in it, and evaporate it, there will crystallize out of it a white, transparent

---

[1] Galvanoscope, *named for Galvani*, one of the early discoverers in electricity.

WHY THE HYDROGEN APPEARS, ETC. 185

solid in needle-like crystals. This substance is a compound of zinc and sulphuric acid, and is called *zinc sulphate*. The solution of the zinc is the result of a chemical action between the zinc and the acid. Hydrogen is another product of the action. The water serves as a solvent of the zinc sulphate. The chemist symbolizes sulphuric acid thus, H_2SO_4; zinc, Zn. He describes the change that occurs by saying that the zinc replaces the hydrogen H_2 in the acid (in other words, the hydrogen is set free from the combination), while the SO_4 part of the acid unites with the zinc, and forms zinc sulphate, $ZnSO_4$. But we have also discovered another important result of the operation; namely, that *electric energy is developed by the chemical action between the liquid and the zinc*.

Is the electric energy created out of nothing? We have already become familiar with the fact (§ 105, page 140) that chemical potential energy in a lump of coal may be converted into kinetic energy, as is constantly done in the steam engine. Similarly, we might burn zinc to make steam. Coal and zinc, then, possess a power to enter into new combinations; this power is usually called *chemical energy*, or *chemism*. It exists in a potential condition, until it is aroused from this dormant state by bringing together suitable substances. When chemical energy becomes kinetic, it may be transformed into mechanical energy, as when a cannon-ball is set in motion by the burning of gunpowder; or it may be changed into heat, as in the ordinary burning of fuel; or into *both heat and electric energy*, as in the burning of zinc in the battery.

§ 159. **Why the hydrogen appears at the copper plate.** — When zinc dissolves in sulphuric acid, hydrogen is liberated, and ordinarily rises at once in bubbles; but in the voltaic cell it rises from the copper, yet no bubbles are seen to move through the liquid between the plates. As a plausible but imperfect explanation of these phenomena, the well-known hypothesis of Grotthuss was offered. It assumes what many chemists believe, that at the instant that a substance is liberated from a compound it possesses unusual readiness to enter into combination with other molecules.

Let the circles 1, 2, 3, etc. (Fig. 125), represent a series of molecules of H_2SO_4 connecting the two plates. At the instant the circuit is closed the SO_4 of molecule 1 unites with a molecule of zinc, setting free its H_2; this instantly unites with the SO_4 of molecule 2, forming a new molecule, 1', of H_2SO_4, and setting free the H_2 of molecule 2. This H_2 unites with the SO_4 of molecule 3, forming molecule 2'. This decomposition and recomposition continues till the H_2 of molecule 6 is set free. This H_2 unites with other molecules of hydrogen, and finally rises in a bubble to the surface; so the molecule of hydrogen that escapes is not the molecule that was first set free at the zinc plate.

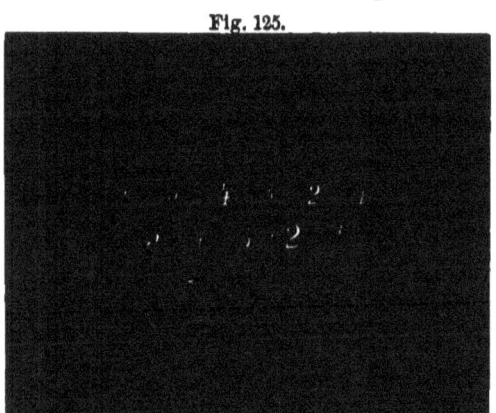

Fig. 125.

§ **160. Electro-chemical series.** — If two plates of zinc were used in a cell, instead of a zinc and a copper, we should have a tendency to opposite currents, which would neutralize each other; or, stated differently, there would be no difference of potential between the two plates, and so no current. It is, therefore, important that only one of the metals should be acted upon. *The greater the disparity between the two solid elements, with reference to the action of the liquid on them, the greater the difference in potential; hence, the greater the current.* In the following *electro-chemical series* the substances are so arranged that the most electro-positive, or those most affected by dilute sulphuric acid, are at the beginning, while those most electro-negative, or those least affected by the acid, are at the end. The arrow indicates the direction of the current through the liquid.

+ Zinc. | Iron. | Tin. | Lead. | Copper. | Silver. | Platinum. | Carbon. −

It will be seen that zinc and platinum are the two metals best adapted to give a strong current.

The essential parts of any galvanic cell in the ordinary form are a liquid and two different solids, one of which is more readily acted upon chemically by the liquid than the other.

§ 161. **Importance of amalgamating the zinc.** — All commercial zinc contains impurities, such as carbon, iron, etc. Figure 126 represents a zinc element having on its surface a particle of iron a, purposely magnified. If such a plate is immersed in dilute sulphuric acid, the particles of iron with the zinc will form numerous voltaic circuits, and a transfer of electricity along the surface will take place. This coasting trade, as it were, between the zinc and the impurities on its surface, diverts so much from the regular battery current, and thereby weakens it. In addition to this, it occasions a great waste of chemicals, because, when the regular circuit is broken, this *local action*, as it is called, still continues. If pure zinc were available, no local action would occur at any time, and there would be no consumption of chemicals, except at times when the circuit is closed. If mercury is rubbed over the surface of the zinc, after the latter has been dipped in acid to clean its surface, the mercury dissolves a portion of the zinc, forming with it a semi-liquid amalgam which covers up its impurities, and the amalgamated zinc then comports itself like pure zinc.

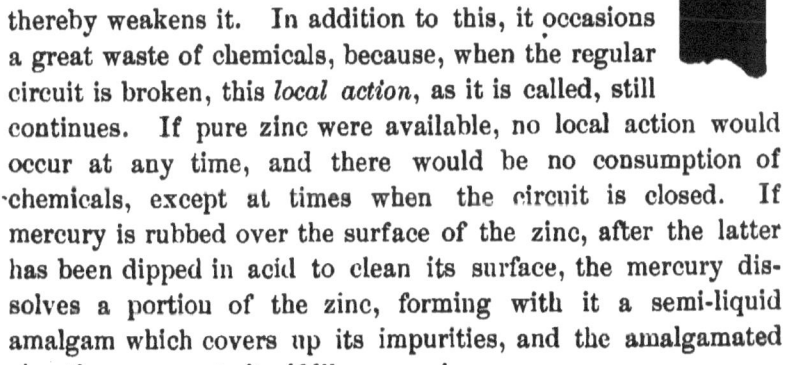

Fig. 126.

XXVIII. VARIOUS BATTERIES.

§ 162. Polarization of plates. — When the zinc and copper elements are first placed in the dilute acid, a very good current of electricity is produced; but the current soon becomes feeble. The cause is easily discovered. The liberated hydrogen adheres very strongly to the copper, as there is nothing for it to unite with chemically; and therefore the plate is very soon visibly covered with bubbles, which may be scraped off with a feather or swab, but only to have the same thing repeated. This coating of bubbles impedes, to a considerable extent, the flow of electricity, and diminishes the current. Besides, a plate coated with hydrogen is more strongly electro-positive than usual, and so, as the coating slowly forms, the difference of potential between the two plates becomes less and less; the current, therefore, must become weaker and weaker as the coating thickens. This action is usually called *polarization of the plates*. Very many methods have been devised for remedying these evils. They are all included in two classes: *mechanical* and *chemical* methods.

Fig. 127.

§ 163. Smee battery. — The Smee battery (Fig. 127) is an example of the former class. A silver plate, or sometimes a lead plate, is coated with a fine, powdery deposit of platinum, which gives the surface a rough character, so that the hydrogen will not readily adhere to it. Dilute sulphuric acid is used in this battery. This plate is suspended between two zinc plates, but not allowed to touch them.

A very effective battery may be constructed by arranging that the copper plate may revolve in the liquid, so that the hydrogen may be removed by friction between the plate and liquid. But this necessitates a constant force to keep the plate in motion.

No mechanical method can wholly prevent the collection of hydrogen on the electro-negative plate. This can only be completely accomplished by furnishing some chemical with which the hydrogen, as soon as liberated, may go into combination.

§ 164. **Grenet battery.** — In the Grenet or bottle battery the hydrogen is disposed of by chemical action. The chemical action is quite complex, and will therefore be omitted. The liquid used is a mixture of potassium bichromate and sulphuric acid dissolved in water. The zinc plate Z (Fig. 128) is suspended between two carbon plates, C, C. The carbons remain in the liquid all the time. (Carbon is now largely used in batteries for the electro-negative plate.)

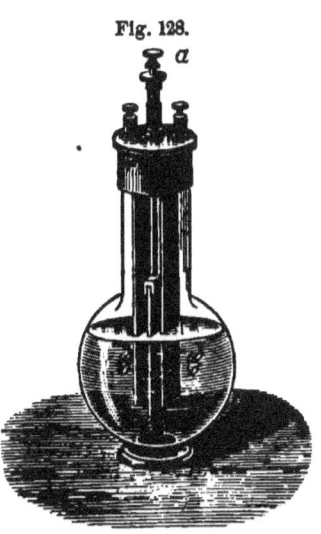

Fig. 128.

This battery gives a very energetic current for a short time, but the liquid soon becomes exhausted. It is a very convenient battery, as, when not in use, we have only to draw the zinc out of the liquid by the brass stem a, and, on pushing the zinc back into the liquid, action commences immediately. It is well to allow the battery to "rest" occasionally by withdrawing the zinc from the liquid for a short time. With one Grenet cell nearly every experiment described in this book can be performed.

§ 165. **Bunsen's and Grove's batteries.** — There is, also, besides the *single-fluid* batteries, a large number of *two-fluid* batteries. The zinc is immersed in the liquid to be decomposed, which most frequently is dilute sulphuric acid, and the conducting plate is surrounded with a liquid which can be decomposed by hydrogen. The two liquids are usually separated by a porous partition of unglazed earthenware, which prevents the liquids from mingling, except very slowly, but does not prevent the passage of hydrogen or electricity. Bunsen's battery (Fig. 129) has a bar of carbon immersed in strong nitric acid contained in a porous cup. This cup is then placed in another vessel containing the dilute sulphuric acid; and immersed in the same liquid is a hollow, cylindrical plate of zinc, which nearly surrounds the porous cup. The hydrogen traverses, by composition and recomposition, the sulphuric acid, passes through the porous partition, and immediately enters into chemical action with the nitric acid, so that none reaches the carbon. There are produced by this action, water — which in time dilutes the acid — and orange-colored fumes of nitric oxide, which rise from the battery. These fumes are very offensive, corrosive, and poisonous. If the nitric acid is first saturated with nitrate of ammonium, the acid will last longer without dilution, and the fumes are almost entirely prevented. Strong sulphuric acid will not answer in any battery. Usually, to one part of sulphuric acid about 12 parts by weight or 20 by volume of water are added to dissolve the sulphate of zinc formed.

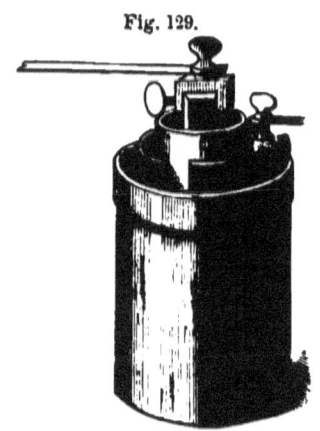

Fig. 129.

Grove used a strip of platinum instead of the carbon rod in his battery. When carbon is used for the negative plate, a solution of bichromate of potassium is frequently substituted for nitric acid, and thereby the disagreeable fumes are avoided. Bunsen's and Grove's batteries are unequalled for powerful and constant currents, and are the best for ordinary lecture-room experiments; but they require frequent attention, and are expensive, so that they are little used for work of long duration.

§ **166. Gravity battery.** — The battery principally used in this country for telegraphing is called the *gravity battery*. A copper plate C, Figure 130, is placed on the bottom of a vessel and covered with crystals of copper sulphate (blue vitriol), and the whole covered with water. As the vitriol dissolves, its specific gravity causes it to remain at the bottom, in contact with the copper plate. The zinc plate Z is suspended in the clear liquid above. To start the action quickly, a teaspoonful of common salt or zinc sulphate is dissolved in the water. As the chemical action proceeds, the vitriol is decomposed, its sulphuric acid constituent unites with the zinc, forming soluble zinc sulphate, and the copper constituent is deposited in a metallic state on the copper plate. The zinc does not require amalgamation.

Fig. 130.

XXIX. EFFECTS PRODUCED BY ELECTRICITY.

§ **167. Heating effect.** — **Experiment 1.** Introduce between the electrodes of a Bunsen or Grenet cell a piece of platinum wire A, Figure 131, about 6^{cm} long and in size about No. 30. The platinum wire becomes white hot.

Experiment 2. Stretch the platinum wire over a gas-burner, turn on the gas, and light it by the heat of the wire.

Experiment 3. Strew lycopodium powder over a tuft of cotton-wool, and ignite it with the heated wire.

Experiment 4. Connect the battery wires (Fig. 131) with a galvanometer (see page 198) G, as in the figure; the needle is deflected. Remove the platinum wire, and close the circuit; the needle is deflected more than before.

What transformations of energy took place in the above experiments?

§ 168. Luminous effect. — We have already seen one illustration of this effect in the glowing of the white-hot platinum wire.

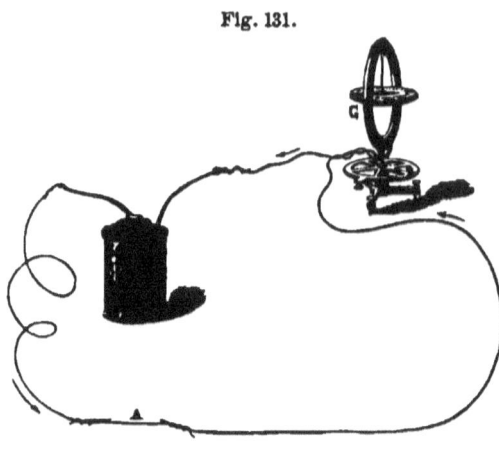

Fig. 131.

Experiment. Attach one pole of the battery to a file (Fig. 132), and pass the other pole over its rough surface. The file forms part of the circuit; and as the wire passes over it, the circuit is rapidly made and broken, and each break causes a spark at the point where the circuit is broken. The shower of sparks that flies from the file is due to red-hot particles of iron that are projected into the air.

Fig. 132.

Fig. 133.

§ 169. Chemical effect. — **Experiment 1.** Steep some leaves of purple cabbage; the infusion has a deep purple color. Dissolve a little caustic soda, and pour a few drops of the solution into a portion of the infusion, and the purple will be changed to a green. Caustic soda is an alkali, and cabbage infusion is turned green only by alkalies. Pour a few drops of dilute sulphuric acid into another portion of the infusion, and the purple will be changed to a red. Only acids turn purple cabbage infusions red. Now prepare a concentrated solution of sodium sulphate. Color the solution with a portion of the purple cabbage infusion, and partly fill a V-shaped glass tube (Fig. 133) with this liquid. Employ a battery of two Grove or Grenet cells connected in series. (See p. 208.) Attach to the poles of the battery-wires two narrow strips of platinum, and place one of these strips in each branch of the tube, a little way apart, so that the current will be obliged to traverse a part of the liquid. Close the circuit; bubbles of gas are immediately disengaged from the platinum strips;

soon the liquid around the −pole is turned green, while that around the +pole is turned red. Evidently decomposition of the sodium sulphate has taken place; an acid and an alkali are the results.

The current which is maintained by chemical action in the battery is capable of doing chemical work outside the battery. A substance that may be decomposed by electricity is called an *electrolyte*, and the process *electrolysis*.[1] *The electrolyte must be a compound substance, and in a liquid state*, either by solution or fusion. A large number of substances are composed, like sodium sulphate, of an acid, and either an alkali or some other substance that will neutralize an acid. Any substance that will neutralize an acid is called a *base*, and a compound of an acid and a base is called a *salt*. When a salt is electrolyzed, the base always appears at the −pole, and the acid at the +pole.

Experiment 2. Prepare a solution of the salt copper sulphate, and subject it to electrolysis, as in the last experiment; copper collects on the −platinum, and sulphuric acid and oxygen at the +platinum. Remove the platinum strips, and introduce the copper terminals; copper is now deposited on the −pole as before, but the +pole wastes away.

The chemical symbol for copper sulphate is $CuSO_4$. By electrolysis it is separated into Cu and SO_4. When a copper +pole is used, the SO_4 immediately unites with a molecule of the copper (Cu) of this pole, and forms a new molecule of copper sulphate ($CuSO_4$), which is dissolved by the water. This accounts for the wasting away of the +pole. The solution does not lose its strength, for as fast as a molecule of copper sulphate is decomposed, another is formed. But when platinum poles are used, the SO_4 does not combine with the platinum, but enters into chemical action with the water. The SO_4 combines with the hydrogen of the water, forming sulphuric acid, and the oxygen of the water is set free. ($SO_4 + H_2O = H_2SO_4 + O$.)

[1] Electrolysis, *a loosening by electricity.*

The liberation of the oxygen is the result of a secondary chemical action, subsequent to the electrolytic action.

Experiment 3. Prepare a solution of tin chloride, by dissolving scraps of granulated tin in hot hydrochloric acid. Add a little water. Electrolyze this salt in solution, using platinum poles. A beautiful growth of tin crystals will shoot out from the −pole and spread towards the +pole, bearing a strong resemblance to vegetable growth; hence it is called the "tin tree."

In a similar manner, silver and lead trees may be prepared from their salts, silver nitrate and lead acetate. Each metal has its own peculiar form of growth; and sometimes the same metal, particularly silver, exhibits different forms, according to the strength of the solution and the power of the current. In Figure 134, A represents a silver tree deposited from a weak solution of silver nitrate, and B a tree formed from a still weaker solution of the same.

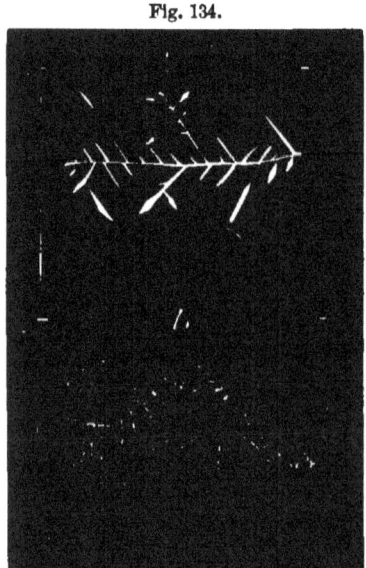

Fig. 134.

Experiment 4. Remove the bottom of a glass bottle having a wide mouth, fit a cork to the mouth, and pass two insulated wires through the cork, terminating in platinum strips (Fig. 135). Fill two test-tubes and part of the inverted bottle with dilute sulphuric acid, and invert the tubes over the platinum poles. The circuit is thus closed through the liquid. Bubbles of gas immediately rise from the poles and displace the liquid in the tubes. About twice as much gas collects over the −pole as over the +pole. Thrust a lighted splinter into each of the gases: the former burns; the latter causes the splinter to burn much more rapidly than it burned in the air. This indicates that the former is hydrogen gas and the latter oxygen gas.

Since pure water is an almost perfect non-conductor of electricity (page 203), the probable explanation of this action is very closely like that already given (page 185) for the action in the simple cell. The sulphuric acid is decomposed; H_2SO_4 becomes $H_2 + SO_4$; then $SO_4 + H_2O$ becomes $H_2SO_4 + O$. It is certain that water is ultimately decomposed, for no sulphuric acid is lost. This electrolysis shows that water is composed of two parts by volume of hydrogen to one part of oxygen. (Why ought not copper poles to be used in this experiment? Ascertain, by inserting a galvanometer in the circuit, whether the current is weakened by performing the work of electrolysis.)

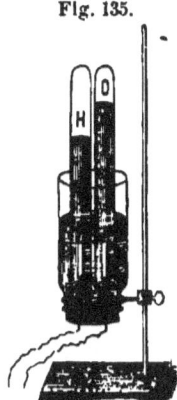

Fig. 135.

When the poles of a strong battery are applied for some time to a person's skin, blisters appear under the poles. The serous fluid that comes from the vesicles under the positive pole is acid; the fluid in the vesicles under the negative pole is alkaline.

§ **170. Physiological effect.** — Experiment. Place the tip of the tongue between the two poles of a single cell, so that the tongue may form part of the circuit; a stinging sensation is felt, accompanied by a peculiar acrid taste.

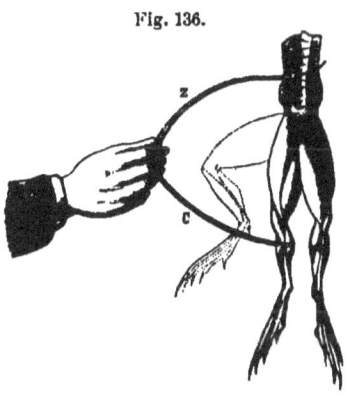

Fig. 136.

When a battery is known not to be very powerful, the tongue serves as a very convenient galvanoscope, to determine whether the circuit is in working condition, and approximately the strength of the current. If the crural nerve (a white cord next the backbone) of a frog, recently killed, is laid bare, and one of the poles of a battery is applied to it, on touching a naked muscle of a leg with the other pole, the muscles are instantly convulsed and the leg drawn up, as represented by the dotted lines in Figure 136. The same convulsion occurs at the instant the circuit is broken. By touching the nerve with a piece of zinc, and the muscle with a copper wire, as represented in Figure 136, similar convulsions occur, on bringing the free ends of the metals in contact, and on their separation. The cause is obvious; for the two metals and the moisture of the flesh furnish all the essentials of a voltaic element.

The irritability of nerves and muscles begins to diminish after death, and sooner or later disappears. It disappears much sooner in warm than in cold-blooded animals. In the limb of a frog that is properly protected, and kept at a cool temperature, it may remain for two, three, or even four weeks. If one pole is armed with a soft sponge, wet with salt water, and pressed firmly on the closed eyelid, while the other is applied at the back of the neck, or held in the hand, making and breaking the circuit will cause a sensation of light of various colors.

§ 171. **Magnetic effect.**—Experiment. Obtain an insulated[1] copper wire, wind twenty or more turns around a rod of well-annealed iron, 10^{cm} long and about 1^{cm} in diameter, and close the circuit. Bring a nail (Fig. 137), or other piece of iron, near the rod. The rod attracts the nail with much force, and this nail will attract other nails. The rod has acquired all the properties of a magnet, as will be seen hereafter. But the instant the circuit is broken, the iron loses its magnetic force, and the nails drop.

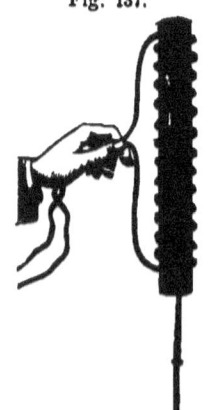

Fig. 137.

The more times the wire is wound around the rod, within a certain limit, the more powerfully is it magnetized. This arrangement is called an *electro-magnet*, because it is a magnet produced by electricity. The rod of iron is called its *core*, and the coil of wire the *helix*.

---

[1] Insulated, *covered with cotton or silk*, to prevent electricity from passing from one section of wire to another in contact with it, without passing through the whole length of the wire.

In order to take advantage of the attraction of both ends or poles of the magnet, the rod is most frequently bent in a U-shape (A, Fig. 138), and then it is called a horse-shoe magnet. Sometimes two iron rods are used, connected by a rectangular piece of iron, as a, in B of Figure 138. The method

Fig. 138.

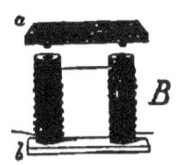

of winding is such that if the iron core of the horse-shoe were straightened, or the two spools were placed together, end to end, one would appear as a continuation of the other. A piece of soft iron, b, placed across the ends, and attracted by them, is called an *armature*. The piece of iron a is called a *back armature*.

XXX. ELECTRICAL MEASUREMENTS.

The wonderful developments of electrical science in recent years are almost wholly due to a better understanding of what electrical measurements can and ought to be made, and how to make them. Most of this increased knowledge has been gained since the first Atlantic cable failed in 1858. Let us learn how to make some of them.

§ 172. Strength of current. — It is evident that the thermal and luminous effects of electrical discharges, electro-chemical decomposition, the deflection of the magnetic needle, the magnetization of iron, and even physiological effects, or any external manifestation, may be employed to detect the presence of an electric current, in a circuit however extended. It is also obvious that *the magnitude of these effects may serve to measure the strength of the current*. Now, as the quantity of water that passes through a given pipe in a minute or an hour indicates the strength of the current, so *by the strength of an electric current is meant the quantity of electricity that passes through an electrical conductor in a unit of time*.

§ 173. Voltameter.

— The quantity of electricity that passes any cross section of any conductor in the same circuit, however long, is, unless there is a leakage at some point, necessarily the same. We may, therefore, introduce a platinum wire into any part of the circuit, and measure the strength of a current by the temperature to which the wire is raised; or we may decompose water and collect the gases resulting therefrom; *the strength of current is measured by the quantity of gas liberated in a unit of time.* The latter arrangement, called a *voltameter*, is easily constructed sufficiently accurate for many purposes, and should be constructed and used by every pupil.

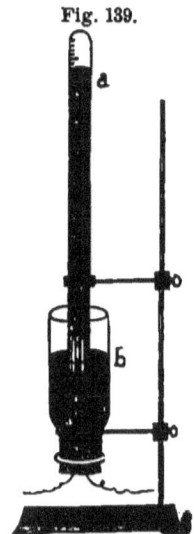

Fig. 139.

In Figure 139, a is a glass tube 50^{cm} long and 3^{cm} in diameter (a much shorter tube will answer; for example, a large sized test-tube), closed at one end, and graduated in cubic centimeters (this may be done by means of a paper scale pasted on one side of the tube); b is a bottomless glass bottle of about 1 liter capacity. Through the stopper of the bottle pass two insulated wires, terminating in platinum strips, which are introduced a little way into the tube. The tube is filled with water slightly acidulated with sulphuric acid, and its orifice is immersed in the same kind of liquid, which partly fills the bottle. When the wires are connected with a battery of two or more cells in series (see page 208), the gas arising from the decomposition of the water will collect in the top of the tube and displace the liquid.

§ 174. Galvanometer.

— The instrument in most common use for measuring current strength is the magnetic needle, which, besides its ordinary use as a *galvanoscope*, performs the still more important office of a *galvanometer*. The simple magnetic needle, used as already described, answers tolerably well when the currents are strong, but it is not sensitive enough to be sensibly moved by very weak currents. If two equal currents, flowing in the same direction, are placed one above and the other below a magnetic needle, they tend to produce opposite deflections, and to neutralize one another's effect, so that no deflection occurs. Evidently, if they flow in opposite directions, they tend to produce a deflection in the same direction, and the result is a deflection twice as great as that produced by a single current. The same result is accomplished if the same current is made to pass both above and below a needle, as in A, Figure 140. If the wire were carried four times around the needle, as

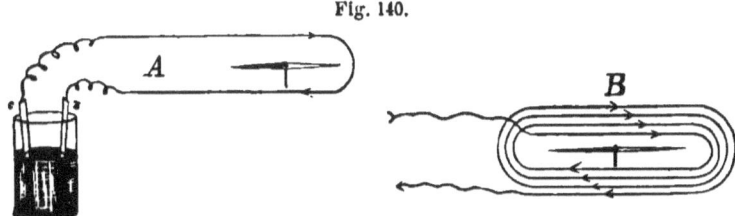

Fig. 140.

in B, the influence of the current on the needle would be about four times that of a single turn. Very sensitive galvanometers, constructed on this principle, often with thousands of turns of wire, are sometimes called *long-coil galvanometers*, in distinction from those having few turns, which are called *short-coil galvanometers*.

§ 175. **Tangent galvanometer.** — The arrangement described above is more commonly used as a galvanoscope than a galvanometer, though it may be so calibrated as to answer the latter purpose. The law connecting the current strength with the deflection of the needle of this galvanometer is not known; but in another form, called the *tangent galvanometer*, the relation is expressed in a simple tangent of the angle of deflection. This apparatus is constructed on the principle that *the strength of currents are proportional to the tangents of the angles of deflection*, when the needle is very short in comparison with the diameter of a circle described by a current circulating around it.

200 ELECTRICITY AND MAGNETISM.

A magnetic needle, about 2.5cm long, is suspended freely by an untwisted thread n, Figure 141, in the center of a copper hoop a, about 30cm in diameter, which terminates in the wires ww'; and these are connected with the battery whose current is to be measured. A circular card-board cc, containing a circle divided to degrees to indicate the extent of deflection, is placed beneath the needle. The ring being placed so that it is parallel with the needle, the needle points to 0° on the scale. When a current passes through the ring a, the needle is deflected. The tangents of the angles of deflection may be found by

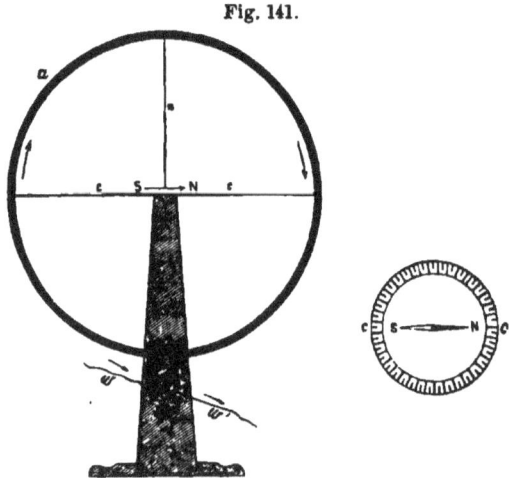

Fig. 141.

reference to a Table of Natural Tangents in Section D of the Appendix, and the relative strengths of currents may be determined by the law given above. The construction of a very simple galvanometer that may be used as a tangent galvanometer, and which will answer all requirements of this book, may be found in Section E of the Appendix.

§ 176. Experiments in measurements. — Inasmuch as the magnitude of the effects that can be produced by an electric current, or the amount of work that can be done by it, depends upon the strength of the current, it is of the utmost importance to understand the principles by which it is regulated. A few experiments will make this apparent. Provide four coils or spools of insulated wire. Mark the coils A, B, C, and D. Let A contain 100 ft. (about 1 lb.) of No. 16 copper wire; B and C respectively 100 ft. and 50 ft. of No. 30 copper wire; and D 50 ft. of No. 30 German silver wire.

Experiment 1. Place a galvanometer G and coil A in the same voltaic circuit, connected as shown in Figure 142. Note the number of degrees the needle is deflected. Next substitute coil B for A, and note the deflection. The deflection is less than before, showing that *of two wires of the same material and equal length, the larger transmits, from the same source, the stronger current.*

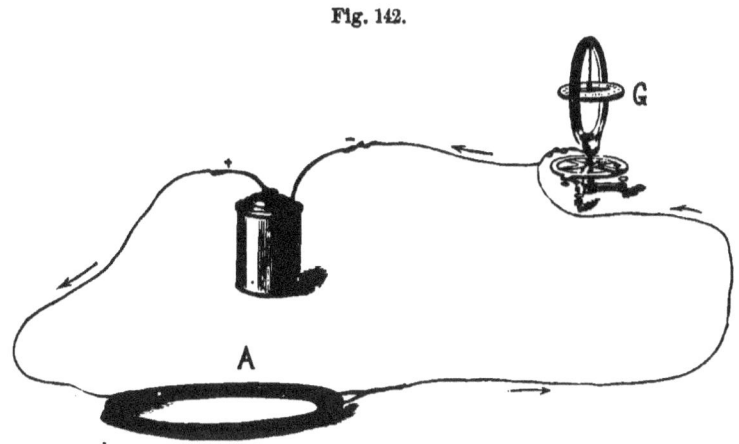

Fig. 142.

Experiment 2. Place coil C in the circuit with B, and compare the deflection with that produced when B alone was in the circuit. The deflection is less than before. (Why?) Take B out, and leave C in the circuit. The deflection is greater than when B alone was in the circuit. *Other things being the same, the shorter wire transmits, from the same source, the stronger current.*

Experiment 3. Introduce D in the place of C, and compare the strengths of the currents in these two wires. The *copper wire transmits, from the same source, a stronger current than the German silver wire of the same length and size.*

Experiment 4. Compare the currents furnished by a Grove or Bunsen, and a Smee or a gravity cell, when the same coil, for instance B, is in the circuit. *The Grove or Bunsen cell gives the stronger current.*

§ 177. On what strength of current depends. — It appears that the strength of the current varies not only with the size, *length*, and *kind of conductor*, but also with the *kind of battery* used. These will be considered consecutively. It is evident that all conductors do not allow the current to pass with equal facility; in other words, some conductors offer more *resistance* to the passage of a current than others. The larger conductor offers less resistance than the smaller. It is found by experiment that (1) *the strength of currents varies directly as the areas of the cross sections of the conductors, or the squares of the diameters of cylindrical conductors*, inasmuch as areas vary as the squares of their diameters. (2) *It varies inversely as the length of the conductor*, i.e., if a wire one mile long offers a certain amount of resistance, a wire two miles long will offer twice as much resistance. (3) *It varies inversely as the specific resistances of the substances used for conductors*. The conducting power of a substance is the reciprocal of its resistance. Hence, if we know the conducting power of any wire, we know that the $resistance = \dfrac{1}{conductivity}$; or the $conductivity = \dfrac{1}{resistance}$.

Resistance is expressed in units called *ohms*[1] (see § 181). The student can easily provide himself with a standard having approximately a resistance of one ohm, by obtaining 40 feet of No. 24 ordinary copper wire 0.5^{mm} in diameter.

§ **178. Formula for resistance.** — Having found that *resistance varies directly as the length, and inversely as the squares of the diameters of conductors*, we may include all its laws in the formula
$$R = K \frac{l}{d^2};$$
in which R = the resistance, l = the length, and d the diameter of a cylindrical conductor. K is a constant, such that when the material of the wire is known and the denomination in which l and d are expressed, a value of K taken from a table may be

---

[1] Ohm, *from the name of a German savant, Dr. G. S. Ohm*, who first enunciated the laws which determine the strength of currents.

FORMULA FOR INTERNAL RESISTANCE.

substituted in the equation, and thus enable us to find the value of R in ohms. Thus let it be required to find R of 1000 ft. of copper wire 0.1 inch in diameter. The table gives the value of K as 9.72 when the length of the wire is measured in feet, and its diameter in thousandths of an inch; since 0.1 inch equals 100 thousandths, $R = 9.72 \times \dfrac{1000}{100^2} = 0.972$ ohm.

In the following table are given the relative resistances of several substances, and the values of K in the above equation when l is expressed in feet and d in thousandths of an inch.

REFERENCE TABLE OF RELATIVE RESISTANCES, ETC.

		Rel. Resist.	K.
Silver	@ 0° C.	1.00	9.15
Copper	"	1.06	9.72
Zinc	"	3.74	34.2
Platinum	"	6.02	55.1
Iron	"	6.46	59.1
German silver	"	13.91	127.3
Mercury	"	63.24	578.6

		Rel. Resist.
Nitric acid — commercial	@ 15 to 28 C.	1,100,000
Sulphuric acid, 1 to 12 parts water	"	2,000,000
Common salt — saturated sol.	"	3,200,000
Sulphate copper	" "	18,000,000
Distilled water		not less than 10,000,000,000
Glass	@ 200° C.	15,000,000,000,000
Gutta percha	@ 0° C.	5,000,000,000,000,000,000

The resistance of metals increases slowly as the temperature rises; but that of liquids and the other poor conductors in the second list decreases very rapidly with a rise in temperature. The resistance of ordinary impure metals is often much higher than that given in the table.

§ **179. Formula for internal resistance.** — Resistance in a voltaic circuit may be divided, for convenience, into two parts; viz., *internal resistance* (r), which the current encounters in passing through the liquid between the two plates in the cell, and *external resistance* (R), which it suffers in the remainder of its path. The internal resistance is governed by the same laws as the external resistance. In this case

$$r = K \frac{\text{distance of the plates apart } (l)}{\text{areas of the plates submerged } (d^2)}.$$

QUESTIONS AND PROBLEMS.

1. What length of copper wire will have the same resistance as a mile of iron wire of the same diameter?

2. How can you reduce the resistance of an iron wire to that of a copper wire of the same length?

3. About how much is the conductivity of water affected by adding a little sulphuric acid?

4. How many times greater is the resistance of dilute sulphuric acid than that of copper?

5. Upon what does the resistance offered by the liquid part of a circuit depend, and how may it be diminished?

6. What is the resistance of 500 ft. of copper wire .014 inch in diameter (No. 30 B.W. gauge)? *Ans.* 24.7 + ohms.

7. What length of copper wire .006 inch in diameter (No. 38) will offer a resistance of 1 ohm?

8. What is the resistance of 16 yards of German silver wire (No. 30) .014 inch in diameter?

9. What is the resistance of 1 mile of iron telegraph wire, the usual size being .175 inch in diameter?

10. Express in ohms the resistance of 1 mile of copper wire, 0.06 inch in diameter? *Ans.* $9.72 \times \frac{5280}{60^2} = 14.256$ ohms.

§ **180. Electro-motive force.** — The experiments described in § 151 show that electricity constantly flows in a closed circuit containing a voltaic cell; hence the cell has the power of setting electricity in motion, or an *electro-motive force* (usually abbreviated E.M.F.). Again, Exp. 4, § 176, proves that a Grove cell, in a circuit of a given resistance, sets in motion a greater quantity of electricity than a Smee or gravity cell; hence we say that the E.M.F. of a Grove cell is greater than that of the other two kinds mentioned. It has been found that *E.M.F. depends*

OHM'S LAW. 205

solely upon the nature and condition of the substances which form the battery, and is, consequently, quite independent of the size of the plates and their distance apart. The unit employed in the measurement of E.M.F. is called a *volt*.[1] It is about the E.M.F. of a current generated by one gravity cell. The following table exhibits the electro-motive force in volts of different cells: —

TABLE OF ELECTRO-MOTIVE FORCES.

Gravity or Daniell	0.98 to 1.08 volts.
Bunsen and Grove	1.75 to 1.95 "
Leclanchè, at first	1.48 to 1.60 "
Grenet "	1.80 to 2.3 "
Smee	.65 .... "

The E.M.F. of the last three decreases considerably if the circuit is closed for a few minutes. These numbers signify, for instance, that it will require 195 Smee cells to give the same current in a circuit (of high resistance) as would be given by 65 Grove cells.

§ 181. Ohm's Law. — The law which expresses the strength of the current, and is the basis of most mathematical calculations on currents, is expressed in the formula known as *Ohm's Law*. Calling the current C, the E.M.F. simply E, and the whole resistance in the circuit R, the formula expressing the law is

$$C = \frac{E}{R}.$$

In words, this means that the *strength of the current is equal to the electro-motive force of the battery, divided by the resistance of the circuit;* i.e., C will be greater or less as E is greater or less, but will be less when R is greater, and greater when R is less. The above relation $\frac{E}{R}$, when the external resistance is considered separately from the internal, must be converted thus: calling the former R, and the latter r, the expression becomes

$$C = \frac{E}{R + r}.$$

[1] Volt, *from the name Volta.*

For single cells in ordinary use the value of r will usually be between .5 and 2 ohms.

The unit of current strength, called an *ampère*, is the current flowing in a conductor having a resistance of 1 ohm, between the ends of which a difference of potential of 1 volt is maintained; or it is a current of 1 *coulomb* per second. A coulomb is the amount of electricity conveyed in 1 second by a current of 1 ampère.

If a cell has $E = 1$ volt, and $r = 1$ ohm, and the connecting wire is short and stout, so that R may be disregarded, then the current has a value of 1 ampère. But if the connecting wire has a resistance R, equal to 1 ohm, then

$$C = \frac{E}{R+r} = \frac{1}{1+1} = \frac{1}{2} = .5 \text{ ampère.}$$

§ 182. Summary of electrical measurements. — Just as we express an amount of money in the denomination *dollars*, or a mass of coal in the denomination *pounds*, we express electrical

Potential, P (commonly, difference of P).... in volts.
Electro-motive force, E .... " volts.
Resistance, R............................. " ohms.
Strength of current, C..................... " ampères.
Quantity of electricity..................... " coulombs.

The following will give some idea of the magnitude of the denominations. A gravity cell produces a difference of potential or an electro-motive force (for these are only different ways of viewing the same quantity) of nearly 1 volt. To produce a spark 1^{mm} long requires from 3,000 to 4,000 volts. A No. 16 ordinary copper wire 250 ft. long (diameter .051 inch, weight 2 lbs.) has a resistance of about 1 ohm. About 150 ft. of copper wire 1^{mm} in diameter has a resistance of 1 ohm. An ordinary Grove cell may have an internal resistance of $\frac{1}{4}$ ohm; this cell will send through 125 ft. of No. 16 copper wire a current whose strength is 1 ampère.

ARRANGEMENT OF BATTERIES. 207

PROBLEMS.

1. What current will be obtained from a gravity cell when $E = 1$, $r = 2$ ohms, and $R = 10$ ohms?
2. What current may be got from a gravity cell whose internal resistance is 3 ohms, and external resistance is 3 ohms?
3. What current will a Grove cell furnish, having the same internal and external resistances as the last?

§ **183. Arrangement of batteries.** — The internal resistance may be diminished by placing the plates as near to each other as practicable, and by employing large plates, and thereby increasing the size of the liquid conductor. But it is not always convenient to employ very large plates, or we may have occasion to employ a battery for certain purposes, as we shall see presently, in which large cells would be of little or no advantage. The same result that can be produced by a single pair of large plates, may be obtained by connecting the similar plates of several pairs in separate cells, thereby practically reducing several pairs to one pair having an area equal to the sum of the areas of the several pairs. Figure 143 illustrates a method of connecting cells for the purpose of reducing the internal resistance. This is called arranging cells *parallel*, *in multiple arc*, or *abreast*.

Fig. 143.

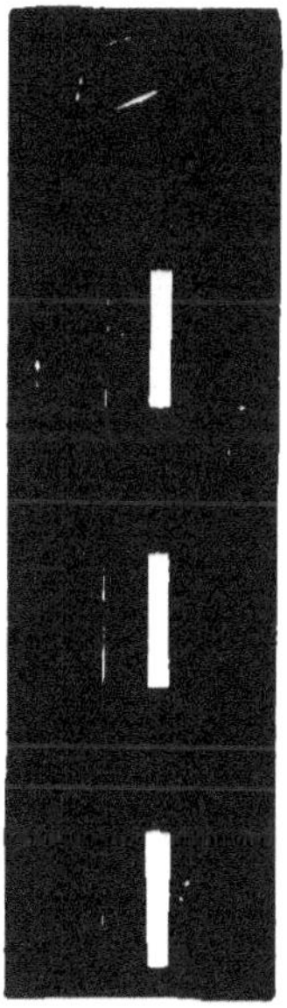

This arrangement is very effectual in increasing the current-strength when the internal resistance is the principal one to be overcome. For instance, call the electro-motive force (E) of a single cell 1 volt, its internal resistance 5 ohms, and let the plates be connected by a short, thick wire, whose resistance may be regarded as nothing; then $C = \dfrac{E}{r} = \dfrac{1}{5} = .2$ ampère. Now connect 10 similar cells abreast. The size of the liquid conductor being increased tenfold, the internal resistance is one-tenth as large, and we have $C = \dfrac{E}{r} = 1 \div \tfrac{5}{10} = 2$ ampères. So that, *when there is no external resistance, the current increases as the size of the plates is increased.* The same is approximately true in case the external resistance is very small in comparison with the internal resistance.

Again, let $E = 1$, $r = 5$ ohms, as above, but the external resistance $R = 200$ ohms; then $C = \dfrac{1}{5 + 200} = .0048+$ ampère. If 10 pairs are connected abreast, $C = \dfrac{1}{\frac{1}{2} + 200} = .0049+$ ampère. In this case, the current is scarcely affected by increasing the number of cells abreast. The question then arises, what can be done to increase the current when the external resistance is necessarily large; as, for instance, when a long telegraph wire is used. In this case R, in Ohm's formula, is unalterable, and lessening r has little effect; so there remains only one way, viz., to increase E, the electro-motive force. How may this be done? If the current from a cell, instead of passing immediately out of the cell on its journey, is made to pass through another cell first, one might naturally expect that either the two cells would counteract one another in the circuit, or that they would double the E.M.F. Experiment shows that the latter result is the true one, and that *the E.M.F. is exactly proportional to the number of cells connected in series.* Cells so connected as to increase the electro-motive force are said to be joined in *series* or *tandem*. The method of connecting the cells for this purpose is shown in Figure 144.

Fig. 144.

It will be seen that in the multiple arc (Fig. 143) all the zinc plates are connected with one another, and all the copper plates with one another. In the tandem arrangement the zinc of one cell is connected with the copper of the next throughout.

ARRANGEMENT OF BATTERIES.

In the last example given above, let us see what would be the effect of connecting the 10 cells in series. In this case E is increased tenfold; and, as the current is obliged to pass through the liquid of 10 cells instead of one, the internal resistance will also be increased tenfold; hence, $C = \dfrac{1 \times 10}{(5 \times 10) + 200} = .0400$ ampère, more than eight times as much as before. Thus it appears that, *when the external resistance is large in comparison with the internal resistance, the current may be largely increased by multiplying the cells in series;* in other words, by forming a *battery of great electro-motive force.* In long telegraph lines the battery is made up of hundreds of cells joined in series. Large cells are used simply because the fluids last longer, and so the cells need less attention.

Fig. 145.

Fig. 146.

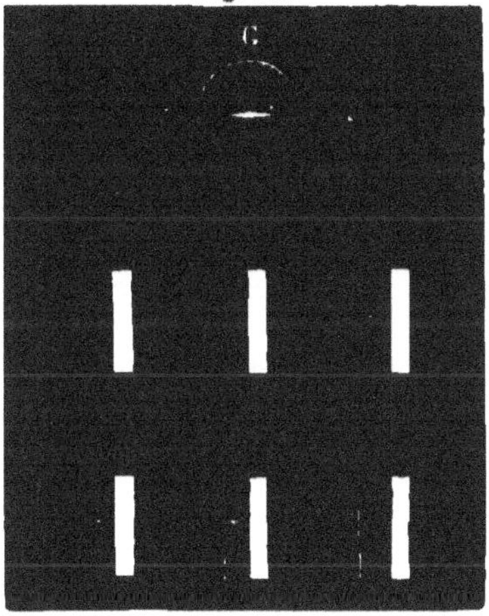

Sometimes a combination of the two arrangements gives a stronger current than either alone. The cells may be grouped together in pairs (as in Figure 145), or in triplets (as in Figure 146), so as to increase the electro-motive force; then the several groups may be connected abreast, to reduce the internal resistance.

PROBLEMS.

1. Suppose that the cells in the last example above be so increased in size that r in each $= .5$ ohm, what current will be got from the battery?

210   ELECTRICITY AND MAGNETISM.

2. What current is there on a telegraph wire 100 miles long, when it is in circuit with 40 Grove cells, the internal resistance of each cell being .5 ohm, the resistance of the wire 15 ohms to the mile, and the resistance of the earth connections 100 ohms?

3. An electric bell, whose resistance is .5 ohm, is found to require a current of .02 ampère to ring it. How many gravity cells will it require, if the circuit consists of an iron wire 1 mile long, having a diameter of .165 in. (= 4.29ᵐᵐ), disregarding the resistance of the cells?

4. Which is the better arrangement (*i.e.*, abreast or tandem) of 210 gravity cells, each of 3 ohms resistance, against an external resistance of 10 ohms, and what will be the current with each?

5. In a battery of 10 cells connected in series, ten times as much zinc and acid are consumed as in 1 cell. Show how about nine-tenths of the chemicals may be wasted.

6. In a certain circuit a battery of 40 gravity cells, each of 3 ohms resistance, is used. The cells are arranged in two groups of 20 cells each in series, and the two groups are so connected as to diminish the internal resistance. If $R = 120$ ohms, what current will be obtained?

7. Devise various arrangements of 30 cells in which $r = .8$ ohm. Which arrangement is best when $R = 10$ ohms? when $R = 30$ ohms?

§ 184. General conclusions.

A perfect battery would have the following qualities : —

1. *It would have a high electro-motive force.*
2. *Its E.M.F. would be constant, whether used to produce strong or weak currents.*
3. *It would have a small and constant internal resistance.*

No battery fulfils perfectly all these conditions; so, in practice, the use to which the battery is to be put, its first cost, and the trouble and expense of keeping it in order, determine which of these qualities shall be given up. The question, What is the best battery? is discussed in the Appendix, Section F.

A current from a single cell, traversing a short, thick wire, will produce as large a deflection of a magnetic needle as a current from 500 cells connected in series. On the other hand, a message has been transmitted through the Atlantic cable, whose resistance is about 7,000 ohms, by a current generated in a lady's thimble, and the signals produced were as distinct as those that would be produced by a cell of several square feet. In this case, the quantity of electricity that passed depended chiefly upon the E.M.F. of the battery, and not upon its internal resistance. (How, in the former case, can the current be increased? How could it have been increased in the latter case?)

The same strength of current that would fuse an inch of platinum wire would fuse a mile of the same wire. But while one cell would fuse the inch of wire, it would require the E.M.F. of many hundred to maintain the same strength of current in the mile of wire.

A battery of three cells arranged abreast will fuse a certain length of platinum wire, but will not be felt by a person holding the poles in his hands; while a battery of 100 cells in series will not fuse the same wire, but will produce quite a shock.

The power of an electro-magnet depends largely upon the number of times a given current circulates around its core, and upon the nearness of the current to the core; for compactness, and to keep the current near the core, a fine wire must then be used. But a long, thin wire would offer large resistance, that might so reduce the current as to more than offset the advantage that would otherwise be gained. Hence, *when there is little other resistance in a circuit, a large wire with few turns will give the strongest electro-magnet.* But if an electro-magnet is to be used in a circuit with other large resistance, then the introduction of a helix of many turns of fine wire would add little more resistance comparatively, so the strength of the current would be reduced but little, while a great gain would be made in the effect on the core. For the same reason, a galvanometer intended to be used in circuits where there is little resistance, should contain only a few turns of large wire; but if it is to be used with large resistance, it should contain a long, fine wire. *Electro-magnets and galvanometers should be adapted to the circuits in which they are used.*

QUESTIONS.

1. A bell in Washington is to be rung by the action of an electro-magnet. The current used comes from a battery in New York. How should the electro-magnet be constructed?

2. If you wished to measure the current, by the introduction of a galvanometer into the circuit in Washington, how should it be made?

3. Would it require a different galvanometer if it were to be introduced into the same circuit in New York, only a few feet from the battery?

4. Provided there were no leakages, how would the deflections at the two places compare?

XXXI. MAGNETS AND MAGNETISM.

§ 185. Permanent and temporary magnets. — One of the most familiar pieces of physical apparatus is a *magnet*. We know how it can pick up bits of iron and steel. By the aid of a small instrument already studied, we may make a pair of small magnets, and study their actions and laws.

Experiment. Take the electro-magnet, described on page 196, and a couple of sewing needles or larger steel rods. Apply these needles, one at a time, to one end of the electro-magnet, and draw them several times across it from end to end, always in the same direction, and not rubbing back and forth. Repeat the operation with an iron wire of the same size; both the wire and the steel are attracted by the electro-magnet, but the iron wire more strongly. Observe that both, while in contact with the electro-magnet, possess the power of attracting bits of iron; but, on removing them, the steel is found to retain the property it had, while the iron does not.

Both of them exerted that peculiar force called *magnetic force*, or possessed the property called *magnetism;* that is, both were magnets; but, as the steel retains its power, it is called a *permanent* magnet in distinction from a *temporary* magnet, like the iron wire or the electro-magnet itself. The quality of steel by which it at first resists the power of magnets, and resists the escape of magnetism which it has once acquired, is called *coer-*

cive force. The harder steel is, the greater is its coercive force. Hence, highly tempered steel is used for permanent magnets. Hardened iron possesses some coercive force; hence, the cores of electro-magnets should be made of the *softest* iron, that they may acquire and part with magnetism instantaneously.

§ 186. **Law of magnets.**—Experiment 1. Suspend the two magnets, each in a horizontal position, by threads that will not untwist, and several feet distant from each other. When they come to rest, notice that they have taken up a direction nearly north and south. Tie a thread on the end of each that points to the north.

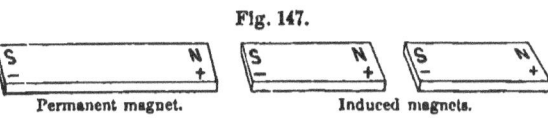

Fig. 147.

Permanent magnet.   Induced magnets.

This end, or *pole*, as it is usually called, we will speak of as the N-end, +, or *marked* end or pole, while the other is the *unmarked*, —, or S-end or pole.

Experiment 2. Bring the marked end of one of the magnets near to the unmarked end of the other; they attract one another. Next bring the marked end of one near to the marked end of the other; they repel one another. Bring the unmarked ends near one another; they repel one another.

We discover the following law of magnets: *Like poles repel, unlike poles attract one another.*

§ 187. **Magnetic transparency and induction.**—Experiment. Interpose a piece of glass, paper, or wood-shaving between the two magnets. These substances are not themselves perceptibly affected by the magnets, nor do they in the least affect the attraction or repulsion between the two magnets.

Substances that are not susceptible to magnetism are, like glass, paper, and wood, *magnetically transparent.* When a magnet causes another body, in contact with it or in its neighborhood, to become a magnet, it is said to *induce* magnetism in that body, *i.e.*, it *influences it to be like itself.* As attraction, and never repulsion, occurs between a magnet and an unmagnetized piece of iron or steel, it must be that the magnetism induced in the latter is such that opposite poles are adjacent; that is, a N or +pole induces a S or —pole next itself, as shown in Figure 147.

§ **188.** Polarity. — Experiment 1. Strew iron filings on a flat surface, and lay a bar magnet on them. On raising the magnet, it is found that large tufts of filings cling to the poles, as in Figure 148, especially to the edges; but the tufts diminish regularly in size from either pole towards the center, where none are found.

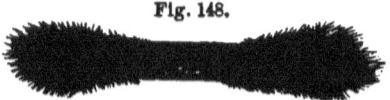
Fig. 148.

Magnetic attraction is greatest at the poles, and diminishes towards the center, where it is nothing, or the center of the bar is neutral. The dual character of the magnet, as exhibited in its opposite extremities, is called *polarity*, and magnetism is styled a *polar force*. If a magnet is broken at its neutral line, as in Exp. 1, p. 28, it is found that equal and opposite polarities exist where there is ordinarily no evidence of them.

Fig. 149.

Experiment 2. Place a copper wire, through which a very strong current of electricity is passing, in a heap of iron filings, — then raise the wire; filings cling to the wire somewhat as they do to a magnet, as shown in Figure 149.

This experiment, and those with the electro-magnet and the deflection of the magnetic needle by an electric current, and a multitude of others that the pupil will meet with, cannot fail to convince him that *an intimate relation exists between electricity and magnetism*, which, though differing in many of their properties, yet alike in many, and almost invariably accompanying one another, and constantly merging one into the other, appear as if they were only different manifestations of one and the same agent.

§ 189. Attraction and repulsion between currents.—

Let us study still further the properties of the current.

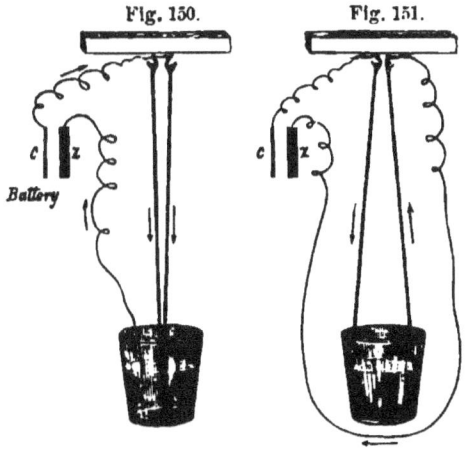

Fig. 150. Fig. 151.

Experiment 1. Suspend two copper wires (Fig. 150), each 50^{cm} long, and about 5^{mm} apart, with their lower extremities dipping about 2^{mm} into mercury, so as to move with little resistance either toward or from each other. In Figure 150 the current divides itself and flows down both wires to the liquid, so that that part of the circuit presents parallel currents flowing in the same direction. Figure 151 is the same apparatus, with the connections so made that the current flows down one wire and up the other, and we have an example of parallel currents flowing in opposite directions. In the former case the wires mutually attract one another. In the latter there is mutual repulsion.

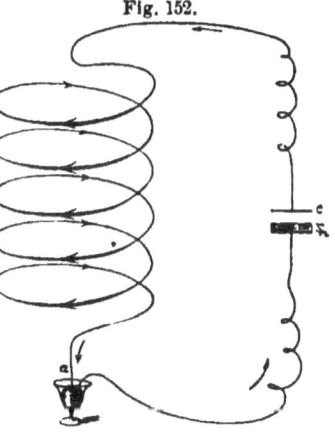

Fig. 152.

Hence, the First Law of Currents: *Parallel currents in the same direction attract one another; parallel currents in opposite directions repel one another.*

An interesting illustration of the former part of this law can be arranged as in Figure 152. A battery wire is bent in the form of a spiral coil. At a the wire is broken, and one end dips just below the surface of mercury in a glass, while the other end is placed in the same liquid at a little distance from the first. When the circuit is closed the current will be parallel with itself, and will flow in the same direction in all parts of the coil that are adjacent. The attraction that follows will cause the coil to contract and lift one pole out of the mercury and break the circuit. The circuit broken, the attraction ceases, and the coil is drawn down again by the force of gravity, and closes the circuit again; and thus constant vibratory motion is produced in the coil.

Experiment 2. Prepare apparatus as represented in Figure 153.

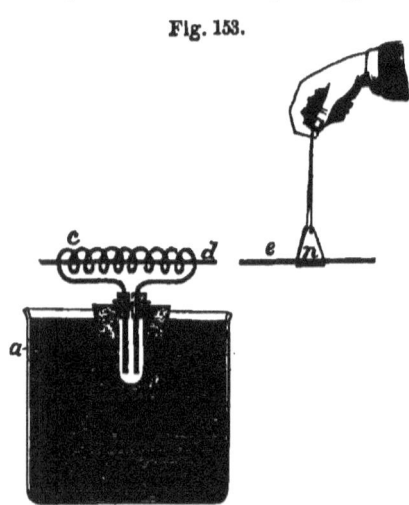

Fig. 153.

Fig. 154.   Fig. 155.

Through a cork a, 8^{cm} in diameter and 5^{cm} thick, cut a circular hole about 4^{cm} in diameter, and insert a glass test-tube b, about 6^{cm} long, that will just fit in the hole. Take an (No. 20) insulated copper wire about 260^{cm} long, wind the central portion into a coil c, 12^{cm} long and 15^{mm} in diameter, with turns about 3^{mm} apart, leaving about 12^{cm} at both extremities unwound. To these extremities solder strips of copper and amalgamated zinc about 3^{cm} long, and as wide as the interior of the test-tube will admit, and allow them to be separated about 5^{mm}. Insert them in the tube, and cover with dilute sulphuric acid. In the center of the coil lay a No. 16 soft iron wire d, and float the whole in a vessel of water. The apparatus constitutes a small floating battery and electro-magnet. Bring one end of a permanent magnet, or a short piece of soft iron wire e, suspended in a paper stirrup n, near to one of the poles of the core of the electro-magnet, and prove by experiment that the coil and its core behave in every respect like a magnet.

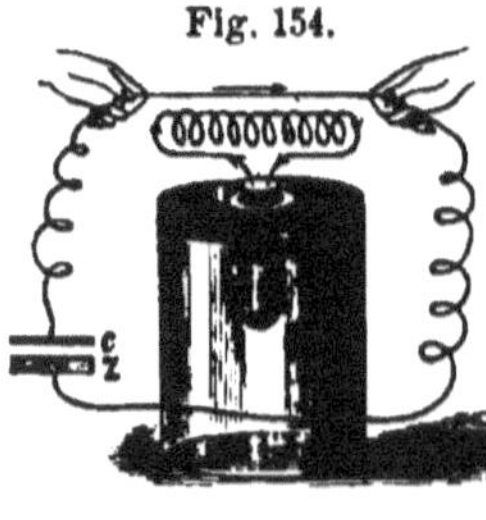

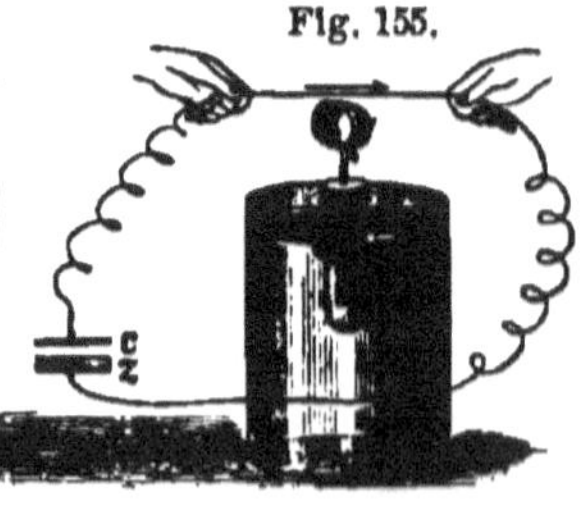

Experiment 3. Remove the iron wire from the floating electro-

magnet, and bring a separate battery wire over and parallel with the helix, as in Figure 154. In this position the two currents flow in planes at right angles to one another. Immediately the coil turns and tends to take a position at right angles to the wire above, so that the two currents may flow in parallel planes and in the same direction, as in Figure 155.

Hence, the Second Law of Currents : *Angular currents tend to become parallel and flow in the same direction.*

Observe that the action of the helix in the last experiment is analogous to the deflection of a magnetic needle by an electric current.

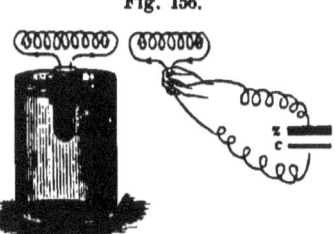

Fig. 156.

Experiment 4. Place opposite one end of the floating helix a second helix, Figure 156, in such a manner that the currents in the two helices may have the same direction. The two poles of the helices attract one another in conformity to the First Law of Currents. Reverse the poles of the helix in your hand so that the currents will flow in opposite directions, though still parallel; they repel one another. (Why?)

The two helices appear to be polarized like two magnets, and for many purposes may be considered as magnets. Observe that at one pole of each helix the current revolves in the direction that the hands of a watch move, and at the opposite pole it revolves in a direction contrary to the movement of the hands of a watch. Bring the north pole of a bar-magnet near that pole of the helix where the motion of the current corresponds to the movement of the hands of a watch. They attract one another; but if the same pole of the helix is approached by the south pole of the magnet, repulsion follows. Hence, that is the south pole of a helix where the current corresponds to the motion of the hands of a watch, (S), and that is the north pole where the current is in the reverse direction, (N). But the important lesson derived from these latter experiments is, that *helices through which currents are flowing behave toward one another, or toward a magnet, in many respects as if they were magnets.*

§ 190. Ampère's theory.

—The facts which we have just studied led Ampère about sixty years ago to devise a theory which furnished a connecting link between magnetism and electricity. It assumed that around every molecule of iron, steel, or other magnetizable substance, electric currents circulate continuously, and thus every molecule becomes a magnet. According to the theory, in an unmagnetized bar these currents lie in all possible planes, and, having no unity of direction, they neutralize one another, and so their effect as a system is zero. But if a current of electricity or a magnet is brought near, the effect of the induction is to turn the currents into parallel planes, and in the same direction, in conformity to the Second Law of Currents. If the coercive force is strong enough, this parallelism will be attained on the removal of the inducing cause, and a permanent magnet is the result.

Intensity of magnetization depends on the degree of parallelism, and the latter depends on the strength of the influencing magnet. When these currents have become quite parallel, the body has received all the magnetism that it is capable of receiving, and is said to be *saturated*. Although the currents really circulate around the individual molecules, yet the resultant of these forces is essentially the same as if the currents circulated around the body as a whole. Figure 157 represents sections of a cylindrical magnet, and the included circles the circulation of the several currents around the molecules lying in these sections. It will be seen that the currents at the contiguous sides of any two of these circles move in opposite directions, and therefore must neutralize one another; while the currents that pass next the circumference of the magnet are not so affected.

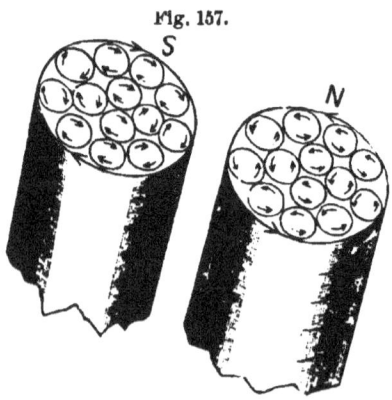

Fig. 157.

AMPÈRE'S THEORY. 219

The hypothetical currents that circulate around a magnetic molecule we shall call *Ampèrian currents*, to distinguish them from the known current that traverses the helix. In strict accordance with this theory, the poles of the electro-magnet are determined by the direction of the current in the helix. The inductive influence of the electric current causes the Ampèrian currents to take the same direction with itself, as represented in Figure 158.

Fig. 158.

However well adapted this theory may be to explain most of the known phenomena of magnetism, it should be borne in mind that physicists of this generation value the theory rather as a help to the imagination and memory, than as a true statement of the facts. It is nearer the truth to say that the molecules are polarized *as if* currents were circulating around them; of the actual existence of such currents we know nothing. So also of the real nature of polarity we know little or nothing.

EXERCISES AND QUESTIONS.

1. Regarding your lead-pencil as a rod of iron, and a string as an electric wire, tie a knot at the end where the current is supposed to enter it, and wind the string spirally around your pencil, so as to make the point of the pencil a south pole.

2. What would be the effect of reversing the current?

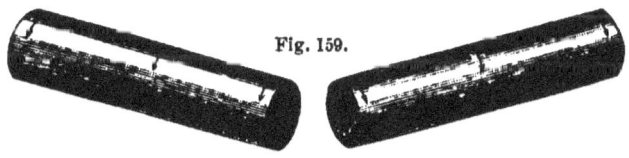

Fig. 159.

3. Take two tin mustard-boxes, and paint arrows around them, also on the ends, all turned in the same direction, to represent Ampèrian currents, as in Figure 159. Imagine each to be a magnet, determine

which is the north and which is the south pole of each, and mark them accordingly with the letters N and S.

4. (*a*) Place the two south poles near one another, and ascertain why they should repel one another. (*b*) Do the same with the two north poles.

5. Let a north and a south pole face one another, and show why they should attract one another.

6. Stretch a string in a northerly and southerly direction, and suspend one of the boxes as a magnetic needle over and parallel with the string, with its north pole pointing north; then imagine a current to enter the string at its southern extremity, and determine its effects on the needle.

7. Why is a magnetic needle deflected by an electric current?

8. Why is the direction of the deflection dependent on the direction of the current?

§ 191. **The earth a great magnet.** — Experiment 1. Magnetize a cambric needle. Suspend it by a fine thread attached to its middle over a magnet, and midway between its poles. The needle, however placed, immediately takes a position parallel with the magnet. The magnet exerts a *directive* influence on the needle. Remove the magnet, and the needle takes a northerly and southerly direction.

If you carry the needle all over your town or State, it will still maintain this direction. Something, like the magnet, exerts a directive influence on the magnetic needle.

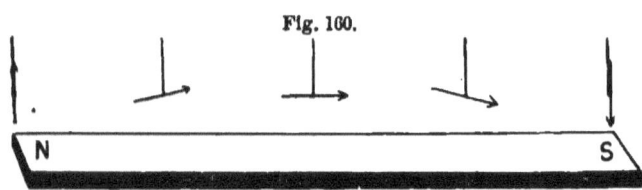

Fig. 100.

Experiment 2. Place the needle once more in its original position over the magnet, and gradually move it from the middle towards one pole of the magnet; the needle ceases to be horizontal. At either side of the center it *dips;* if it is nearer the N-pole of the bar, the S-pole dips, and conversely, as shown in Figure 160. If the needle is properly supported, the dip increases till at the poles the inclination is 90°.

If a magnetic needle freely suspended is carried to different parts of the earth's surface, it will dip as it approaches the polar regions, and is only horizontal at or near the earth's

THE EARTH A MAGNET. 221

equator. A common compass needle must have the S-end loaded to keep it horizontal. Like effects are commonly attributed to like causes. These phenomena are just what we should expect if (as is very improbable) a huge magnet were thrust through the axis of rotation of the earth, as represented in Figure 161, — having its N-pole near the S geographical pole, and its S-pole near the N geographical pole; or if (as is more probable) the earth itself is a magnet.

Fig. 161.

Experiment 3. Magnetize a circular steel disk, so that its poles may be at the extremities of one of its diameters. Place it beneath a plate of glass. Sift over the glass fine iron-filings, as in Exp. 2, p. 28. Gently tap the glass a few times, so as to agitate the filings. Once in motion, they arrange themselves in lines radiating from either pole, forming graceful curves from pole to pole, as represented in Figure 162. These represent what are called *lines of magnetic force*. They represent the resultants of the combined action of the two poles. Now carry the little magnetized cambric needle around the disk. It follows the lines of magnetic force as mapped out by the filings, always assuming a position tangent to the magnetic curve, as shown in Figure 162.

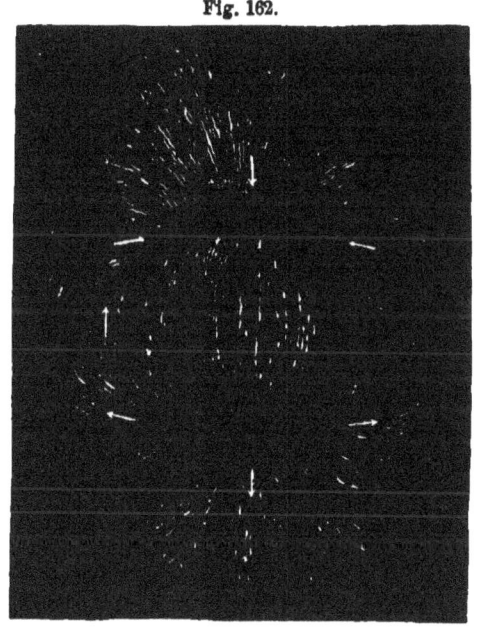

Fig. 162.

It is evident that the space around a magnet is the seat of a peculiar influence; this space, extending as far as the magnet exerts any effect, is called the *magnetic field*. The last experiment presents a true exhibition, on a small scale, of what the earth does on a large one, and thereby presents one of many phenomena which lead to the conclusion that *the earth is a magnet*.

§ **192. Magnetic poles of the earth.** — It will be seen that there are two points where the needle points directly to the center of the disk. A point was found on the western coast of Boothia, by Sir James Ross, in the year 1831, where the dipping needle lacked only one-sixtieth of a degree of pointing directly to the earth's center. The same voyager subsequently reached a point in Victoria Land where the opposite pole of the needle lacked only 1° 20' of pointing to the earth's center.

It will be seen that, if we call that end of a magnetic needle which points north the N-pole, we must call that magnetic pole of the earth which is in the northern hemisphere the S-pole, and vice versa. (See Fig. 162.) Hence, to avoid confusion, many careful writers abstain from the use of the terms *north* and *south poles*, and substitute for them the terms *positive* and *negative*, or *marked* and *unmarked poles*.

§ **193. Variation of the needle.** — Inasmuch as the magnetic poles of the earth do not coincide with the geographical poles, it follows that the needle does not in most places point due north and south. The angle which the needle makes with the geographical meridian is known as the *angle of declination*. This angle differs at different places.

Experiment. As Columbus found, we can easily find, the declination at any place as follows: Set up two sticks so that a string joining them points to the North Star; the string will lie in the geographical meridian. Place a long magnetic needle over the string; the angle between the needle and the string is the required declination. If great accuracy is required, allowance must be made for the fact that the star is not exactly over the pole, but appears to describe daily around it a circle whose diameter is about 4°.

VARIATION OF THE NEEDLE. 223

Let A (Fig. 163) represent a magnetic pole, and B the North Star; it will be seen that there is a position in which the needle will point due north. A line passing around the earth through the two magnetic poles, connecting those places where the needle points due north, is called a *line of no variation*.

Fig. 163.

On the map, Plate II., it is marked 0. Lines east and west of this line, and approximately parallel with it, represent lines of equal variation. At places in the United States east of this line the needle points west of north, *e.g.*, New England; but most of the States lie west of this line, so in them the needle points east of north.

The magnetic poles are not fixed objects that can be located like an island or cape, but are constantly changing. They appear to swing, somewhat like a pendulum, in an easterly and westerly direction, each swing requiring centuries to complete it. The north magnetic pole is now on its westerly swing. The chart given in this book was only true at the time of observation in 1870. To be true for the present time, each of the lines should be moved westward at the rate of about 1° for every twelve years.

§ 194. Natural magnets. — On the assumption that the earth is a magnet, it would not be strange if magnetizable substances should partake of its magnetic properties by induction. An ore of iron called *lodestone*, composed of a mixture of two oxides of this metal, possesses more or less magnetic power. Such magnets are termed *natural* magnets, to distinguish them from the *artificial* magnets of steel.

§ 195. Cause of the earth's magnetism. — The cause of the earth's magnetism is not known. The theory that it is an electro-magnet in virtue of currents flowing around it near its surface, from east to west, explains all the effects that it produces on the magnetic needle. But what sustains these electric currents? There are many things that point to the sun as the source of the earth's magnetism. Those who adopt this theory generally regard the terrestrial currents as *thermo-electric*.

A single instance must suffice to illustrate the intimate relation tha certainly exists between the sun's condition and the earth's mag netism. In 1859 two observers remote from each other saw simul taneously a bright spot break out on the face of the sun, whose duration was only five minutes. Exactly at this time there was a general dis turbance of magnetic needles, and telegraph wires all over the world were traversed with so-called *earth currents*. Telegraphers received shocks, and an apparatus in Norway was set on fire. These phenomena were quickly followed by auroral displays. Sometimes telegraphs are worked by earth currents alone, without any battery in the circuit.

§ 196. **General remarks on magnets and magnetism** — Artificial magnets, including permanent magnets and electro magnets, are usually made in the shape either of a straight bar or of the letter U, called the *horse-shoe*, according to the use made of them. If we wish, as in the experiments already de scribed, to use but a single pole, it is desirable to have the other as far away as possible; then obviously the bar-magnet is most convenient. But if the magnet is to be used for lifting or hold ing weights, the horse-shoe form is far better, because the attraction of both poles is conveniently available and because their combined power is more than twice that of a single pole. This is due to the reflex influence of the poles on one another through the armature. Magnets, when not in use, ough always to be protected by armatures (A, Fig. 164) of soft iron; for, notwithstanding the coercive power of steel, they slowly part with their magne tism. But when an armature is used, the opposite poles of the magnet and armature being in contac with one another, *i.e.*, N with S, they serve to bind one another's magnetism.

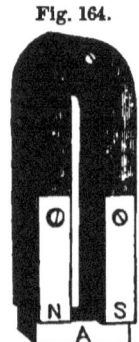

Fig. 164.

Thin bars of steel can be more thoroughly magnetized than

Plate II.

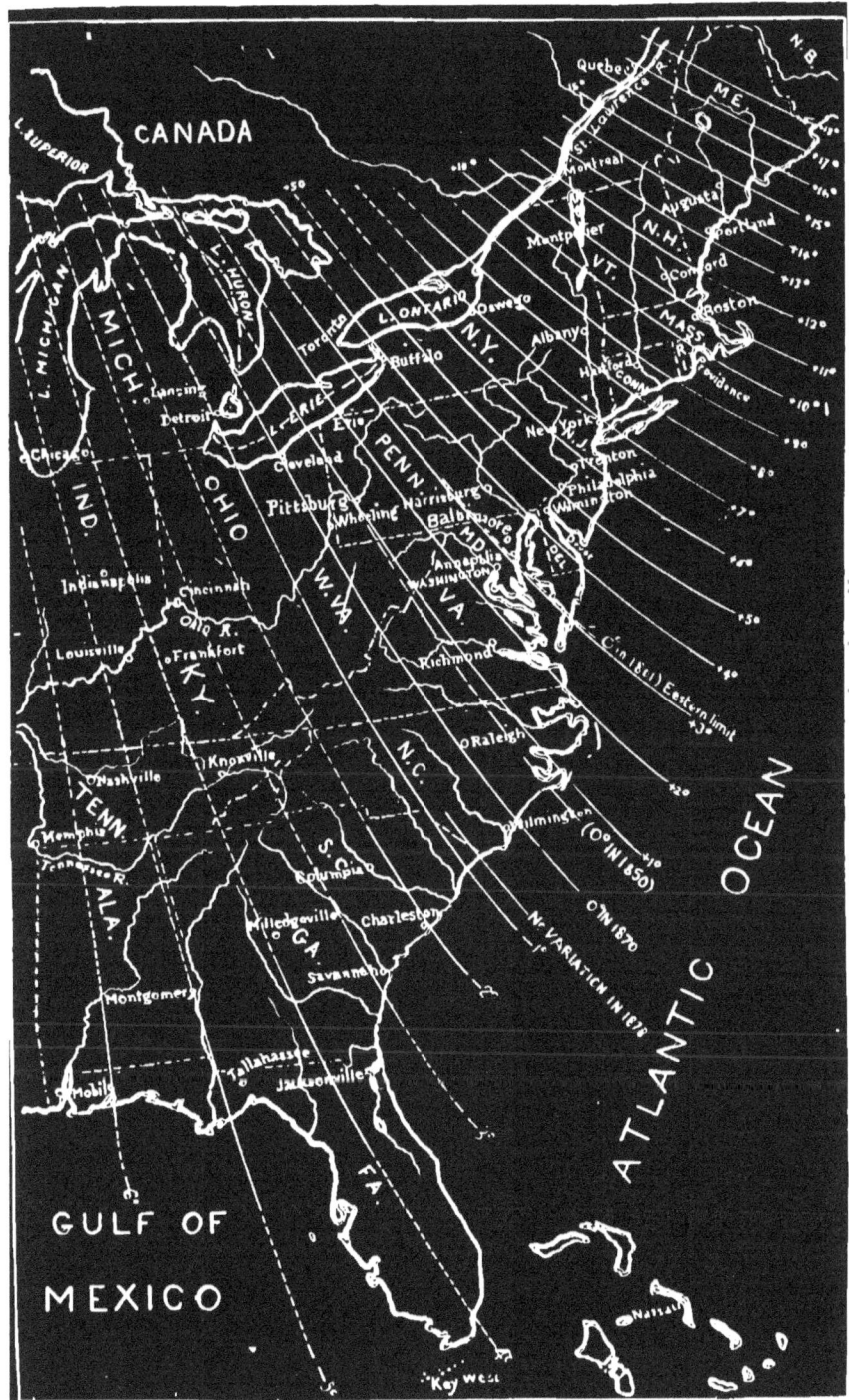

thick ones. Hence, if several thin bars (Fig. 164) are laid side by side, with their corresponding poles turned in the same direction, and then screwed together, a very powerful magnet is the result. This is called a *compound magnet*. In any magnet the outer layers are far more strongly magnetized than the central ones; so a steel tube makes very nearly as strong a magnet as a rod of the same diameter, and is much lighter than the latter.

§ 197. **Diamagnetism.** — Besides iron and steel, many other substances, and possibly all substances, both in the liquid and gaseous, as well as in the solid state, are more or less susceptible to magnetic influence. Conspicuous among these are nickel and cobalt. But this influence is not always of the same kind. A small bar of bismuth suspended between the poles of a powerful electro-magnet, instead of being attracted is repelled by the poles of the magnet, as shown by its taking a position with its longest axis at right angles to a direct line between the poles. Substances which behave in this manner are called *diamagnetic*, and they are said to place themselves *equatorially* between the poles. Substances that place themselves *axially* between the poles, as iron and nickel, are called *paramagnetic*, or simply magnetic.

Paramagnetic liquids placed in a watch-glass between the poles become heaped up at the poles and depressed in the center, while the opposite phenomena occur with diamagnetic liquids. The magnetic behavior of gases may be learned by inflating soap-bubbles with them, and noting the direction of their distension. Alcohol, water, nitrogen, and carbonic acid are diamagnetic. Oxygen is paramagnetic. The only substances whose magnetic properties can be shown without extraordinary apparatus are iron and its compounds.

§ 198. **Magnets not sources of energy.** — Perpetual-motion seekers are easily led into the error of supposing that in the magnet they have an inexhaustible supply of energy; but a very little study will serve to exhibit the character of the error. If, for instance, we bring a piece of iron near a magnet, it is attracted, and, if allowed to move up to the magnet, this force of attraction will do a certain amount of work. Take now another piece of iron similar to the first; this also will be attracted, and a certain amount of work will be performed, but a less amount than that done in the first instance. Continue the operation until the magnet no longer attracts; then the magnet has done a definite amount of work, and lost the power of doing more. To restore it to its original condition, we must remove all the pieces of iron; this will require an expenditure of external work exactly equal to that originally performed by the magnet.

XXXII. MAGNETO-ELECTRIC AND CURRENT INDUCTION.

§ **199.** Introductory experiments.—**Experiment 1.** Connect a helix with a delicate galvanometer (Fig. 165), and quickly thrust a magnetized steel rod into the coil. A deflection of the needle shows that a current of electricity at that instant traverses the wire. But

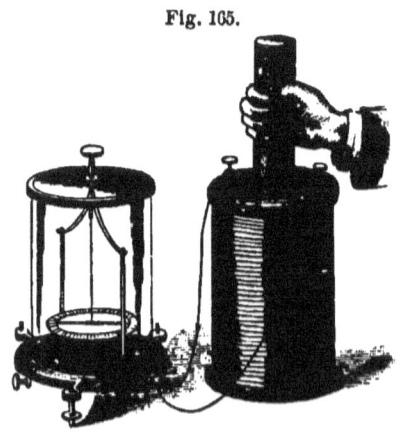

Fig. 105.

the needle, after a few oscillations, assumes its original position. This shows that the current was only momentary. Quickly remove the magnet; again the wire is traversed by a current, but this time in an opposite direction to the first, as shown by an opposite deflection. Repeat the experiment, and notice that when the magnet approaches the coil the induced current runs in the opposite direction to the Ampèrian currents, as represented in Figure 166. But when the magnet is withdrawn, the induced current takes the same direction with the Ampèrian currents, as in Figure 167. In the former case, the repulsion due to opposite currents must act as a resistance to the force that brings them

MAGNETO AND DYNAMO MACHINES. 227

together. Likewise, in the latter case, the attraction due to currents flowing in the same direction must resist the force that separates them. Hence, *the energy shown by the electric current has been generated at the expense of mechanical energy.*

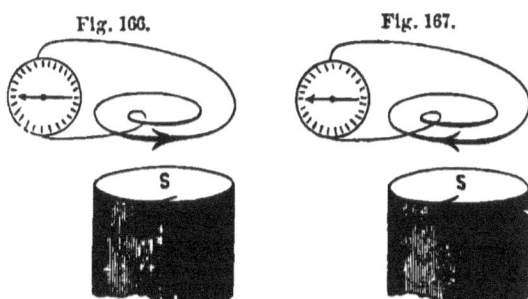

Fig. 166. Fig. 167.

Experiment 2. Place within the coil a core of soft iron. Wave back and forth, over one extremity of the core, one of the poles of a powerful bar-magnet. The needle of the galvanometer is violently agitated, being deflected in one direction at each approach, and in the opposite direction at each departure. Now repeat the experiments with the opposite pole of the magnet. The effect is, as we should expect, to reverse all the currents.

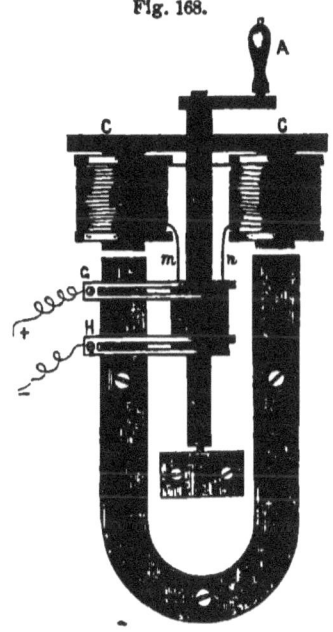

Fig. 168.

§ 200. Magneto and dynamo machines. — If the permanent magnet is stationary, and the electromagnet is moved back and forth, the result is the same as when the magnet was moved and the electromagnet was stationary. Machines constructed for the purpose of generating electric currents in this manner are called *magneto-electrical machines.*

Figure 168 will give a general idea of the construction of the simpler kinds of magneto machines. N S is a permanent compound horse-shoe magnet. E E are coils containing cores of soft iron connected by the back-armature C C, the whole constituting a sort of an armature to the permanent magnet. The brass axle D D is rigidly connected with the back-armature C C, so that when the axle is rotated by means of the crank A, both helices are carried around with it. Now, suppose the crank to be turned; during the first quarter of a revolution a separation of poles occurs, and currents of electricity are established in both helices. The wire that constitutes the helix is wound in opposite directions around the two cores, so that the two currents may not flow in opposite directions through the wire, and thereby neutralize one another, but may have a common direction, and thereby produce a current of double the electro-motive force that would be produced in a single helix. During the second quarter-revolution the poles approach one another, and the effect would be to reverse the current; but the polarity of the cores also change as they are now brought under the influence of the poles which they are approaching, and this double change leaves the current to flow in the same direction as it did before. At the end of a half revolution there is a reversal of current, as the poles do not change at this point. The result would be that during every revolution there would be a current half of the time in one direction, and half of the time in the opposite direction. In order to secure a constant current in one direction, a current-reverser I, or *commutator*, as it is called, attached to the axle, is so arranged that the current is reversed at the end of each half revolution, and is then conducted away by the wires G H.

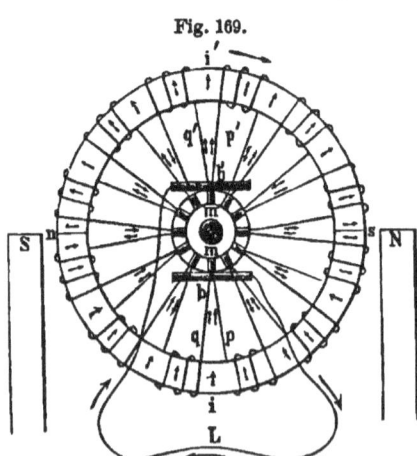

Fig. 169.

Each of the two currents produced in a single revolution has a maximum point, or point of greatest intensity, when the cores are nearest the poles of the magnet; and a minimum point, or point of least intensity, when they are farthest from the poles. Between these two points the current is constantly growing or diminishing. It is apparent that such a machine gives not only an intermittent current, but one that resembles a succession of waves or a stream produced by the strokes of a pump, alternately rising and sinking. But for most purposes for which electricity is employed, it is important that the current should be continuous and uniform. Figure 169 will serve to illustrate the principle by which this is secured in the widely-known Gramme machine.

MAGNETO AND DYNAMO MACHINES. 229

The armature *ns* consists of a ring composed of a bundle of soft iron wires (better shown in Fig. 169 *a*) surrounded by what is virtually an endless coil of wire. The wire, however, is put on in separate coils, the in-wire of one united to the out-wire of the next, and from each junction a branch wire is led to a copper plate on the axis of rotation *mm*. A horse-shoe magnet NS (only a portion of which is shown in the cut) is so placed that one-half of the ring is under the influence of the N-pole, and the other half under that of the S-pole. Suppose the ring to rotate in the direction of the arrow; then every point of the iron core, as it comes opposite a given point of the magnet, will successively become a pole of opposite name, while the points *i* and *i'* are the neutral points. If we imagine the core to be divided at the points *n* and *s*, we have two semicircular magnets whose north poles and whose south poles respectively face one another. In the two mutually-facing poles on either side, the Amperian currents must be in opposite directions. Now an attentive study of this ideal diagram in the light of what you have previously learned respecting the generation of induced currents will enable you to see that as the ring armature rotates, the corresponding advance of the induced poles of the ring will induce currents in the wire in such a manner that all the coils which at any given moment are in the semicircle next one of the magnet poles — say the North — are traversed by a current of one direction. Similarly, the semicircle formed by the coils immediately approaching, or immediately receding from the South pole are at the same time traversed by a current of the opposite direction. The result is that currents in the lower half tend *toward* the point *m* on the axis, and in the upper half *from* point *m'*. So long as the leading out-wires from these points are open, these currents have no outlet, and consequently oppose and neutralize one another. But if the points *m* and *m'* are connected by a wire L, we shall have a *constant and non-alternating current* flowing through the wire from *m* to *m'*. The contact at these points is made by means of brushes of thick wire provided as springs. These press on the contact pieces and make practically a constant connection with the two halves of the circuit.

Inasmuch as an electro-magnet may be made a much more powerful magnet than a permanent magnet, it is now extensively used as the inducing or the so-called *field magnet*. Such a machine is called a *dynamo-electrical machine*, or often more briefly a *dynamo*. Figure 169 *b* represents such a machine. EE is the stationary field magnet, A the moving armature, and N and S large pole-pieces brought as near as practicable to the armature and partially encircling it. When the machine is at rest, there are no currents; but when the armature is in motion, the residual magnetism (a small portion of the magnetism which soft iron always retains after it has been magnetized) induces at first a weak current in the wire of the armature; but as a portion of this current is carried by means of a shunt wire l through the coil of the field magnet, and magnetizes the core more and more strongly, the current in both the shunt l and the main wire L quickly reaches its maximum.

By the kind permission of the United States Electric Lighting Company we introduce a cut, Fig. 169 c, of the popular American dynamo called the Weston. It will be seen that in this machine two powerful field magnets are placed one on either side of the revolving armature. Power originated in a steam engine is communicated to the dynamo by means of a belt passing over the circumference of the wheel W and causes the armature, which is on the axle of this wheel, to revolve.

§ **201. Current induction.** — If, in the original experiment in magneto-electric induction (p. 226), a helix connected with a battery is substituted for the permanent magnet, precisely the same results are obtained as with the magnet. Indeed, we ought to expect the same results. (Why?) The wire A, Figure 170,

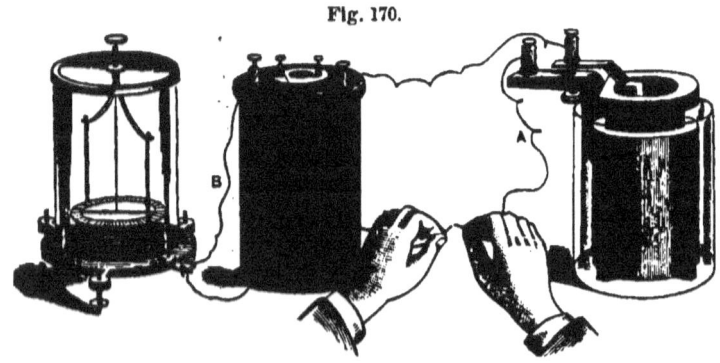

Fig. 170.

through which the battery current circulates, is known in this case as the *primary wire*, and the battery current, the *primary* or *inducing current*. The wire B, through which the induced currents circulate, is called the *secondary wire*, and the currents that traverse this wire are frequently called *secondary currents*.

Fig. 169 c.

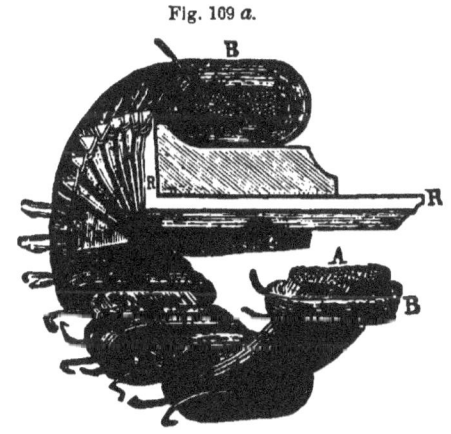

Fig. 169 a.

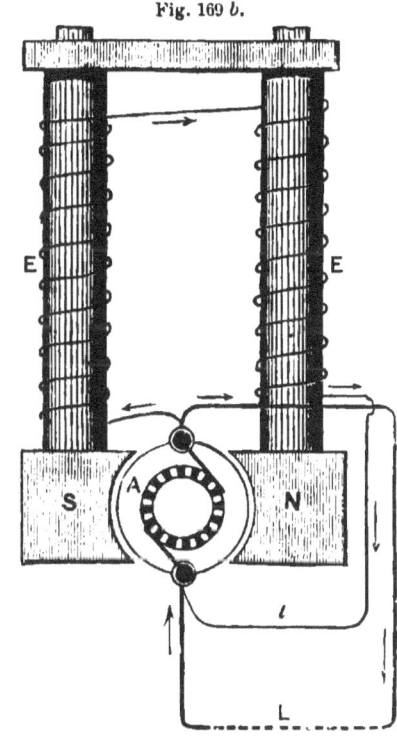

Fig. 169 b.

It will be observed that in all these experiments we have a relative motion between a conductor and an inducing body (magnet or current-bearing conductor) ; also, that electricity flows only during the continuance of the relative motion. Delicate measurements have proved that the total quantity of induced electricity transmitted in the conductor depends on the total quantity of change in relative motion, and not at all on the time occupied in this change. Hence, it is evident, that the more rapid the change, the more intense must be the momentary current; *i.e.*, the greater must be the quantity of electricity flowing at the moment. Combining this statement with Ohm's law, remembering that the resistance of the secondary circuit is constant, we derive the following very important law: *In any induced current, the E.M.F. at any instant is proportional to the rapidity of the relative change at that instant.*

If, instead of bobbing the primary helix in and out of the secondary helix, the former is allowed to remain stationary within the latter, it is found that making and breaking (Fig. 170) the primary current, *i.e.*, starting and stopping a primary current, induces currents in the secondary wire. Indeed this process is evidently much the same thing (and in theory exactly the same thing), as moving the primary conductor, with unbroken circuit, from an infinite distance where its action would be zero, into the secondary circuit, the whole change occupying a very brief time. A reversal of the process would evidently correspond to breaking the primary circuit. The results obtained in Exp. 1, § 197, enable us at once to predict the direction of an induced current, and we may formulate the two cases thus: —

(*a*) *When the primary current is approached, or a current originated in the primary circuit, the induced current has an opposite direction to that of the primary.*

(*b*) *At a departure from the primary current, or when a current in a primary circuit is stopped, the induced current has the same direction as the primary.*

§ **202. Extra current.** — The conclusions of the preceding section admit of an important extension. We learned that *any* electric or magnetic disturbance in the neighborhood of a conductor gives rise to a current. But it is evident that in a single wire every portion must be considered as a neighboring conductor with respect to every other portion; consequently there can be no change of electrical condition in a wire without accompanying induction phenomena. If we suddenly close a circuit the current does not abruptly assume its final intensity, because there is a current induced in the same circuit whose direction is such as to retard the change from zero. So, too, if a closed circuit be suddenly broken, there is a current induced in the direction of the primary current which retards the change to zero. These induced currents are often, for the sake of distinction, called *extra currents*. Of course, if all parts of the conductor are kept close together by winding the wire into a helix or a spiral, the effect is much increased. It is evident that the direct extra current must produce a much greater effect than the indirect, because the former is added to the primary, while the latter is subtracted. This is the cause of the bright spark on breaking a strong current, and also of physiological effects or shocks experienced on breaking the primary circuit in the experiment illustrated by Figure 170. If a soft iron core is introduced into a helix, the extra currents are vastly increased by the action of the magnetic changes. (Why?)

§ 203. Induction coils. — If a core of iron, or, still better, a bundle of wires (A A, Fig. 171), is inserted in the primary coil, it is evident that it will be magnetized and demagnetized every time the primary is made and broken. The starting and cessation of Ampèrian currents in the core in the same direction as the primary current, and simultaneous with the commencement and ending of the primary current, greatly intensifies the secondary current. To save the trouble of making and breaking by hand, as in Figure 171, the core is also utilized in the construction of an automatic make-and-break piece. A soft iron hammer b is connected with the steel spring c, which is in turn connected with one of the terminals of the primary wire.

The hammer presses against the point of a screw d, and thus, through the screw, closes the circuit. But when a current passes through the primary wire, the core becomes magnetized, draws the hammer away from the screw, and breaks the circuit.

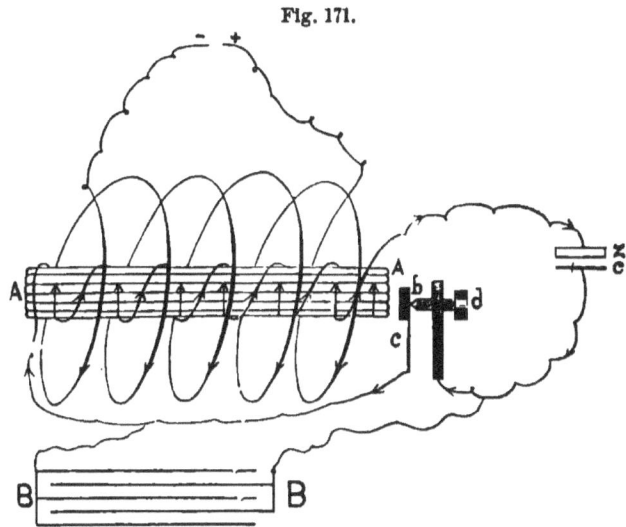

Fig. 171.

The circuit broken, the core loses its magnetism, and the hammer springs back and closes the circuit again. Thus the spring and hammer vibrate, and open and close the primary circuit with great rapidity. An instrument made on these principles is called an *induction coil*.

§ **204. Ruhmkorff's coil.**—This instrument has the important addition, to the parts already explained, of a *condenser* B B. This consists of two sets of layers of tinfoil separated by paraffine paper; the layers are connected alternately with one and the other pole of the battery, as the figure shows, so that they serve as a sort of expansion of the primary wire. When the circuit is broken, the extra current would jump across at b, and would vaporize the points of contact, and form a bridge with the vapor of metal that would prolong the time of breaking. But, when the condenser is attached, the extra current finds an escape into it easier than to jump across at *b*, so the vaporizing of the contact is avoided, and the time of breaking being much shortened, the secondary is much more intense.

The primary helices of induction coils consist of comparatively few turns of coarse insulated wire; but the secondary helices contain many turns of very fine wire, insulated with great care. The secondary current is, at breaking, as we ought to expect from the extreme rapidity with which the primary circuit is broken, distinguished from the primary, or galvanic current, by its vastly greater tension, or power to overcome resistances. A coil constructed for Mr. Spottiswoode of London has two hundred and eighty miles of wire in its secondary coil. With five Grove cells this coil gives a secondary spark forty-two inches long, and perforates glass three inches thick. Many brilliant experiments may be performed with these coils which will be indicated in connection with frictional machines.

XXXIII. THERMO-ELECTRICITY.

§ **205.** So far in our experiments we have obtained a current of electricity by using the potential energy due to the chemical affinity of zinc and sulphuric acid, or by expending mechanical energy; can we not also get a current directly from the molecular energy that we know as heat?

Experiment. Insert in one screw cup of a sensitive galvanometer an iron wire, and in the other cup a copper, or, better, a German silver wire. Twist the other ends of the wire together, and heat them at their junction in a flame; a deflection of the needle shows that a current of electricity is traversing the wire. Place a piece of ice at their junction. A deflection in the opposite direction shows that a current now traverses the wire in the opposite direction.

These currents are named, from their origin, *thermo-electric*. The apparatus required for the generation of these currents is very simple, consisting merely of bars of two different metals joined at one extremity, and some means of raising or lowering their temperature at their junction, or of raising the temperature at one extremity of *the pair* and lowering it at the other; for the electro-motive force, and consequently the strength of the current, is nearly proportional to the difference in temperature of the two extremities of the pair. The strength of the current is also dependent, as in the voltaic pair, on the thermo-electro-motive force of the metals employed. The following *thermo-electric series* is so arranged that if the temperatures of both junctions are near the ordinary temperatures of the air, those metals farthest removed from each other give the strongest current when combined; and the current passes, when heated at their junction, from the one first named to that succeeding it. The arrows indicate the direction of the current at the heated and cold ends respectively. At high temperatures the current may be reversed.

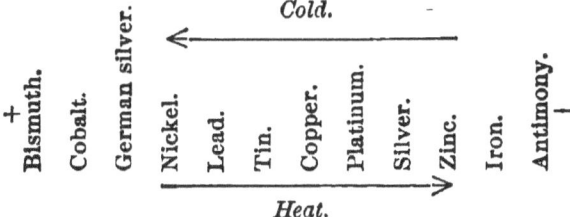

§ **206. Thermo-electric batteries and thermo-pile.**—
The electro-motive force of the thermo-electric pair is very small in comparison with that of the voltaic pair; hence the greater necessity of combining a large number of pairs with one another in series. This is done on the same principle, and in the same manner, that voltaic pairs are united, viz., by joining the +metal of one pair to the —metal of another. Figure 172 represents such an arrangement. The light bars are bismuth, and the dark ones antimony. If the source of heat is strong and near, by either conduction or convection one face may be

Fig. 172.

heated much hotter than the other, and a current equal to that from an ordinary galvanic cell is often obtained. Instruments constructed on these principles, and used as a source of electricity, are very convenient and efficient for many purposes, especially when a steady current is required with small external resistance ; they are called *thermo-electric batteries*.

If the source of heat is feeble or distant, the feeble current may serve to measure the difference of temperature between the ends of the bars turned toward the heat (as in Figure 172) and the other ends, which are at the temperature of the air. The apparatus, when used for this purpose, is called a *thermo-pile*, or a *thermo-multiplier*. A combination of as many as thirty-six pairs of antimony and bismuth bars, connected with a very sensitive galvanometer, constitutes an exceedingly delicate *thermoscope* and *thermometer*. Quantities of heat, that would not perceptibly expand the mercury in an ordinary thermometer, can, by the use of a thermo-electric pile, be made to produce large deflections of the galvanometer needle. Heat radiated from the body of an insect several inches from the pile may cause a sensible deflection.

XXXIV. FRICTIONAL ELECTRICITY.

§ 207. Mechanical energy transformed into electrification. — **Experiment.** Prepare an insulated stool (see § 214) by placing a square board on four *dry* and *clean* glass tumblers, used as legs. Let a person whom we will call John stand on this stool, and let a second person, James, strike John a few times with a cat's fur. Then let James bring a knuckle of a finger near to some part of John's person, for instance a knuckle of his hand, or his chin or nose; an electric spark will pass between the two, and both will experience a slight shock. The length of the spark shows that the electricity is urged by a high E.M.F., like the induced currents of the magneto-machine and induction coil.

Fig. 173.

As mechanical energy is transformed into a kind of molecular motion, or internal energy, called heat, when one hammers an anvil, so in this case a portion of the motion of the fur at each stroke is transformed into another phase of internal energy known as *electrification*. Electricity made apparent in this manner is called *frictional electricity*, because the electrification is developed by friction between two surfaces.

Fig. 174.

§ 208. Electroscope. — **Experiment.** Suspend in a loop, tied in a white silk thread, a strip of gold foil 20^{cm} long and 15^{mm} wide, so that the two vertical portions may be near each other. After John has been struck a few times with the fur, let him bring a finger gradually near the upper extremity of the foil; the two portions of the foil gradually diverge, as in Figure 174, indicating the presence of an unusual force in his body.

We have already found that this force is due to electricity. Bodies in this state are said to be *charged* with electricity, or simply *electrified*. Such electrification in a person is often manifested by a divergence of hair on his head. Any arrangement, like that of the foil just described, intended to detect the presence of electrification, is called an *electroscope*. One of the most common and useful electroscopes consists of one or two pith-balls, made from the pith of elder or sunflower, suspended by silk thread. If an electroscope is brought near to either pole of a secondary wire of an induction coil, a similar electrification is manifested by the poles. Likewise, by means of very delicate electroscopes, the poles of a galvanic battery, or of a thermo-battery, are found to be feebly electrified.

§ 209. Attractions and repulsions. — **Experiment 1.** Poise a flat wooden ruler on an inverted bottle or flask, having a round bottom, as in Figure 175. Draw a rubber comb two or three times through your hair, or rub it with a woolen cloth, and place it near one end of the ruler; instantly the ruler moves toward the comb.

Fig. 175.

Experiment 2. Hold the comb over a handful of bits of tissue paper; the papers quickly jump to the comb, stick to it for an instant, and then leap energetically from the comb. The papers are first attracted to the comb, but in a short time acquire some of its electrification, and then are repelled.

§ 210. Two states of electricity. — It is quite apparent that we are now dealing with a very different class of electrical phenomena from any that we have previously observed. It is also quite as obvious that we are dealing with electricity in a very different state or condition from that in which we have before studied it. Hitherto we have studied only those phenomena produced by electricity

TWO STATES OF ELECTRICITY. 239

when in motion; and, inasmuch as when in that state its energy is expended in work, or transformed into some other form of energy as rapidly as it is generated, there was no such thing as an *accumulation* of electricity. In our late experiments there is wanting anything like a current; but, on the other hand, we find that electricity in this new state may accumulate, be stored up, and remain in a quiescent state for an indefinite time. In the latter state it is incapable of affecting a magnetic needle, magnetizing, generating heat, illuminating, producing decomposition, or giving shocks. But in this state of apparent repose it may attract and afterwards repel light bodies in the vicinity of the body in which it resides. These attractions and repulsions are quite distinct from the attractions and repulsions which occur between parallel currents.

This state of electricity is called *static*, in distinction from the current state, which is often called *dynamic*. We have seen that, under certain conditions, electricity may change from one state to the other, as when the electricity which had accumulated in the boy on the insulated stool passed to the other boy, producing, in its current state, both illuminating and physiological effects; and again, when a current is broken, the current ceases, but electricity accumulates in the wire (see § 208). We have also learned that electricity of high potential, such as is most readily developed by friction, exhib-

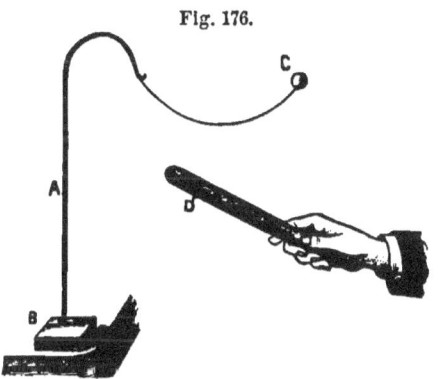

Fig. 176.

its the static phenomena, *i.e.*, attractions and repulsions, most strikingly; but we must be careful to avoid the notion that these are peculiar to electricity so derived.

§ **211. Two kinds of electrification.**—Experiment 1. Bend a small glass tube into the form represented by A, Figure 176, and insert one end in a block of wood B for a base; and suspend from the tube a pith-ball C by a silk thread. Rub a glass rod D with a silk handkerchief, and present it to the ball; attraction at first occurs, followed by repulsion after contact. Now rub a stick of sealing-wax, or a hard-rubber ruler, with flannel, and present it to the ball, which is in a condition such that it is repelled by the electrified glass; it is attracted by the electrified sealing-wax. We are led to suspect that the sealing-wax possesses a different kind of electrification from that of the glass. Let us further test the matter.

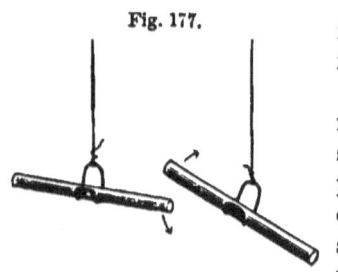

Fig. 177.

Experiment 2. Suspend two glass rods that have each been rubbed with silk in two wire stirrups (Fig. 177), and present them to each other; they repel one another. Suspend two sticks of sealing-wax that have been rubbed with flannel in the same manner; the same result follows. Now, in a like manner, present one of the glass rods and one of the sticks of sealing-wax to each other; they attract one another.

Fig. 178.

Experiment 3. Make a pin hole in each end of a hen's egg, and blow its liquid contents out. Apply, with flour paste, tinfoil smoothly to the surface of the shell, and completely cover it. With a drop of melted sealing-wax attach one end of a silk thread midway between the ends of the shell, so that it may be suspended, as in Figure 178. Repeat the last two experiments with the shell as with the pith-ball; you obtain similar results.

It is evident (1) that there are *two kinds or conditions of electrification;* or, for convenience, we sometimes say *two kinds of electricity;* (2) that they are so related to each other that *like kinds repel and unlike kinds attract one another.* The two kinds are usually distinguished

INDUCTION. 241

from one another by the names *positive* and *negative*, or, more briefly, as +e and −e. The former is, by definition, such as is developed on glass when rubbed with silk, and the latter is the kind developed on sealing-wax when rubbed with flannel. There is no reason, except custom, for calling the one positive rather than the other.

Experiment 4. Once more electrify a stick of sealing-wax with a flannel, and present it to the ball or shell, and after the ball is repelled, bring the surface of the flannel which had electrified the rod near the ball; the ball is attracted by it, showing that the rubber is also electrified and with the opposite kind to that which the sealing-wax possesses. (Which kind of electrification has the flannel?)

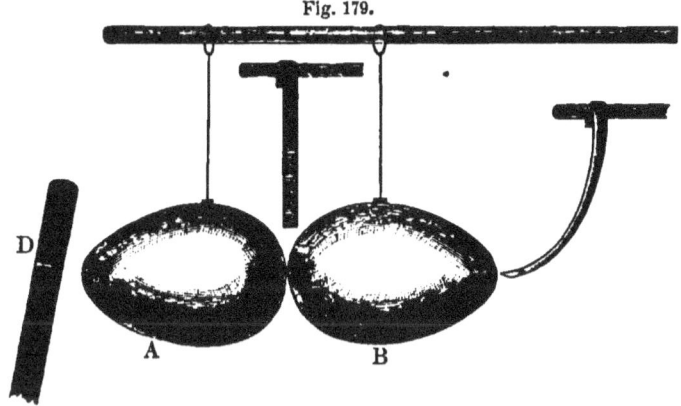

Fig. 179.

One kind of electrification is never developed alone; when two substances are rubbed together, both always become oppositely electrified, and to an equal amount. In general, *whenever any electricity of either sign is developed, an equal quantity of the other is to be found.* (Ascertain the kind of electrification developed on a rubber comb when it is passed through the hair; also the kind developed on a person when whipped with fur, by presenting the bodies whose electrification is to be tested to a body having a known electrification.)

§ **212.** Induction.—**Experiment.** Suspend two egg-shells, prepared as above, so as to touch one another, end to end, as in Figure 179.

Bring near to one end of the shells, but not to touch, a sealing-wax rod excited with flannel, and therefore having $-e$. While the rod is in this position, carry a thin strip of tissue paper, or a pith-ball suspended by a silk thread, along the eggs. The paper is attracted most strongly at the ends; but in the middle, where the shells are in contact, there is very little electrification. Separate B from A about 10^{cm}, while the rod D is still in position. Then place D midway between A and B; the rod repels B and attracts A. It appears that when the two shells touched one another, thereby constituting practically one body, that the shells were oppositely electrified, as represented by the signs $+$ and $-$ in the diagram; and when the two bodies were separated, they retained their opposite charges.

We learn from this experiment that by *induction* we may charge at the same time two bodies, one with $+e$, and the other with $-e$.

§ 213. Discharge. — **Experiment.** Bring the two shells oppositely charged near one another; when near enough they exhibit mutual attraction for one another. On bringing them still nearer, a spark passes between them, their mutual attraction suddenly ceases, and on testing them with an electroscope, it is found that both have lost their electrification, *i.e.*, both have become *discharged*.

When two bodies containing equal amounts of opposite electricities are brought together, both become discharged. During the process of discharge, the electricity which was previously in a condition of rest, or a static state, assumes a condition of motion, or a dynamic state, as is shown by a spark passing between the two bodies when brought near one another. One of the bodies — that positively charged — is at a potential higher than that of the earth, the other being lower. When they are brought sufficiently near, the tendency for the electricity to pass from the region of higher potential becomes strong enough to penetrate the insulating air and establish a condition of equilibrium. In this particular case the result is zero potential or no electrification; but in general both bodies would be left at a like condition of electrification, its sign depending upon the sign of that electricity which was in excess.

We may now understand how it is that an electrified body

attracts to itself light bodies in its vicinity. For example, a stick of sealing-wax, excited with $-e$, brought near a pith ball, induces $+e$ next itself, and repels $-e$ to its farthest side; then, of course, attraction follows. There is the same attraction between heavy bodies, but usually not sufficient to produce motion.

§ 214. Insulation.—Experiment. Bring again the electrified sealing-wax near one end of one of the shells; the shell becomes polarized, that is, the opposite ends become oppositely electrified. Touch the shell with the finger. Through your body the negative charge is driven to the earth, while the positive charge remains in proximity to the rod. (Explain.) Remove the finger, and afterwards remove the rod; test the shell, and you will find that it is charged with electricity. (Is it $-e$ or $+e$?) Touch this shell with the other shell, then separate them. Test them, and you find that they have the same kind of electrification. It is evident that the first shell became electrified by *induction* and the last shell by *conduction*. Touch with the finger one of the shells; it loses its electrification.

When you touch the shell with your finger, the electric charge diffuses itself through your body and the earth. It is evident that the electricity could not traverse the silk thread, otherwise we could not have charged the shell. Substances which do not allow electricity to pass readily through them are called *non-conductors* or *insulators*. A body that is to receive a permanent charge of electricity must be insulated, *i.e.*, have no connection with the earth through a conducting substance. Some of the best insulating substances are *dry air*, *ebonite*, *shellac*, *resins*, *glass* (free from lead, *e.g.*, common bottle glass), *silks*, and *furs*. In experiments with electricity in the statical state, the E. M. F. is in general so much greater than when a galvanic battery is the source of electricity, that substances — such as dry wood, for instance — which are practically good insulators in the latter case are not so regarded in the former. Moisture injures the insulation of bodies; hence experiments succeed best on dry, cold days of winter, when moisture of the air is least liable to be condensed on the surfaces of apparatus, *especially if it is kept warm*.

244   ELECTRICITY AND MAGNETISM.

§ 215. Electrification confined to the external surface.
— **Experiment 1.** Place a tin fruit-can on a clean, dry glass tumbler (Fig. 180). Fasten a circular disk *a* of tin 15^{mm} in diameter to one end of a rod of sealing-wax. Charge the can heavily with electricity from an electrical machine (see p. 245). Through an orifice *c* in the can introduce the disk, and touch the interior surface of the can. Withdraw the disk, and present it to an electroscope. It shows no electrification. Now touch the exterior surface of the can with the disk, and present it to the electroscope; it is found to be electrified.

Fig. 180.

Experiment 2. Attach to the can a gold foil, or double pith-ball electroscope, and put into the can a few feet of metal chain. Fasten the outer end to a rod of glass, or some other insulator, and charge the can till the leaves of the electroscope diverge widely. Then draw up the chain by the glass rod; the leaves come together somewhat. Drop the chain into the can; the leaves separate again, showing that the charge had not been lost.

These experiments show (1) that no electricity can be found inside of a hollow-charged body; or, roughly stated, *electricity at rest resides on the exterior surfaces of bodies;* (2) that *when the exterior surface of an electrified body is increased without increasing its mass or the charge, the amount of electricity at any point is diminished.*

§ 216. Electrical potential. — We have seen that the passage of electricity from point to point sometimes causes a spark; so, conversely, the spark indicates the passage of electricity. The passage of a current from one shell to the other (Fig. 179) might be proved, and its direction determined, by connecting the shells by wires joined to a suitable galvanometer. The current would flow from A charged with $+e$, to B charged with $-e$, thus showing that A had a higher potential than B. A body charged with $+e$ is understood to be one that has a higher potential than that of the earth, and a body charged with $-e$ is one that has a lower potential than that of the earth, the potential of the earth being regarded for convenience as zero.

With a very sensitive electroscope it can be shown that the wires connected with the plates of a galvanic battery are at different potentials when the circuit is broken. But the difference of potential is so small, compared with the difference produced by friction, that a thousand gravity cells in series give a spark only about $\frac{1}{100}$ of an inch long.

XXXV. ELECTRICAL MACHINES. — CONDENSERS, ETC.

§ 217. If, then, for any purpose we wish electricity of high potential, we must use an enormous number of cells, an induction coil, or, more cheaply and conveniently, an electrical machine depending either on friction or on the induction of a charge of electricity. Brief descriptions of a few machines will now be given, followed by a series of experiments that may be performed with them.

Fig. 181.

§ 218. Plate machine. — It consists of a positive or prime conductor A, a negative conductor B, a glass plate C, a rubber D, made of two cushions of leather covered with an amalgam, four insulating supports E, F, G, and H, a silk insulating bag I, and a brass chain K, used to connect either conductor with the earth. An extension of the prime conductor L consists of two combs, one on either side of the plate; their pointed teeth are turned toward the plate. M is a pith-ball electroscope.

When the plate is turned in the direction indicated by the arrow, it passes between the rubbers, and the friction generates $+e$ on the plate and $-e$ on the rubber. The electrified portion of the plate then passing through the silk bag comes opposite the comb, when it polarizes the prime conductor, attracting $-e$ and repelling $+e$. But the $-e$ escapes from the points of the comb (see § 226) to the plate and neutralizes the $+e$ of the plate, and thereby leaves the conductor charged with $+e$. If both conductors are insulated at the same time, the mutual attraction of the two kinds of electricity would prevent their becoming heavily charged, so one of the conductors is always connected to earth by a chain. If $+e$ is wanted, A is insulated; if $-e$ is wanted, B is insulated.

Fig. 182.

§ 219. **Electrophorus.** — *Experiment.* On a circular disk of sheet-iron or tin 26^{cm} in diameter cement a circular disk of vulcanite 22^{cm} in diameter. To the center of another circular disk of tin 18^{cm} in diameter (Fig. 182) apply with heat one end of a stick of sealing-wax for a handle. Strike the surface of the vulcanite a few times with a cat's fur or a fox-tail; it will become electrified with $-e$. Then place the tin disk on the vulcanite; $-e$ of the vulcanite will polarize the disk, inducing $+e$ on its lower surface and $-e$ on its upper surface. Now place a finger on the disk. The $-e$ will escape through your body to the earth, but the $+e$ will remain

Fig. 183.

on the disk, *bound* by the $-e$ of the vulcanite. Finally, raise the disk by its insulating handle. Removed from the influence of the $-e$ on the vulcanite, the $+e$ of the disk is now free, and if a knuckle of one of your hands (Fig. 183) is brought near it, a bright spark will pass from it to your hand, and it will become discharged.

The disk may be charged and discharged in the same manner a great number of times without again whipping the vulcanite with the fur. A Leyden jar (page 250) may be charged with this apparatus in a few minutes. (Is the disk charged by conduction or induction? What are the proofs?)

§ 220. Continuous electrophorus. — Various methods have been adopted for developing electricity continuously from the electrophorus, and more rapidly and with less manipulation than can be done with the apparatus above described. Figure 184, from which the supporting parts are omitted for the sake of simplicity, will serve to illustrate the general principle of such machines.

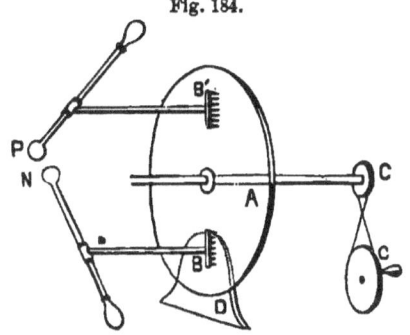

Fig. 184.

A is a vulcanite or glass wheel, which can be rotated by means of a system of wheels C C. About 2^{cm} back of A is a vulcanite sector D, which serves as an inducer, or the same purpose as the vulcanite disk of the electrophorus. Opposite D and in front of A is a metallic comb B, which is connected with the conductor N. Let $-e$ be excited on D with a cat-skin. Then the conducting system N B will be polarized by its influence; $-e$ will be driven to its farthest extremity N, and $+e$ drawn to the points of the comb B. Then, since electricity escapes readily from points, the $+e$ will leap from B to A, drawn off from the points by the $-e$ of D. But vulcanite being a non-conductor, only that portion of the surface of A will be charged with $+e$ that is directly opposite the comb B. The conductor B N, being deprived of its $+e$, is left charged with free $-e$. Now, rotate the disk A. When it has accomplished half a revolution, that portion of it which is charged with $+e$ comes opposite another comb B', which is also connected with a conductor P. The conductor B'P becomes polarized. Its $-e$ passes off from the points of B' to the disk A, and discharges the $+e$ on this disk, while the conductor PB' is left charged with free $+e$. It is evident that a constant rotation of A would cause it to be constantly charged with $+e$ at its lower part by the influence of the sector D, and constantly discharged at its upper part, while the conductor BN is constantly receiving a charge of $-e$ in consequence of the loss of its $+e$; and for a similar reason the conductor B'P is constantly receiving a charge of $+e$. With a rapid rotation of the disk the two conductors will be so rapidly and highly electrified, the one with $-e$ and the other with $+e$, that under the influence of their mutual attraction almost an incessant flow of sparks will pass between them, even when the extremities, P and N, are several inches apart.

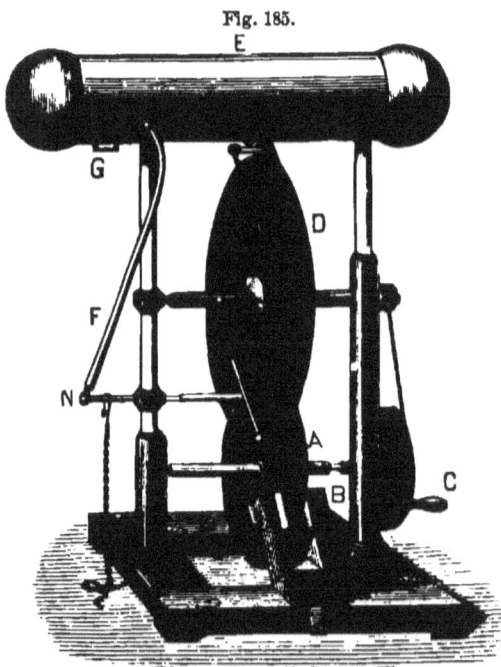

Fig. 185.

§ 221. Carré machine. — The sector D is liable to lose its charge suddenly, and loses it very quickly after the operation of the machine ceases; so that, to begin again, it is necessary to recharge the sector. In the Carré machine this difficulty is avoided.

A circular plate of glass A, which serves as the inducer, passes between two leather cushions B, as it is rotated by means of the crank C; and thus by friction, occasioned by rubbing against the cushions, the plate is kept constantly and highly electrified during the operation of the machine. By means of a system of pulleys, motion is communicated from the shaft of the plate A to the shaft of an ebonite disk D. The glass plate A becomes electrified with $+e$. The method by which the plate D and the conductors F and E become charged is like that of the machine last described. (Explain.) The cylinder E, called the *prime conductor*, has a large area, and is therefore capable of receiving a large charge. The conductor F is so jointed at N that it may be brought near or carried away from E. During operation the conductor F should be connected by a chain, or other good conductor, with the earth. When in operation, if F is brought within two or three inches of E, sparks will pass between them so rapidly as to present the appearance of a constant and continuous line of light. If F is removed to quite a distance from E, the charge will so accumulate in E that sparks five or six inches in length may be drawn from it with the fist. If a Leyden jar (§ 223) is suspended by its knob from the loop G, and its outside coating connected with F, the discharge will become so intense as to produce a report nearly as loud as that of a pistol.

§ 222. Condenser. — A very important adjunct to an electrical machine is a *condenser* of some kind, by means of which a large quantity can be collected on a small surface.

Experiment. Let a person stand on an insulated stool (p. 237), and place one hand on the prime conductor of a machine. Let the other open hand press against a plate of glass or disk of vulcanite, held on the open hand of a second person standing on the floor. After a few turns of the machine, let the hand that has been on the prime conductor grasp the free hand of the second person. Quite a shock will be felt by both. Or the connection may be made through a group of persons having hold of one another's hands, when the whole company may receive a shock.

It is evident that by this process an unusual quantity of electricity had collected previous to the discharge. The explanation is simple. The hand of the first person, charged with $+e$, acts by induction through the glass upon the second person, attracting $-e$ to the surface of the glass with which his hand is in contact, and repelling $+e$ to the earth. Thus, through their mutual attraction, the two kinds of electricity become, as it were, heaped up opposite each other, and yet are prevented, by the insulating glass, from uniting.

§ 223. **Leyden jar.** — The most convenient form of condenser is the *Leyden jar*. Coat a green glass quart fruit-jar (Fig. 186), within and without, for about two-thirds its hight, with tin foil, using flour paste. Close the mouth with a cork saturated with hot paraffine. Through the cork pass a stout brass wire till it touches the inner foil. Cast a lead bullet a on the exposed end of the wire. Clean, warm, and varnish the exposed glass surface of the jar, and when *thoroughly dry* it is ready for use.

Fig. 186.

The jar may be charged by connecting one of its coatings with the +conductor, and the other with the −conductor of an electrical machine, or by connecting one of the coatings with one of the conductors, and the other with the earth. Or it may be charged by connecting the outside coating with one of the poles of the secondary coil of an induction coil, and bringing the other pole near to the ball leading from the inner coating. To discharge the jar, connect the outer coating with the knob of the jar. To avoid a shock in so doing, prepare a discharger as follows: Through the cork of a bottle (*e.g.*, a soda-water bottle, Figure 187) pass a stout brass semicircular wire. Cast on each of its ends a lead bullet. Use the bottle as an insulating handle.

Fig. 187.

The effects are greater in proportion to the number and size of the jars in electrical connection. Let any number of jars (Fig. 188) be placed on a sheet of tin foil, by which their outer coatings are connected. Connect also their inner coatings with one another by a wire running around their projecting

ELECTRICITY NOT IN THE COATINGS.   251

rods. The several jars are by this means practically converted into one large jar. This combination of jars is called a *Leyden battery*. A strip of gold leaf placed on the glass slip a may be fused, and even volatilized, by a battery discharge passed through it; cards and slips of glass may be perforated, and gas or ether ignited.

§ **224. Electricity not in the coatings.** — If the two persons in the experiment (p. 249) both remove their hands from the glass plate, after they have been charged, and grasp one another's hands, they experience little or no shock. But if they

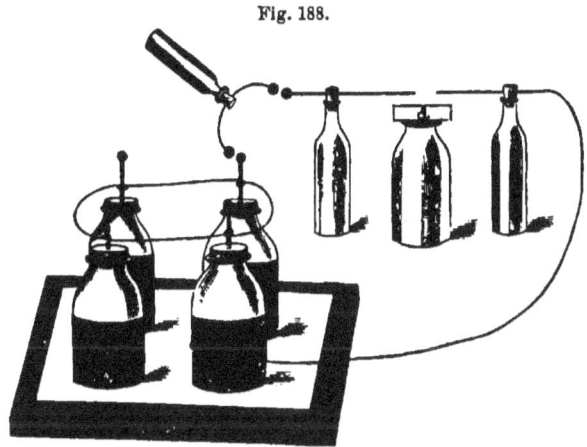

Fig. 188.

replace their hands on the glass, and grasp one another's hands, they receive a shock. This shows that electricity was not on their bodies, but on the surface of the glass. The coatings of a Leyden jar serve the purpose of conductors to spread electricity on the glass at the time of charging, and to allow its escape from all parts of its electrified surface at the time of a discharge.

QUESTIONS.

1. An insulated jar cannot receive a great charge. Why?
2. If points are attached to the outer coating of an insulated jar, it can receive a much larger charge. Why?

3. If the knob of a second jar be held near the outer coating of an insulated jar, sparks will pass from the coating to the knob, and both jars will be charged. Suppose that the inner coating of the first jar is charged with $+e$, what kind of electricity will each of the other coatings have?

§ 225. **Attractions and repulsions.** — **Experiment 1.** Support a plate of window glass (Fig. 189) about 5^{cm} from a table. Rub its upper surface with a silk handkerchief, and place pith-balls, or bits of tissue paper, on the table beneath the glass. They will dance up and down between the plate and table in a lively manner. (Explain.)

Fig. 189.

Experiment 2. Place a handful of bits of tissue paper on a tin disk supported by a prime conductor of an electrical machine. The papers become excited, are repelled into the air, and fall on all sides, giving the appearance of a miniature snow-storm.

Experiment 3. Apply one end of a discharger to the conductor of a machine, and the other end to the inner surface of a glass tumbler, and charge the interior with electricity, and then place it over some pith-balls, or images cut from pith; a ludicrous dance will be kept up for several minutes.

The *electric whirl* consists of a cap of metal resting upon a pointed wire, which serves as a pivot. The cap has pointed wires branching out from it, like the spokes of a wheel, and bent near their ends at right angles, and all turned in the same direction, as shown in Figure 190. When this apparatus is placed upon the conductor of a machine, the air particles around the highly electrified points become excited, and are repelled, producing a current of air issuing from the points. The reaction causes the wheel to revolve in the opposite direction, as indicated by the arrows in the figure. A candle flame placed near the point of a rod attached to a conductor will be extinguished.

Fig. 190.

§ 226. **Effect of points.** — We might reasonably expect that a current of excited air-particles issuing from points on an excited conductor would serve to carry away with them electricity from the conductor; in other words, to discharge it. Do they produce this result?

Experiment. While the electrical machine is in operation, hold your knuckle near the conductor; a succession of sparks will pass from the conductor to your hand. Now place several points on the conductor, and again present your knuckle as before; either no sparks will pass to your knuckle, or, at most, very feeble ones, and in a few seconds after the operation of generating electricity ceases, the conductor will be found completely discharged, although it is thoroughly insulated. It is apparent that the electricity escaped from the points.

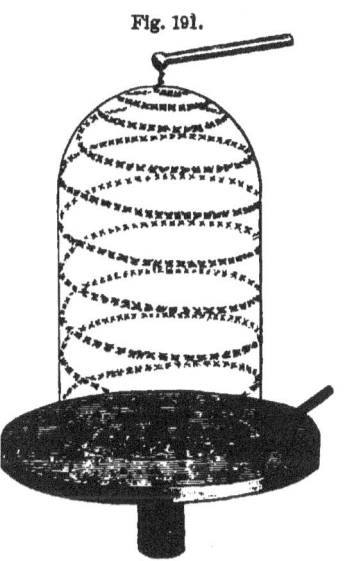

Fig. 191.

We conclude, therefore, that *the effect of points on an electrified insulated body is greatly to facilitate the discharge of its electricity.*

§ 227. **Luminous effects.** — Figure 191 represents a glass shade having circular bits of tinfoil pasted spirally around it, from top to bottom, and about 1^{mm} apart. If the poles of an induction coil, or the conductors of an electrical machine, are connected, one with each extremity of this spiral line, an intermittent line of light will be produced in the path of the current by the sparks which appear at the intervals between the bits of foil. All experiments illustrating luminous effects should be performed in a dark room.

Beautiful luminous effects may be produced with apparatus arranged as follows: Apply to one surface of a mica disk (Fig. 192), about 15×10^{cm}, a sheet of silver leaf or tin foil, 8×5^{cm}. Place two pointed poles of an inductorium, or a Carré machine, within 1^{cm} of the foil, and as far apart as the power of the machine will admit. Sparks will leap from the poles to the foil, and travel in tortuous branching lines between the poles.

254   ELECTRICITY AND MAGNETISM.

If air is partially exhausted from the glass tube used in illustrating the law of falling bodies (Fig. 87, page 106), and the poles of a coil or machine are applied to the opposite extremities of the tube, sparks of electricity passing through the rarefied air spread out in sheets of bluish white light resembling the auroral lights, hence this tube has received the name *Aurora tube*.

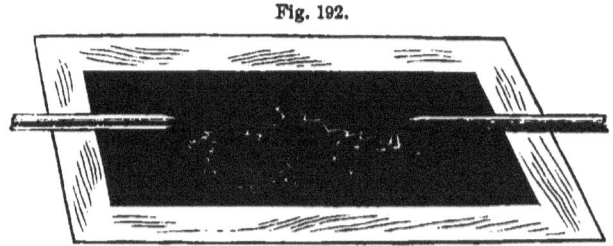
Fig. 192.

If a circular disk is divided into black and white sectors, as in Figure 193, and rotated very rapidly in ordinary daylight, the colors blend, and the disk appears of a uniform gray color.

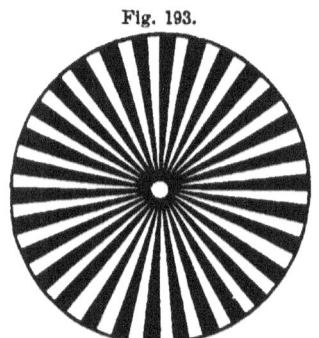
Fig. 193.

But if the disk is illuminated in a darkened room by the electric spark, each sector appears separate, and the disk appears to be at rest. A railroad train in rapid motion, and even its wheels, appear to be at rest when illuminated on a dark night by a flash of lightning. This shows that the duration of an electric spark must be very brief, inasmuch as it fails to illuminate these objects in two successive positions.

The remarkable beauty and brilliancy of the discharge is perhaps best exhibited by means of the well known *Geissler's tubes*. These tubes contain highly rarefied vapors and gases of various kinds. Platinum wires are sealed into the glass at each end to conduct the electric current through the glass. The light, instead of appearing, as in the Aurora tube, like a stream pouring from pole to pole, is often striated or divided transversely into luminous sections, with alternating darker sections, as shown in Figure 194. These striæ vary in shape and color

Fig. 194.

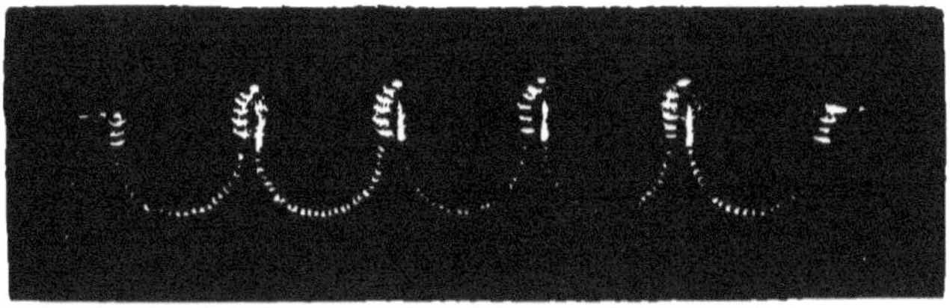

with the degree of the vacuum, and the kind of gas or vapor through which the discharge passes. Experiments with these tubes succeed best when used with the induction coil.

§ 228. Lightning. — Certain clouds which have formed very rapidly are highly charged with electricity, usually positively charged. The surface of the earth and objects thereon immediately beneath the cloud are charged inductively with the opposite kind of electricity. The cloud and the earth correspond to the coatings, and the intervening air to the glass of a huge Leyden jar. The charge in the earth and that in the cloud hold one another prisoner by their mutual attraction, until, as the charges accumulate, the attraction becomes great enough to overcome the resistance of the intervening air, when a discharge takes place. It is the accumulation of induced electricity on elevated objects, such as buildings and trees, that offers an attraction for the opposite electricity of the cloud, and renders them especially liable to be struck by lightning.

§ 229. Lightning-rods. — The flash will pass along the line of least resistance. A *good* lightning conductor offers a peaceful means of communication between the earth and a cloud; it leads the electricity of the earth gently up toward the cloud, and allows it to combine with its opposite without disturbance, thereby so far discharging the cloud as to prevent a lightning stroke; or, if the tension is too great to be thus quietly disposed of, the flash strikes downward, and is led harmlessly to the earth by the conductor. *An ill-constructed lightning-rod may be worse than none.* A good rod should be made of good conducting material, so large that it will not be melted, and free from loose joints. The lower end should be buried in earth that is always moist, and the upper end should terminate in several sharp points.

§ **230. General observations.** — Although the E. M. F. of the frictional machine is enormous, still, the current which it can produce is always small on account of its very great internal resistance. That this resistance must be almost immeasurably greater than that of any galvanic battery is evident when we consider that a part of the circuit is always through the air, for instance in the plate machine, that part between the plate and the comb. Any source of electricity that cannot yield a strong current is, ordinarily, of little value, inasmuch as the amount of work that can be done by electricity is proportional to the square of the strength of the current. The frictional machine is, therefore, of little practical value, except as a source of amusement, and a convenient means of investigating a certain class of electrical phenomena.

By an ingenious application of the principle illustrated by the disk of black and white sectors (Fig. 193), it has been ascertained that the duration of the spark is often times less than the millionth part of a second, and the velocity of the electric discharge from a Leyden jar through a short, thick copper wire is 280,000 miles in a second.

The phenomena of electricity in a statical state are limited to those of attractions and repulsions. Heating, luminous, magnetic, physiological, chemical, and mechanical effects can be produced by electricity in the dynamical state only. In the former state it seeks the surface; in the latter it travels through the body. Statical and dynamical phenomena are rarely coexistent. One must cease before the other begins.

Much is known of electricity, its nature, its laws, and its capacity for work; much remains to be learned. The question, *What is electricity?* is so far unanswered. But we may reason as follows concerning it, and the conclusion answers all practical purposes. For example, the energy of the chemical combination of coal and oxygen in the furnace is transformed into heat, heat works an engine, the engine rotates a coil of wire in a magnetic field, the motion of this coil in the vicinity of a magnet induces currents of electricity in the wire, these currents produce an arc, and thereby heat and light. So the energy of the coal is transformed into heat and light, through the intermediate agency of electricity. Hence, it is certain that this intermediate agency, this so-called *electricity*, whatever it is, *may receive and impart energy.*

§ 231. Transformation of energy. — We have found that *every contrivance for the development of electric energy is simply a machine for the transformation of some other form of energy into electric energy.* In the voltaic battery the chemical potential energy of the combustibles is transformed into the kinetic energy of the electric current. With the magneto and frictional machines, mechanical energy is transformed into electric energy. In the thermopile, heat is changed directly into electric energy. By means of an induction coil, a strong current of small E.M.F. is transformed into a momentary weak current of great E.M.F. The kinetic or current form of electricity may, under suitable conditions, be converted into the potential or static state, and vice versa. Not only are these various forms of energy transformable into electric energy, but electric energy may be changed into any one of these. Thus electric energy may be transformed into heat, magnetism, light, chemical action, and mechanical motion. These forms of energy are all interchangeable; as in fact, all known forms of energy are mutually convertible.

EXERCISES.

1. A spark of fire applied to a mass of soap bubbles, filled with a mixture of oxygen and hydrogen gases generated by a galvanic current, produces a powerful explosion. Commencing with the battery, state the transformations of energy, in order, to the final result.

2. The needle of a galvanometer connected with a thermopile is deflected. Trace the transformations of energy concerned.

3. A steam engine and a dynamo machine furnish an electric light. State all the transformations of energy necessary.

XXXVI. USEFUL APPLICATIONS OF ELECTRICITY.

§ 232. **Medical and surgical operations.** —Currents from an induction coil have great E.M.F., like frictional electricity, and so can pass through the poorly conducting tissues of the human body and produce violent muscular contractions. Currents induced by a single voltaic cell, through the mediation of an induction coil, may produce agonizing convulsions. A voltaic current has a similar effect at the instants of making and breaking the circuit (why?); but by beginning with a mild current, and slowly and gradually increasing its strength, a current from two hundred cells has been passed through a person with impunity. (Explain.) The physiological effect produced by an induced current at its negative pole is more violent than at the positive pole. In this way we may readily distinguish one pole from the other by simply holding one in each hand. The gradual current produces a benumbing influence, or insensibility to pain. A to-and-fro motion of the current produces a muscular agitation of the part through which it is sent, the tonic and stimulating effects of which are similar to those of muscular exercise. The galvanic current also exerts a powerful electrolytic effect on the system. On this principle it has been successfully employed in reducing tumors, etc.

A platinum wire heated by a galvanic current is used like a knife in surgical operations. The former has the advantage over the latter in that it sears the extremities of the blood vessels and thereby prevents hemorrhage. Enough has been said to show that a medical practitioner who can apply the laws of electricity has at his command a powerful therapeutic agent; but except in experienced hands it is likely to prove useless, if not positively dangerous.

§ 233. **Electric light.** — If the terminals of wires from a powerful magneto machine or galvanic battery are brought together, and then separated 1 or 2 millimeters, the current does not cease to flow, but volatilizes a portion of the terminals. The vapor formed becomes a conductor of high resistance, and remaining at a very high temperature produces intense light. The light rivals that of the sun both in intensity and purity. The heat is so great that it fuses the most refractory substances, including even the diamond. Metal terminals quickly melt and drop off like tallow, and thereby become so far separated that the electro-motive force is no longer sufficient for the increased resistance, and the light is extinguished. Hence, pencils of carbon (prepared from coke deposited in the distillation of coal inside of gas retorts), which are less fusible, are used for terminals. For simple experi-

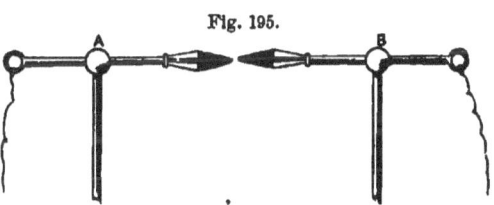

Fig. 195.

ments, these pencils may be held in forceps (Fig. 195) at the ends of two brass rods, to which the battery-wires are attached. These rods slide in brass heads A and B, supported by insulating pillars, so that the distance between the carbon points may be regulated.

§ 234. **Voltaic arc.** — The light is too intense to admit of examination with the naked eye; but if an image of the terminals is thrown on a screen by means of a lens, or a pin-hole in a card (see page 330), an arch-shaped light is seen extending from pole to pole, as shown in Figure 196. This light has received the name of the *voltaic arc*. The larger portion of the light, however, emanates from the tips of the two carbon terminals, which are heated to an intense whiteness, but some from the arc. The +pole is hotter than the −pole, as is shown by its glowing longer after the current is stopped. The carbon of the +pole becomes volatilized, and the light-giving particles are transported from the +pole to the −pole, forming a bridge of luminous vapor between the poles. What we see is not electricity, but *luminous matter*. Neither light nor a current can exist without matter, as may be shown by trying to pass a current between two metallic poles, a little way apart, in a charcoal vacuum (page 56); no spark can be produced.

Fig. 196.

§ 235. **Electric lamp.** — It is apparent that the +pole is subject to a wasting away, and the −pole to a slight accession of matter. At the point of the former a conical-shaped cavity is formed, while around the point of the latter warty protuberances appear. When, in consequence of the wearing away of the +pole, the distance between the two pencils becomes too great for the electric current to span, the light goes out. Numerous self-acting regulators for maintaining a uniform distance between the poles have been devised. Such an arrangement is called an *electric lamp*. In some, the carbons are moved by clock-work, which requires winding up occasionally; in others, the movement of the carbons is accomplished automatically by the action of the current itself.

§ 236. **Electric candle.** — The "Jablochkoff Candle" obviates all necessity for regulators. In this candle, instead of the carbons pointing toward each other, they are placed side by side, a and b (Fig. 197), separated by a thin insulating septum, *c*, of kaolin. The current passes up one carbon, across the space between the points, and down the other. In its passage between the points it forms the luminous arc. The heat of the arc fuses and volatilizes the kaolin, and it wastes slowly away like the wick of a candle; hence its name.

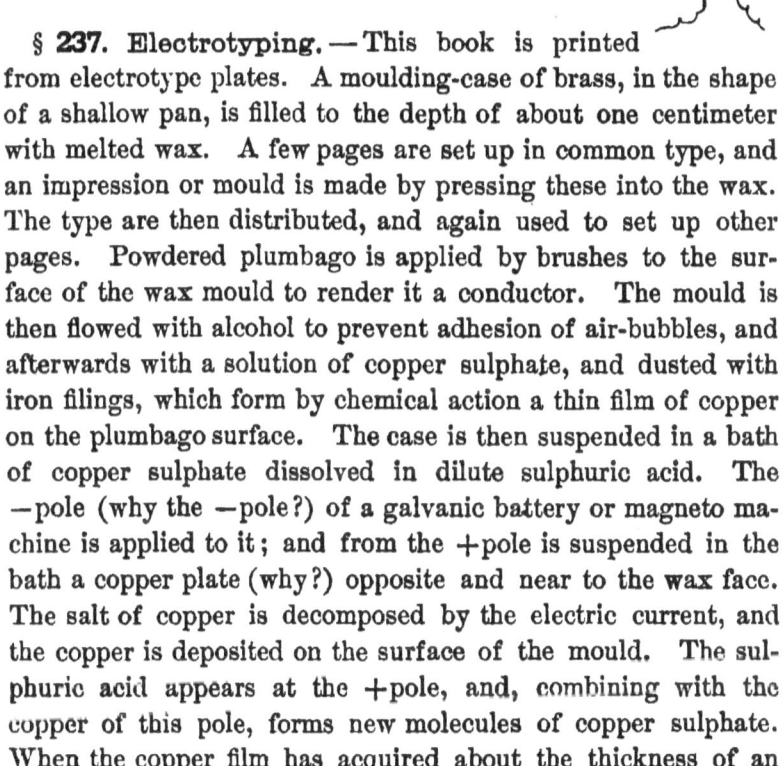

Fig. 197.

The electric light is of the purest white. In it the most delicate colors retain their noonday purity of tint, while a gas light appears of a sickly yellow hue in comparison.

§ **237. Electrotyping.** — This book is printed from electrotype plates. A moulding-case of brass, in the shape of a shallow pan, is filled to the depth of about one centimeter with melted wax. A few pages are set up in common type, and an impression or mould is made by pressing these into the wax. The type are then distributed, and again used to set up other pages. Powdered plumbago is applied by brushes to the surface of the wax mould to render it a conductor. The mould is then flowed with alcohol to prevent adhesion of air-bubbles, and afterwards with a solution of copper sulphate, and dusted with iron filings, which form by chemical action a thin film of copper on the plumbago surface. The case is then suspended in a bath of copper sulphate dissolved in dilute sulphuric acid. The −pole (why the −pole?) of a galvanic battery or magneto machine is applied to it; and from the +pole is suspended in the bath a copper plate (why?) opposite and near to the wax face. The salt of copper is decomposed by the electric current, and the copper is deposited on the surface of the mould. The sulphuric acid appears at the +pole, and, combining with the copper of this pole, forms new molecules of copper sulphate. When the copper film has acquired about the thickness of an ordinary visiting card, it is removed from the mould. This shell shows distinctly every line of the types or engraving. It is then backed with melted type-metal to give firmness to the

Let B, Figure 1, Plate III., represent the message-sender, or operator's key; Y, the message-receiver. It may be seen that the circuit is broken at B. Let the operator press his finger on the knob of the key. He closes the circuit, and the electric current instantly fills the wire from Boston to New York. It magnetizes a; a draws down the lever b, and presses the point of a style on a strip of paper c that is drawn over a roller. The operator ceases to press upon the key, the circuit is broken, and instantly b is raised from the paper by a spiral spring d. Let the operator press upon the key only for an instant, or long enough to count one, a simple *dot* or indentation will be made in the paper. But if he presses upon the key long enough to count three, the point of the style will remain in contact with the paper the same length of time; and, as the paper is drawn along beneath the point, a short straight line is produced. This short line is called a *dash*. These dots and dashes constitute the *alphabet of telegraphy*. For instance, a part of a message, "man is in," is represented as printed in telegraphic characters on the strip of paper. The Roman letters above interpret their meaning.

§ 241. Sounder. — If the strip of paper is removed, and the style is allowed to strike the metallic roller, a sharp click is heard. Again, when the lever is drawn up by the spiral spring, it strikes a screw point above (not represented in the figure), and another click, differing slightly in sound from the first, is heard. A listener is able to distinguish dots from dashes by the length of the intervals of time that elapse between these two sounds. Operators generally read by ear, giving heed to the clicking sounds produced by the strokes of a little hammer. A receiver so used is called a *sounder*, a common form of which is represented in the lower central part of Plate III.

§ 242. Relay and repeater. — The strength of the current is diminished, of course, as the line is extended and the number of instruments in the circuit is increased. Hence, a current that would move a single sounder audibly, on a short line, would not move many sounders on a long line with sufficient force to render the message audible. Resort is had to *relays* and *repeaters*. The principle on which they remove this diffi-

Plate III.

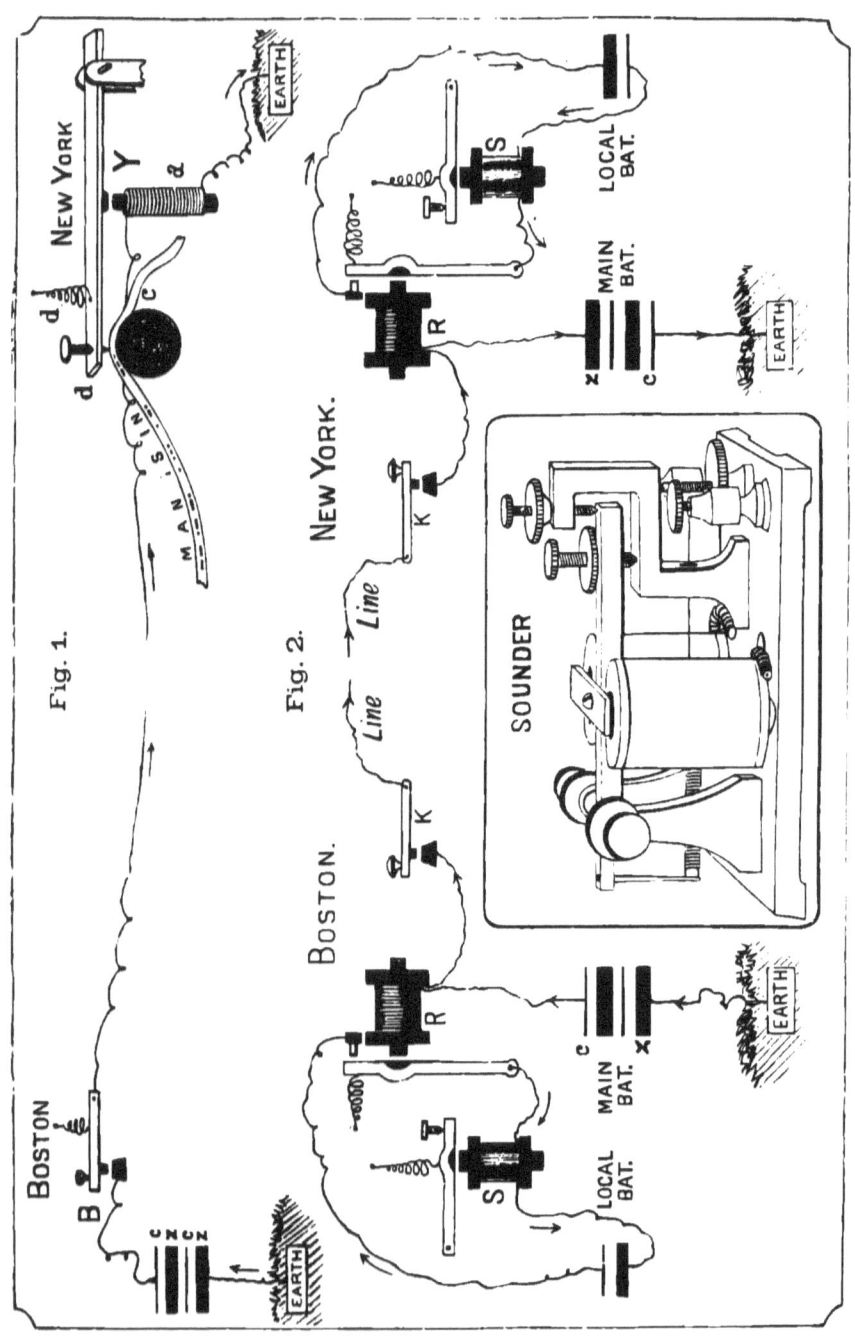

culty may, perhaps, be best explained by analogy. In days gone by, posts for couriers were stationed a day's journey apart. At each post were a courier and a horse at all hours ready to start. The courier, bearing a dispatch, rode all day, and at night reached a post where fresh horses were saddled ready for the next stage of the journey; he himself was exhausted, his force was nearly spent, but he could awaken a courier who was stationed there and deliver the despatches to him, and he with fresh strength instantly took up the journey. In a similar manner we picture to ourselves the electric current arriving at a station so nearly exhausted that it cannot deliver intelligible signals, yet it may still have strength to wake up another battery and set in motion a fresh current which shall receive and announce audibly, or carry forward the message which the exhausted current has just strength to whisper.

In Figure 2, Plate III., the letter R represents a relay and S a sounder. Suppose a weak current arrives at New York from Boston, and has sufficient strength to attract the armature of the relay at that station. This, as may be seen by examination of the diagram, will close another short circuit, called the local circuit, and send a current from a local battery located in the same office, through the sounder at that station. The sounder, being operated by a battery in a circuit of only a few feet in length, delivers the message audibly. If it is desired that the message should go beyond New York, — for instance, to Philadelphia, — then we have only to suppose the local line at New York to be lengthened so as to extend to Philadelphia, and a powerful line battery to be substituted for the small local; then the message that leaves Boston will be shifted from one circuit to the other at New York, and be delivered in Philadelphia without the intervention of any operator on the route. In this case a relay is called a *repeater*. The electro-magnets in relays are wound with long and thin wire, while those of sounders are wound with short, large wire. (Explain. The main battery consists of many cells; how should they be connected? It may be located at either terminus, but it is generally split in halves, and one half placed at each terminus; how should the two halves be connected?)

In the diagram the circuit is represented as open at both keys. When the line is not in use, the circuit ought always to be left closed, by means of switches connected with the keys (not represented in the diagram), so that, when the line is not "at work," an electric current is constantly traversing the wire. Sending a message, consequently, consists in interrupting this current by means of a key. Suppose that Boston wishes to communicate with New York. He first removes the switch on his key, which breaks the circuit and enables him to control the circuit with his key. He then manipulates his key so as to produce an understood signal, which will attract New York's attention. Every time that Boston presses on his key, every armature in his own office, and in the New York office, and at way-stations, falls. Of course the message may be read at every station on the route.

TELEGRAPHIC ALPHABET.

A	B	C	D	E	F
·—	—···	·· ·	—··	·	·—·
G	H	I	J	K	L
—— ·	····	··	—·—·	—·—	———
M	N	O	P	Q	R
——	—·	· ·	·····	··—·	· ··
S	T	U	V	W	X
···	—	··—	···—	·——	·—··
Y	Z	&	,	?	.
·· ··	····	· ···	·—·—	—··—·	·· —· ··

TELEGRAPHIC FIGURES.

1	2	3	4	5	6
·——·—	··—··	···—·	····—	———	······
	7	8	9	0	
	——··	—····	—··—	———	

§ 243. Fac-simile telegraph. — This is an autographic apparatus by means of which a message may be, practically, transmitted over a wire and appear at a distant terminus in the exact hand-writing of the sender, and ready at once for delivery. The principle on which it operates may be learned from the diagram in Plate I., in which all details of its mechanism are omitted for simplicity of illustration. X is a sheet of tin-foil, on which the message to be sent is written with an ink prepared by dissolving sealing-wax in alcohol. The alcohol quickly evaporates, leaving the lines of sealing-wax adhering to the foil. Y is a sheet of paper moistened with a solution of prussiate of potash. Each of the pens is simply a small, pointed iron needle. Now suppose that both of the pens are moved at the same time and with the

THE ELECTRIC FIRE-ALARM. 267

same rapidity across their respective sheets. Then the electric current, decomposing the prussiate of potash, will cause the needle in New York to trace a continuous blue line on Y, until the needle in Boston reaches a line of sealing-wax on X, when the circuit is broken as it passes over this line. At the same time there is a break in the continuity of the line traced on Y. If, further, each needle is moved down a hair's breadth each time it traverses its respective sheet, then we shall have an exact fac-simile of the writing on the tin-foil produced on the chemically-prepared paper, except that whereas the original is written in dark letters on a light ground, the message is received in light letters on a dark ground. Pen-and-ink sketches of photographs and other pictures may be transmitted in the same way. The pens are not, of course, held and guided by human hands, but by complex machinery. The rigorous exactness requisite in the movements of the two pens is secured by the absolute synchronism in the vibrations of two pendulums, one at each terminus, controlled by the electric current.

Fig. 199.

§ 244. **The electric fire-alarm.** — This is a modification of the electro-magnetic telegraph. Figure 199 will serve to illustrate the general plan of the American system, invented by

Prof. M. G. Farmer, and by him first introduced into Boston in the year 1852.

From some central station wires radiate to every part of the city. At suitable intervals there are inserted in these circuits small cottage-shaped boxes, usually attached to buildings at the corners of streets. On opening one of these boxes, a person who is to give an alarm finds a crank A, which he is directed to "pull down once and let go." This winds the spring H, which sets in motion a train of wheels, and causes a make-and-break wheel C to revolve. This wheel bears upon its circumference notches corresponding to the number of the box. Two terminals of the line are so connected, one with C and the other with a lever b, that when the lever touches the wheel the circuit is closed. But when the wheel revolves, and a notch passes under the lever, the circuit is broken. The effect of breaking the circuit is to demagnetize the electro-magnet F at the central station, and release the armature which is attached to the tongue of a bell. The tongue then being drawn forcibly by the spring G in the opposite direction, produces one stroke on the bell. By pulling the lever down once, the spring is wound up just enough to cause C to revolve three times, and thus the number of the box is struck three times in succession. The watchman at the central station, being thus notified of the existence and locality of the fire, at once and in a similar manner notifies the several fire-engine companies.

XXXVII. TELEPHONE AND MICROPHONE.

§ 245. Bell telephone. — Figure 200 represents a sectional and a perspective view of this instrument. It consists of a steel magnet A, encircled at one extremity by a spool B of very fine insulated wire, the ends of which are connected with the binding screws DD. Immediately in front of the magnet is a thin circular iron disk EE. The whole is enclosed in a wooden or rubber case F. The conical-shaped cavity G serves the purpose of either a mouth-piece or an ear-trumpet. There is no difference between the transmitting and receiving telephone; consequently either instrument may be employed as a transmitter, while the other serves as a receiver. Two magneto telephones in a circuit, are virtually in the relation of a magneto-electric generator and a motor. The transmitter being in itself a diminutive magneto machine, of course no battery is required in the circuit. Connect in circuit two such telephones, and the apparatus is ready for use.

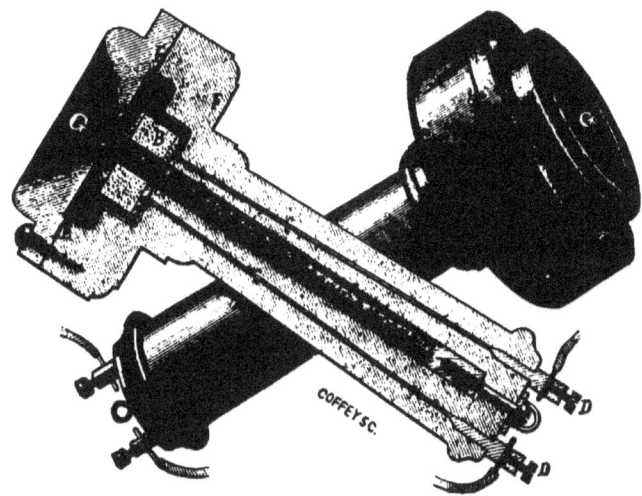

When a person talks to the disk of the transmitter, he throws it into rapid vibration. The disk, being quite close to the magnet, is magnetized by induction; and, as it vibrates, its magnetic power is constantly changing, being strengthened as it approaches the magnet, and enfeebled as it recedes. This fluctuating magnetic force will of course induce currents in alternate directions in the neighboring coil of wire. These currents traverse the whole length of the wire, and so pass through the coil of the distant instrument. When the direction of the arriving current is such as to reënforce the power of the magnet of the receiver, the magnet attracts the iron disk in front of it more strongly than before. If the current is in the opposite direction, the disk is less attracted, and flies back. Hence, whatever movement is imparted to the disk of the transmitting telephone, the disk of the receiving telephone is forced to repeat. The vibrations of the latter disk become sound in the same manner as the vibrations of a tuning fork or the head of a drum.

The above is a description of the original and simplest form of the Bell telephone. It is apparent that the original energy, *i.e.*, that of the voice, applied at the transmitter must, during its successive transformations and especially during its transmission in the form of electric energy through large resistances, become very much enfeebled, so that when it reappears as sound, the sound is quite feeble and frequently inaudible. The first grand improvement on the original consists in introducing a battery into the circuit and so arranging that the voice instead of being obliged to generate currents should be required to act only as a controlling force of a current already

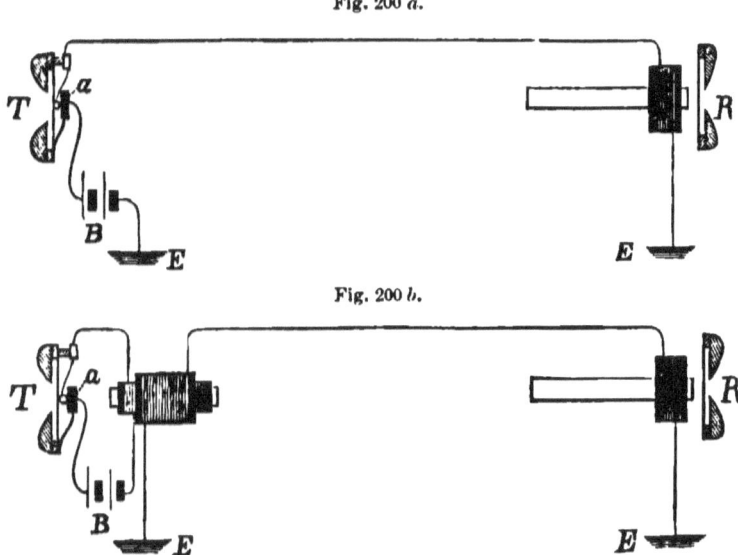

Fig. 200 a.

Fig. 200 b.

generated by the battery. It is evident that only a fluctuating or undulating current can produce the necessary vibrations in the disk of the receiver. The fluctuations are caused by a varying resistance in the circuit. The pupil must have learned by experience ere this that the effect of a loose contact between any two parts of a circuit is to increase the resistance and thereby weaken the current; but the effect of a slight variation in pressure is especially noticeable when either or both of the parts are carbon. Figure 200 a illustrates a simple telephonic circuit in which are included a variable resistance transmitter T, a magneto receiver R, and a battery B. One of the electrodes, a platinum point, touches the center of the transmitter disk; the other electrode, a carbon button a, is pressed by a spring gently against the platinum point. Every vibration of the disk, however minute, causes a variation in the pressure between the two electrodes and a corresponding variation in the circuit resistance. As changes the resistance, so changes the current strength, and so consequently changes the force with which the magnet in the receiver

R pulls its disk. The varying tension between magnet and disk causes the latter to vibrate and reproduce sounds.

The next improvement of considerable importance consists in the adoption of an induction coil, which, we have learned, produces a current of much greater force than is possessed by the original battery current. By its adoption we are able to converse over much longer distances, and since the battery current traverses only a local circuit, as may be seen by reference to Fig. 200 *b*, a single Leclanché cell is generally sufficient to operate it. The currents induced by the fluctuating primary current traverse the line wire and generate sonorous vibrations in the disk of the receiver in the same manner as in the original telephone.

Fig. 201.

§ **246. Microphone.**—In Figure 201, A and B are buttons of carbon; the former is attached to a sounding-board of thin pine wood, the latter to a steel spring C, and both are connected in circuit with a battery and a telephone used as a receiver. The spring presses B against A, and any slight jar will cause a variation in the pressure and corresponding variations in the current strength.

By means of this instrument, called the *microphone*, any *little sounds*, as its name indicates, such as the ticking of a watch or the footfall of an insect, may be reproduced at a considerable distance, and be as audible as though the original sounds were made close to the ear.

CHAPTER V.

SOUND.

The subjects of Sound and Light, which we have now to study, have two important characteristics in common that distinguish them from the subjects already studied. First, each of them affects its peculiar organ of sense, the ear or eye, and very many of the phenomena to be studied under each subject are of importance only to one or the other of these senses; while the most common, and many of the most important applications of heat and electricity have no direct relation to any organ of sense. Second, both sound and light, we shall find originate in vibrating bodies, and reach us only by the intervention of some medium capable of being set in vibration.

Here, as in all other kinds of motion, energy is involved; the ear or eye absorbs energy whenever a sensation is produced, but the amount absorbed is so minute, and the difficulty of measuring it so great, that usually other points better deserve the student's attention.

Let us begin with the study of such vibrations as will neither produce sound. nor in a dark room affect the eye.

XXXVIII. VIBRATION AND WAVES.

§ 247. Vibration.—**Experiment 1.** Repeat the experiment with the pendulum 1^m long, page 111, and note in what respects its motion differs from most other motions.

Experiment 2. Take a pendulum 50^{cm} long, hold it with the string just touching an edge of a table, having the hand about 38^{cm} above the table, and set it vibrating; the ball will be seen to vibrate faster in the portion of its arc that is under the table than in the other portion, and so more vibrations are made in ten seconds than if the string swing freely without touching the table.

DIRECTION OF VIBRATION. 273

Experiment 3. Without the pendulum, move the hand quickly from side to side every two seconds, turning instantly at one side, and waiting at the other till the two seconds are up.

These three motions, though very different, have this in common: the motions in each case occur at *equal intervals of time*. This interval of time is called the *period of vibration*. In Exp. 1 it was two seconds; in Exp. 2, about one second; and in Exp. 3, two seconds. The motion from one side to the other and back is called a *vibration*. If $n =$ the number of vibrations in one second, and $t =$ the period, $t = \dfrac{1}{n}$. The amplitude of pendulum vibrations is assumed to be very small. In Exp. 1 the motion is called a *simple* (or pendular) *vibration;* in the other cases the *vibration* of the ball or the hand is *complex*. Do not confound period with duration of the vibrating state; in Exp. 1 the pendulum may have vibrated ten or one hundred seconds, before coming to rest, but the period was two seconds. Considered mathematically, other periodic (and therefore vibratory) motions are, the movements of the hands of a watch, the regular trips of a stage-coach, etc. *A vibration is a recurrent change of position.*

§ 248. **Direction of vibration.** — A small rod, like a yardstick, fixed at one end, may be set in vibration by pulling the other end to one side; a tree vibrates in the wind; the strings of a piano swing from side to side when vibrating; in all these cases the motion is at right angles to the length of the body, and so the body is bent. These are all cases of *transverse vibrations*.

Experiment. Hang up a spiral spring, or elastic cord, with a weight attached to the lower end; lift the weight, and, dropping it, notice that the cord vibrates, lengthening and shortening rapidly.

The motion of the body is in the direction of its length, and so it is not bent; this is a case of *longitudinal vibration*. Twist the string, and see that it is possible to set up *torsional vibra-*

tions. Compare these kinds of vibration with the kinds of elasticity studied on page 30.

§ 249. Propagation of vibration. — Waves. — Experiment. Take a soft cotton rope, a few meters long, lay it straight on a floor, set one end in vibration by quick movements of the hand. Notice any point in the rope, and see that it is set in vibration; that is, it moves up and down, or laterally from side to side through its original position of rest. Make a single movement of the hand, which is better called a *pulse* than a vibration; it is easy to see that the pulse does not reach all points of the rope at the same time. Send a quick succession of equal pulses along the rope; at any instant different pulses affect different parts of it, and you get more or less perfectly the familiar form that we call a *wave-line*. Notice that any point of the rope only moves up and down, while the *form* of the wave moves on. Vibrate the hand in a longer period, and notice that the distance from crest to crest is longer than before.

§ 250. Wave-length and amplitude. — Imagine an instantaneous photograph taken of the rope along which the waves are passing. It would appear much like the curved line CD, Figure 202. This curve represents what is known as a simple wave-line. The shortest of the similar portions into which a wave-line can be cut is called a *wave-length*, as *wx*, *uv*, or *en*. The greatest distance of any point in a wave from the axis, as *ou*, is called the *amplitude* of the wave.

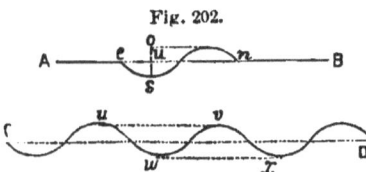
Fig. 202.

§ 251. Reflection of waves. — Interference.[1] **— Experiment 1.** Stretch the rope horizontally between two elevated points, and pluck it with the hand or strike it with a stick near one end, and send along it a single pulse, forming a crest on the rope (A, Fig. 203). This travels to the other end, and there we see it reflected and inverted (B).

[1] See Section G of the Appendix.

Experiment 2. Just at the instant of reflection, start a second crest; these two, the crest and the returning inverted crest or trough (C), are now traveling along the rope in opposite directions, and must meet at some point. This point will be urged upward by the crest and downward by the trough, and so its motion will be due to the difference of the two forces.

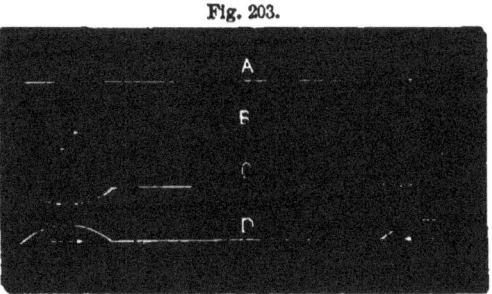

Fig. 203.

Experiment 3. Send along the rope, first a trough, then a crest; now two crests (D) will meet near the middle of the rope, and the motion here will be due to two forces acting in the same direction, and so the resulting crest will be greater than either of the original ones.

This action on a single point of two pulses, or two trains of waves, no matter if from different sources, is termed *interference*. The resulting motion may be greater or less than that due to either pulse alone, or it may be zero.

§ 252. Stationary vibrations, nodes, etc. — **Experiment.** Hold one end of a rubber tube, about 2ᵐ long, while the other is fixed, and send along it a regular succession of equal pulses from the vibrating hand; it will be easy, by varying the tension a little, to obtain a succession of gauzy spindles (Fig. 204) separated by points that are

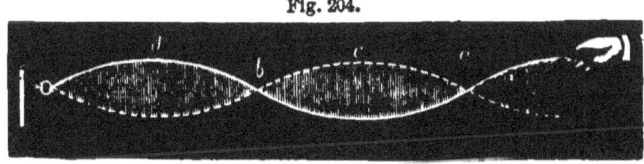

Fig. 204.

nearly or quite at rest. Unlike the earlier experiments, the waves here do not appear to travel along the tube; yet in reality they do traverse it. The deception is caused by stationary points being produced by the interference of the advancing and retreating waves.

This interference of direct and reflected waves gives rise to the important class of so-called *stationary vibrations*. The points of least motion, as a and b, are called *nodes;* the points of greatest motion, c and d, are called *antinodes;* and the portion of the rope between a node and an antinode, as ac, is a *semi-ventral segment*, and ab is a *ventral segment*.

§ **253. Water-waves.** — If you have a long and rather narrow box or trough, nearly filled with water, you can produce much the same effects as with the rope. A crest is caused by thrusting the hand into the water; a trough, by suddenly withdrawing it. Chips floating on the water show that here, as with the rope, it is only a form that advances — not the matter. But more careful observation will show that the chip, when on the crest of the wave, does move forward a little, and when in the trough moves backward a little; thus it does not merely rise and fall, but goes round in a curve that is approximately a circle. After a little practice, you may be able to produce interference and stationary waves; or you may produce them by blowing on the surface of water in a basin, or by tapping the basin. Water-waves furnish important illustrations of the fact that energy may be transmitted by vibration as truly as by the actual transfer of the medium, as in the river's current or the wind.

§ **254. Longitudinal waves.** — **Experiment.** Procure a brass wire wound in the form of a spiral spring,[1] about 4^m long. Attach one end to a cigar box, and fasten the box to a table. Hold the other end H of the spiral firmly in one hand, and with the other hand insert a knife-blade between the turns of the wire, and quickly rake it for a short distance along the spiral toward the box, thereby crowding closer together for a little distance (B, Fig. 205) the turns of wire in front of the hand, and leaving the turns behind pulled wider apart (A) for about an equal distance. The

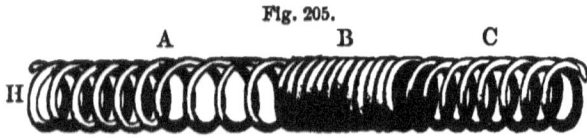

Fig. 205.

---

[1] About 25^m of No. 20 brass spring-wire should be wound with care in a lathe on a spindle 1^{cm} in diameter, as close as possible.

crowded part of the spiral may be called a *condensation*, and the stretched part a *rarefaction*. The condensation, followed by the rarefaction, runs with great velocity through the spiral, strikes the box, producing a sharp, loud blow; is reflected from the box back to the hand, and from the hand again to the box, producing a second blow; and by skilful manipulation three or four blows may be produced in rapid succession. If a piece of twine be tied to some turn of the wire, it will be seen, as each wave passes it, to receive a slight jerking movement forward and backward in the direction of the length of the spiral.

How is energy transmitted through these 4^m of spring so as to deliver the blow on the box? Certainly not by a bodily movement of the spiral as a whole, as might be the case if it were a rigid rod. The movement of the twine shows that the only motion which the coil undergoes is a vibratory movement of its turns. Here, as in the case of water-waves, energy is transmitted through a medium by the transmission of vibrations.

There are two important distinctions between this kind of wave and a liquid wave: the former consists of a condensation and a rarefaction; the latter, of an elevation and a depression; in the former, the vibration of the parts is in the same line with the path of the wave, and hence these are called *longitudinal waves;* in the latter, across its path, and they are therefore *transverse waves.*

A wave cannot be transmitted through an inelastic soft iron spiral. *Elasticity is essential in a medium, that it may transmit waves made up of condensations and rarefactions; and the greater the elasticity, the greater the facility and rapidity with which a medium transmits waves.*

§ 255. **Air as a medium of wave-motion.** — May not air and other gases, which are elastic, serve as media for waves?

Experiment. Place a candle flame at the orifice a of the tube,[1] Figure 206, and strike the table a sharp blow with a book near the orifice b.

---

[1] This tube, which will serve many important purposes, may be made of tin in three parts, A and B, each 2.5^m long, and 10^{cm} in diameter, and a conical-shaped cap C about 30^{cm} long, having an orifice of 3^{cm} diameter. The ends of the three parts should be made slightly tapering, so that they may be put together like a stove-pipe.

Instantly the candle flame is quenched. The body of air in the tube serves as a medium for transmission of motion to the candle.

Was it the motion of a current of air through the tube, as when blown through, or was it the transfer of a vibratory motion? Burn touch-paper[1] at the orifice b, so as to fill this end of the tube with smoke, and repeat the last experiment.

Evidently, if the body of air is moved along through the tube, the smoke will be carried along with it. The candle is blown out as before, but no smoke issues from the orifice a. It is clear that there is no

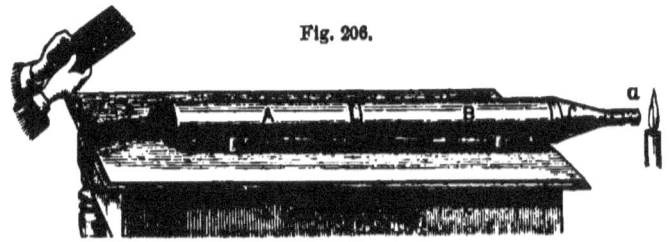

Fig. 206.

translation of material particles from one end to the other, — nothing like the flight of a rifle bullet. The candle flame was struck by something like a *pulse* of air, and not by a *wind*.

§ 256. How a wave is propagated through a medium. — The effect of applying force with the hand to the spiral spring is to produce in a certain section (B, Fig. 205) of the spiral a crowding together of the turns of wire, and at A a separation; but the elasticity of the spiral instantly causes B to expand, the effect of which is to produce a crowding together of the turns of wire in front of it, in the section C, and thus a forward movement of the condensation is made. At the same time, the expansion of B causes a filling up of the rarefaction at A, so that this section is restored to its normal state. This is not all: the folds in the section B do not stop in their swing when they have recovered their original position, but, like a pendulum, swing beyond the position of rest, thus producing a rarefaction at B, where, immediately before, there was a con-

---

[1] To prepare touch-paper, dissolve about a teaspoonful of saltpeter in a half teacupful of hot water, dip unsized paper in the solution, and then allow it to dry. The paper produces much smoke in burning, but no flame.

densation. Thus a forward movement of the rarefaction is made, and thus a pulse or wave is transmitted with uniform velocity through a spiral spring, air, or any elastic medium.

§ 257. **How a wave-line represents a vibration.** — We saw (page 274) that a wave might be caused by a vibration, and readily believe that if the vibration is changed in any way, the wave must show some corresponding change. In fact the wave-line shows at once something about the vibration for several successive periods; so, if we could attach a cord to a vibrating body, or set up vibrations in water by it, and then take an instantaneous photograph of the rapidly-disappearing waves, we might learn much about the nature of the original vibrations.

A much more practical arrangement, which gives a permanent wave-line, is the following

Experiment. Attach, by means of sealing-wax, a bristle or a fine wire to the end of one of the prongs of a tuning-fork, as seen in Figure 207. Set the fork in vibration, and quickly draw the point of the bristle lightly over a smoked glass (A, Fig. 207). A beautiful wavy line will be traced on the glass, each wave corresponding to a vibration of the prong when vibrating as a whole.

Fig. 207.

Next tap the fork, near its stem, on the edge of a table, and trace its vibrations on a smoked glass as before. You will generate the same set of waves that you did before; but, running over these, is another set of waves, of much shorter period, much like No. 3 of Figure 223, page 312, showing that the prong vibrates, not only as a whole, but in parts. The serrated wavy line produced represents the resultant of the combined vibrations, and may be called a *complex wave-line*.

If we imagine the piece of twine on the spiral (page 277) replaced by a bristle pointing downward, and under it a smoked glass drawn at right angles to the length of the spiral, the vibrating bristle will trace a characteristic curve on the glass. We may even conceive a writing point attached to a particle of air and tracing these curves, and thus understand how it can illustrate the nature of the vibration of the air. This method is known as *the graphical method of studying vibrations*.

XXXIX. SOUND-WAVES.

§ 258. How sound originates. — Listen to yonder sounding church-bell. It produces a sensation; it is heard. If the orifices of the ears be stopped by pressing the palms of the hands against them, the sensation, in a great measure, ceases. The ear is, therefore, the organ of sense through which the sensation of hearing is produced. The bell must be the cause of the impression made on the ear. But the bell is at such a distance that it cannot itself act on the ear; yet something must act on the ear, and it must be the bell which causes that something to act.

Commencing at the origin of sound, let the first inquiry be, How does a sounding body differ from a silent body?

Experiments. Strike a bell or a glass bell-jar, and touch the edge with a small cork ball suspended by a thread; you not only *hear* the sound, but, at the same time, you *see* a tremulous motion of the ball, caused by a motion of the bell. Touch the bell gently with a finger, and you feel a tremulous motion. Press the hand against the bell; you stop its vibratory motion, and at that instant the sound ceases. Strike the prongs of a tuning-fork, press the stem against a table, you hear a sound. Touch gently the cheek with the end of one of the prongs; you feel a tickling sensation produced by its minute vibrations. Thrust the ends of the prongs just beneath the surface of water; the water is thrown off in a fine spray on either side of the vibrating fork. Watch the strings of a piano, guitar, or violin, or the tongue of a jews-harp, when sounding. You can *see* that they are in motion.

The difference between a body when sounding and when not sounding is, that when sounding it is in a state of continuous vibration; when not sounding, this vibration is absent, and the parts of the body are at rest among themselves. We conclude that *sound originates in a vibrating body*.

Sounds that proceed from the tuning-fork and the violin string are examples of sound produced by transverse vibrations.

Experiment 1. With one hand grasp at its center a glass tube about 1^m long; lay a damp woolen cloth on the palm of the other hand, and grasp the tube tightly with this hand, and slide it quickly lengthwise the tube. The friction between the cloth and the tube will throw the latter into *longitudinal vibrations*, and a loud, shrill sound will be produced.

Experiment 2. Take a strip of sheet-iron or brass 15^{cm} long and 6^{cm} wide, make a hole near one end, and suspend by a string 1^m long from the hand, and rotate it rapidly about the hand after the manner of a sling. The string will rapidly twist and untwist, and a loud sound will result from the *torsional vibrations*.

§ 259. How sound travels. — How can a bell, sounding at a distance, affect the ear? If the bell while sounding possesses no peculiar property except motion, then it has nothing to communicate to the ear but motion. But motion can be communicated by one body to another at a distance only through some medium.

Does sound require a medium for its communication? If so, what is the medium?

Experiment. Lay a thick tuft of cotton-wool on the plate of an air-pump, and on this, face downward, place a loud-ticking watch, and cover with the receiver. Notice that the receiver, interposed between the watch and your ear, greatly diminishes the sound, or interferes with the passage of *something* to the ear. Take a few strokes of the pump and listen; the sound is more feeble, and continues to grow less and less distinct as the exhaustion progresses, until either no sound can be heard when the ear is placed close to the receiver, or an extremely faint one, as if coming from a great distance. The removal of air from a portion of the space between the watch and your ear destroys the sound, although the watch continues to tick. Let in the air again and the sound is restored.

Thus it appears that *sound cannot travel through a vacuum;* in other words, *without a medium*, and the medium in this case is *air*.

By which of the two methods described on page 278 is motion transmitted from the sounding body through the air? Take an extreme case: A cannon is discharged at a distance of one-fourth of a mile from you. You not only hear the sound, but feel the shock communicated by the air; the windows are shaken by it; at the same time, you easily perceive that it is not the motion of a wind, but the motion of a pulse. It can easily be shown that the pulse travelled at a rate of about 800 miles in an hour, or with nearly the velocity of a rifle ball, whereas the wind of a hurricane seldom exceeds 75 miles an hour. What, think you, would be the result if you were to be struck by a gust of wind of such velocity? Yet the softest whisper travels with very nearly the same speed.

§ 260. **Air-waves.**—Boys amuse themselves by inflating paper bags, and with a quick blow bursting them, producing with each a single loud report. First the air is suddenly and greatly condensed by the blow, the bag is burst; the air now, as suddenly and with equal force, expands, and by its expansion condenses the air for a certain distance all around it, leaving a rarefaction where just before had been a condensation. If many bags were burst at the same spot in rapid succession, the result would be that alternating shells of condensation and rarefaction would be thrown off, all having a common center, enlarging as they advance, like the waves formed by stones dropped into water; only that, in this case, the waves are not like rings, but hollow globes; not circular, but spherical.

As a wave advances, each individual air-particle concerned in its transmission performs a short excursion fro and to in a straight line radiating from the center of the shells or hollow globes. A particle begins to move when the front of the shell of compression touches it, and completes its motion when the back of the next shell of rarefaction leaves it. Accordingly, *an air-wave travels its own length in the time that a particle occupies in going through one complete vibration so as to be ready to start again.*

SOLIDS AND LIQUIDS AS MEDIA OF SOUND. 283

§ 261. **What sound is.** — The term *sound* is sometimes used to denote a sensation, sometimes to denote the external cause of the sensation; it is in this latter sense that the word is used in Physics, and that we have to define it.

If the ear replace the candle in the experiment (page 278), the air pulse produces a loud sound. Conversely, air-waves started by the voice may affect a flame, as shown on page 313. In fact, the relation between the cause of our sensation and a vibration is so uniform, that we may say, *Sound is vibration that may be appreciated by the ear.*

§ 262. **Solids and liquids as media transmitting sound.**
— **Experiment 1.** Lay a watch, with its back downward, on and near to one end of a long board (or table), and cover the watch with loose folds of cloth till its ticking cannot be heard through the air in any direction at a distance equal to the length of the board. Now place the ear in contact with the distant end of the board, and you will hear the ticking of the watch very distinctly.

Experiment 2. Place one end of a long pole on a resonance box (page 296), and apply the stem of a vibrating tuning-fork to the other end; the sound-vibrations will be transmitted through the pole to the box, and a loud sound will be given out by the box, as though that, and not the tuning-fork, were the origin of the sound.

Experiment 3. Place the ear to the earth, and listen to the rumbling of a distant carriage; or put the ear to one end of a long stick of timber, and let some one gently scratch the other end with a pin.

Experiment 4. The following experiment will be found very instructive and satisfactory: Let two persons stand about fifteen rods apart, and one of them strike two pebble-stones together, so as to be scarcely audible to the other. Then, when at the same distance apart, let one of them dive to the bottom of a pond of water, or hold one ear for a few seconds beneath the surface of the water, while the other, extending his hands into the water, strikes the stones together as before. The sound is much more audible than when conveyed by air.

Solids and liquids, as well as gases, transmit sound-vibrations.

XL. VELOCITY OF SOUND.

§ 263. On what velocity of sound depends. — The flash of a gun, however distant, is seen by an observer at the instant it is made. But the report, if the distance is several hundred yards, is heard a little later. If the distance is a mile, an interval of nearly five seconds will occur; so that sound must occupy that time in traveling a mile, or it must travel about 1100 feet in a second, — a velocity somewhat less than that of a rifle ball.

It is apparent that sound must travel more slowly in a dense than in a rare medium, inasmuch as in the former there is a greater mass to be moved; on the other hand (see page 277), it travels faster in the medium that is the most elastic. *Density retards and elasticity increases the velocity of sound*. The relation of velocity to the density and elasticity of gases, as ascertained by careful experiment, is as follows: *the velocity of sound in gases is directly proportional to the square root of their elasticity, and inversely proportional to the square root of their respective densities.*

The velocity of sound in air at 0° C. has been found to be 333^m (1093 ft.) per second. Its velocity increases nearly six-tenths of a meter for each degree centigrade. At the temperature of 16° C. (60° F.) we may reckon the velocity of sound at about 342^m (1125 ft.) per second.

The greater density of solids and liquids, as compared with gases, tends, of course, to diminish the velocity of sound; but their greater elasticity[1] more than compensates for the decrease of velocity occasioned by the increase of density. As a general rule, solids are more elastic than liquids; hence, sound generally travels faster in the former than in the latter. For example, sound travels in water about 4 times as fast as in air; in lead,

---

[1] The question will very pertinently arise here, inasmuch as gases are (page 53) perfectly elastic, How can solids and liquids be regarded as having greater elasticity? It should be understood that while gases completely recover their volume after a compressing force is removed, they do it more sluggishly than solids and liquids.

4 times; in gold, 5 times; in brass, 10 times; in copper, 11 times; in iron, 16 times; in glass, 16 times; in wood, along the fiber, between 10 and 15 times; in wood, across the fiber, between 4 and 6 times.

QUESTIONS.

1. (*a*) If a body of gas is compressed, how is its density or volume affected? (See page 156.) (*b*) How is its elasticity affected? (*c*) How is it affected as regards the velocity with which it will transmit sound?

2. Hydrogen is sixteen times lighter (or rarer) than oxygen under the same pressure. (*a*) In which will sound travel faster? (*b*) Why? (*c*) How many times faster?

3. When sound travels in air with a velocity of 331^m per second, it travels in carbonic acid gas at the rate of 262^m per second. (*a*) Which is the denser gas? (*b*) How many times denser?

4. When a confined body of air is heated, it has its elasticity increased without any change of density. How will this affect transmission of sound?

5. If air is heated and allowed to expand freely, as on a warm summer day, its elasticity is unaffected, but its density is diminished; how will this affect the transmission of sound?

XLI. REFLECTION AND REFRACTION OF SOUND.

§ 264. **Reflection.** — In the experiment with the spiral spring, waves were reflected from the box to the hand, and from the hand to the box. When a sound-wave meets an obstacle in its course, it is reflected; and a sound heard after being thus reflected is often called an *echo*, or *reverberation* when many times reflected, so that the sound becomes nearly continuous.

§ 265. **Sound reflected by concave mirrors.** — Experiment. Place a watch at the focus (page 286) A, Figure 208, of a concave mirror G. At the focus B of another concave mirror H, place the large opening of a small tunnel, and with a rubber connector attach the bent glass tube C to the nose of the tunnel. The extremity D being placed in the ear, the ticking of the watch can be heard very distinctly, as though it were somewhere near the mirror H. Though the mirrors be 5ᵐ apart, the sound will be heard much louder at B than at an intertermediate point E.

How is this explained? Every air-particle in a certain radial line, as Ac, receives and transmits motion in the direction of this line; the last particle strikes the mirror at c, and being perfectly elastic, bounds off in the direction cc' in conformity to the *law of reflection* (page 118), communicating its motion to the particles in this line. At c' a similar reflection gives motion to the air-particles in the line c'B. In consequence of these two reflections, all divergent lines of force, as Ad, Ae, etc., that meet the mirror G, are there rendered parallel, and afterwards rendered convergent at the mirror H. The practical result of the concentration of this scattering force is, that a sound of great intensity is heard at B. The points A and B are called the *foci* of the mirrors. The front of the wave as it leaves A is convex, in passing from G to H it is plane, and from H to B concave. If you fill a large circular tin basin with water, and strike one edge with a knuckle, circular waves with concave fronts will close in on the center, heaping up the water at that point.

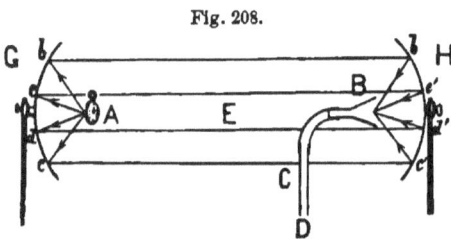

Fig. 208.

Long "whispering-galleries" have been constructed on this principle. Persons stationed at the foci of the concave ends of the long gallery can carry on a conversation in a whisper which persons between cannot hear. A most notable instance was that of the "Ear of Dionysius," in the dungeon of Syracuse. The roof of the prison was so constructed as to transmit through a narrow passage cut in the rock, to the ear of the tyrant, even the whispers of the victims there confined.

The external ear is a sound condenser. The hand held concave behind the ear, by its increased surface, adds to its efficiency. An ear-trumpet, by successive reflections, serves to concentrate, at the small orifice opening into the ear, all the sound-waves that enter at the large end.

§ 266. **Refraction.**— If you place your ear at the small end of a tunnel C (Fig. 209), and listen to the ticking of a watch A,

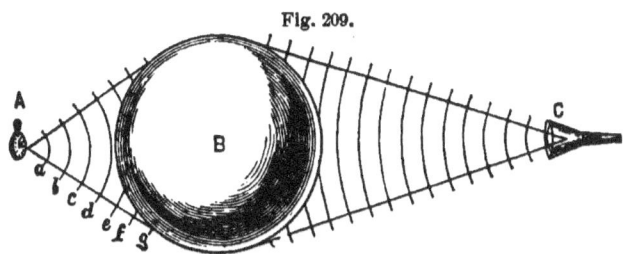

Fig. 209.

4^m distant, and then introduce a collodion balloon B filled with carbonic acid gas between your ear and the watch, and very near the latter, the sound becomes much louder.

The cause is obvious; for, let the curved lines a, b, c, etc., represent sections of sound-waves with convex fronts, and B a spherical body of carbonic acid gas which is denser than air; then it is clear that, owing to the slower progress of the waves in the denser gas, they would become flattened on entering this gas, and the waves of convex fronts may be changed to waves of plane fronts. Again, points at the extremities of the waves, having less distance to travel in the denser gas than points near the center, would emerge first and get in advance, and thus the wave fronts which are plane while wholly in the dense gas, become concave on leaving it. By these changes in the form of the wave fronts, sound energy which was originally becoming diffused through wider and wider space, and therefore becoming less intense as it progressed, is so changed in direction in passing into and out of a medium of greater density, that the energy is finally concentrated at a distant point, as at C, and thereby intensified.

Any change in direction of sound, caused by passing from a medium of a certain density into a medium of different density, is called *refraction*.

XLII. LOUDNESS OF SOUND.

§ 267. Loudness depends on amplitude of vibrations. — Gently tap the prongs of a tuning-fork and dip them into water, — the water is scarcely moved by them; increase the force of the blow, — the vibrations become wider, and the water spray is thrown with greater force and to a greater distance. The same thing occurs when the fork vibrates in air; though we do not see the air-particles as they are batted by the moving fork, yet we feel the effects as a sound sensation, and we judge of their energy by the intensity of the sensation. Loudness of sound is really the measure of a sensation; but as we have no suitable or constant standard of measurement for a sensation, we are compelled to measure rather the intensity of the sound-wave, knowing at the same time that the loudness is not proportional to this intensity; unfortunately the expressions *loudness* and *intensity of sound* are often interchanged. The intensity of a vibration is measured by the energy of the vibrating particle. It is clear that if the amplitude of vibration of a particle is doubled while its period remains constant, its velocity is doubled (or nearly so), and its energy becomes therefore four times as much as at first. Hence, (1) *Measured mechanically, the loudness or intensity of sound is proportional to the square of the amplitude of the vibrations of the sounding body.*

§ 268. Loudness depends upon the density of the medium. — In the experiment with the watch under the receiver of the air-pump (page 281), the sound grew feebler as the air became rarer. Aeronauts are obliged to exert themselves more to make their conversation heard when they reach great hights than when in the denser lower air. Fill a glass bell-jar with hydrogen gas, and place in it a small alarm clock; the sound is exceedingly weak and thin, as compared with the sound when the jar is filled with air. These experiments teach us, (2) *that the intensity of sound depends upon the density of* *the medium in which it is produced.* In a rare medium a vibrating body during a single vibration sets in motion either fewer particles, as in the case of the partially exhausted receiver, or, as in the case of the hydrogen gas, it sets in motion lighter particles than in a dense medium; consequently it parts with its energy more slowly, and the sound is consequently weaker.

(In which ought the vibrations of a body to last longer, — in a dense or in a rare medium? Why?)

§ 269. **Loudness depends on distance.** — It is a matter of every-day observation that the loudness of a sound diminishes very rapidly as the distance from its source to the ear increases. The ear is not, however, able to compare very accurately the loudness of two sounds; for instance, it cannot determine when one sound is just twice as loud as another. This, however, so far as it is affected by distance, can be very accurately determined by calculation. For it is evident that as a sound-wave recedes from its source in an ever-widening sphere, a given amount of energy becomes distributed over an ever-increasing surface; and as a greater number of particles partake of the motion, individual particles receive proportionally less energy; hence it follows, — as a consequence of the geometrical truth, that the surface of a sphere varies as the square of its radius, — that (3) *the intensity of sound varies inversely as the square of the distance from its source.* For example, if two persons, A and B, are respectively 500 and 1000 meters from a gun when it is discharged, the report that reaches A will be four times as loud as the same report when it reaches B.

§ 270. **Speaking tubes.** — Experiment. Place a watch at one end of the long tin tube (Fig. 206), and the ear at the other end. The ticking is heard very loud, as though the watch were close to the ear.

Long tin tubes, called *speaking tubes*, passing through many apartments in a building, enable persons at the distant extremities to carry on conversation in a low tone of voice, while persons in the various rooms through which the tube passes hear nothing. The reason is that the sound-waves which enter the tube are prevented from expanding, consequently the intensity of sound is not affected by distance, except as its energy is wasted by friction of the air against the sides of the tube.

§ 271. Reënforcement of sound. — Observe the difference between the loudness of a sound made in a room, and the same made in the open air. In a room of moderate size the original and reflected sounds are heard simultaneously, one intensifying the other and producing what is called *resonance*. In large rooms, the blending of the two is not perfect, resulting in a sort of blurred sound, which is loud but indistinct.

Experiment. Set a tuning-fork in vibration; you can scarcely hear the sound produced unless it is held near the ear. Press the stem against a table; the sound rings out loud, but seems to proceed from the table. Again set it vibrating, hold it to the ear, and, watch in hand, note the number of seconds it can be heard; then note the time that it can be heard when the stem rests on the table. The vibrations continue longer in the former case than in the latter.

When only the fork vibrates, the prongs presenting little surface cut their way through the air, producing very slight condensations, and consequently sounds of little intensity. When the fork rests upon the table, the vibrations are communicated to the table; the table with its larger surface throws a larger mass of air into vibration, and thus greatly intensifies the sound. But as the sound is rendered more intense, the energy of the vibrating body is sooner exhausted, and the sounds have shorter duration.

In all stringed instruments, like the piano, violin, etc., reënforcement of sound is necessary; here the *sounding board* of thin wood fills more perfectly the place of the table. This sounding board must strengthen all notes within the compass of the instrument. The strings of the piano, guitar, and violin owe as much of their loudness of sound to their elastic sounding boards, as does the fork to the table.

§ 272. Reënforcement by masses of air. — Resonators.

— **Experiment.** Take a glass tube A, Figure 210, 45cm long and 4cm in diameter; thrust one end into a vessel of water, C, and hold over the other end a vibrating tuning-fork B that makes (say) 256 vibrations in a second. (See page 300.) Gradually lower the tube into the water, and when it reaches a certain depth, *i.e.*, when the column of air oc attains a certain length, the sound of the fork becomes very loud; continuing to lower the tube, the sound rapidly dies away. Try other forks that make different numbers of vibrations in a second. The sound of each is intensified, but each requires a length of air-column suited to its particular vibration number.

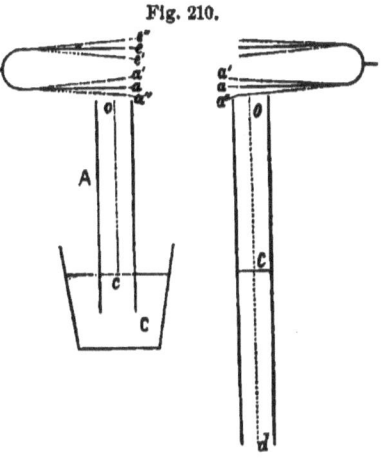

Fig. 210.

Columns of air are thus found to serve as well as sounding boards to strengthen a sound. When so used they are called *resonators;* but unlike the sounding board they can respond loudly to only one tone, or to a few tones of widely different pitch. An important form of resonator is shown on page 309.

How is this reënforcement effected? When the prong a moves from one extremity of its arc a' to the other a'', it sends a condensation down the tube; this condensation striking the surface of the water, is reflected by it up the tube. Now suppose that the front of this reflected condensation should just reach the prong at the instant it is starting on its retreat from a'' to a'; then the reflected condensation will conspire with the condensation formed by the prong in its retreat to make a greater condensation in the air outside the tube. Again the retreat of the prong from a'' to a' produces in its rear a rarefaction, which also runs down the tube, is reflected, and will reach the prong at the instant it is about to return from a' to a'', and to cause a rarefaction in its rear; these two rarefactions moving in the same direction conspire to produce an intensified rarefaction. The original sounds thus combine with their echoes to produce resonance; but this can only happen when the like parts of each wave coincide each with each; for if the tube were somewhat longer or shorter than it is, it is plain that condensations would meet rarefactions in the tube, and tend to destroy one another.

The loudness of sound of all wind instruments is due to the resonance of the air contained within them. A simple vibratory movement at the mouth or orifice of the instrument, scarcely audible in itself, such as the vibration of a reed in reed pipes, or a pulsatory movement of the air produced by the passage of a thin sheet of air over a sharp wooden or metallic edge, as in organ pipes, flutes, and flageolets, or more simply still by the friction of a gentle stream of breath from the lips sent obliquely across the open end of a closed tube, bottle, or pen-case. is sufficient to set the large body of enclosed air in the instrument into vibration, and thus reënforced, the sound becomes audible at long distances.

Fig. 211.

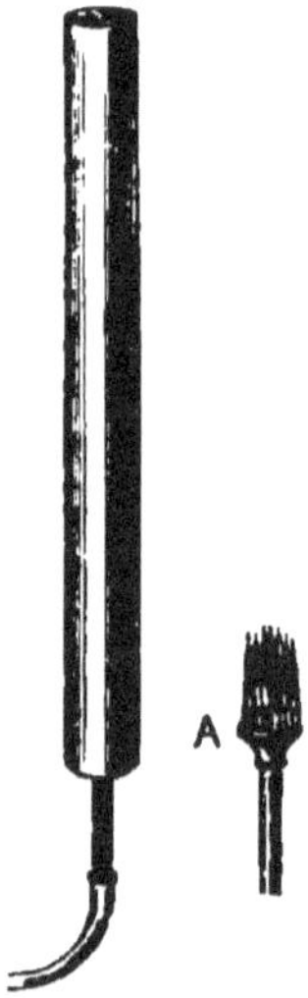

Experiment 2. Attach a rose gas-burner A, Figure 211, to a metal gas tube about 1ᵐ in length, and connect this by a rubber tube with a gas-burner. Light the gas at the rose burner, and you will hear a low rustling noise. Remove the conical cap from the long tin tube (Fig. 206, page 278), support the tube in a vertical position, and gradually raise the burner into the tube; when it reaches a certain point not far up, the body of air in the tube will catch up the vibrations, and give out a deafening sound that will shake the walls and furniture in the room.

§ 273. Measuring wave-lengths and velocity of sound. — Experiments like that described on page 291 enable us readily to measure the wave-length produced by a fork that makes a given number of vibrations in a second, and also to measure the velocity of sound. It is evident that if a condensation generated by the prong of the fork in its forward movement from a' to a'' (Fig. 211) met with no obstacle, its front, meantime, would traverse the distance od, or twice the distance oc; hence the length of the condensation is the distance od. But a condensation is only one-half of a wave, and the passage of the prong from a' to a'' is only one-half of a vibration; consequently the distance od is one-half of a wave-length, and the distance oc is one-fourth of a wave-length. The measured distance of oc in this case is about 33^{cm}; hence the length of wave produced by a C'-fork making 256 vibrations in a second is (33^{cm} × 4 =) 1.32^m. And since a wave from this fork travels 1.32^m in $\frac{1}{256}$ of a second, it will travel in an entire second (1.32 × 256 =) 338^m. The distance oc is modified by temperature. It is also modified by the diameter of the tube. For accuracy about two-thirds of the diameter should be added to the length of the tube to obtain one-fourth of the wave-length. It is evident that the three quantities expressed in the formula

$$\text{wave-length} = \frac{\text{velocity}}{\text{number of vibrations}}$$

bear such a relation to one another that if any two are known the remaining quantity can be computed. It will further be observed that *with a given velocity the wave-length varies inversely as the number of vibrations*, i.e., the greater the number of vibrations per second, the shorter the wave-length.

QUESTIONS.

1. (a) Which produces greater wave-lengths, a fork making 256 vibrations in a second, or one making 512 vibrations in the same time? (b) How many times?

2. Disregarding the diameter of the tube, what number of vibrations does a fork make in a second, whose resonance tube is 22.26^{cm} long, when the temperature is 16° C.?

3. What is the wave-length produced by a fork that makes 384 vibrations in a second, when the temperature is 16° C.?

§ 274. Interference of sound-waves. — Is it possible for two sounds to destroy one another and produce silence?

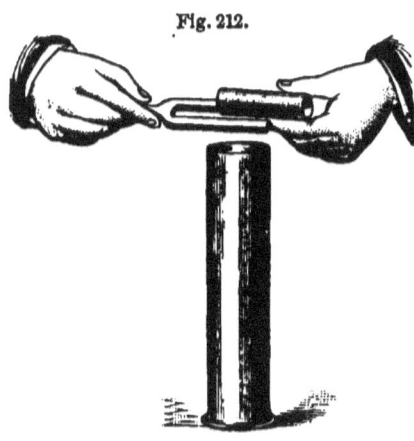

Fig. 212.

Experiment 1. While the fork B (Fig. 210) is vibrating in the position represented in the figure, slowly roll the fork over in the fingers, through a quarter of a revolution, until the two prongs are in the same horizontal plane, with their edges turned toward the opening of the tube. When the fork has accomplished one-eighth of a revolution, and is in an oblique position, as in Figure 212, the reënforcement from the tube entirely disappears, but reappears when the fork completes its quarter of a revolution. Return to the position where there is no resonance, and enclose the prong furthest from the tube, without touching the fork, in a loose roll of paper, so as to prevent the sound-waves produced by that prong from passing into the tube; the resonance resulting from the vibrations of the other prong immediately appears in full force.

Experiment 2. Carry, while sounding, a tuning-fork mounted on a resonance-box (see Fig. 214), from a distance slowly toward a wall of a room; the sound will become wavy, rising and sinking at regular intervals. That is, at certain points the condensations and rarefactions of the waves advancing from the fork will coincide each to each with those of the waves reflected from the wall, and when this is the case the sound is louder. At other points the condensations of the waves issuing from the fork occur at the same places where the rarefactions of the reflected waves occur; and in this case they nearly destroy one another, and a fainter sound is the result.

Thus it appears that *two sounds of the same pitch may unite to form a sound louder or weaker than either alone, or even cause silence, according to their difference of phase and their relative intensities.* When like phases of two sets of sound-waves coincide, *i.e.*, condensation with condensation, and rarefaction with rarefaction, the result is, as we might expect, an intensified sound. In this case the air-particles are subject to the action of two joint forces in the same direction, which tends to quicken their motions. On the other hand, when unlike phases of two sets of sound-waves coincide, *i.e.*, the condensations of one set of waves with the rarefactions of another, as would happen if one set of waves should be half of a wave-length behind the other; the air-particles will be subject to two forces in opposite directions; and the evident result of an equal tendency to the two opposing forces is a state of repose.

§ **275. Forced and sympathetic vibrations.** — Experiment 1. Suspend by a string 1^m long a stone weighing about 2^k. Swing the stone, and learn its rate of vibration; then stop it, and blow gentle puffs of breath against the stone in the same periods. The first few puffs produce no visible effect; but, persevering, the stone will soon move visibly, and after a large number of these feeble impulses, the stone will move through a wide arc, and will require considerable force to stop it.

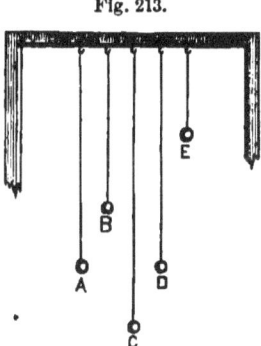

Fig. 213.

Experiment 2. Suspend from a frame several pendulums, A, B, C, etc. (Fig. 213). A and D are each 1^m long, C a little longer, and B and E shorter. Set A in vibration, and slight impulses will be communicated through the frame to D and cause it to vibrate. The vibration-period of D being the same as that of A, all the impulses tend to accumulate motion in D, so that it soon vibrates through arcs as large as those of A. On the other hand, C, B, and E, having different rates of vibration from that of A, will at first acquire a slight motion, but soon their vibrations will be in opposition to those of A, and then the impulses received from A will tend to destroy the slight motion they had previously acquired.

Experiment 3. Hang up, a few feet distant from the pendulum of Exp. 1, a bullet or shot by a shorter string, and connect the bullet and stone by a tight thread. Set the stone swinging, and the bullet must vibrate in the same period, although its natural time of vibration is shorter.

Experiment 4. Press down gently one of the keys of a piano so as to raise the damper without making any sound, and then sing loudly into the instrument the corresponding note. The string corresponding to this note will be thrown into vibrations that can be heard for several seconds after the voice ceases. If another note be sung, this string will not respond audibly.

Raise the dampers from all the strings of the piano by pressing the foot on the right-hand pedal, and sing strongly some note into the piano. Although all the strings are free to vibrate, only those that correspond to the note you sing (*i.e.*, those that are capable of making the same number of vibrations per second as are produced by your voice), will respond loudly.

Experiment 5. Take two forks, A and B (Fig. 214), tuned exactly in unison, and mounted on resonance-boxes, and place them from three to ten meters apart. Fasten, by a bit of sealing-wax, a thread to a thin piece of glass 12^{mm} square (glass used for microscopic mountings is the best, or a piece of photographic tintype plate will answer well), and suspend so as to touch a corner of one of the prongs of the fork B. Set the fork A in vibration by drawing a resined bass-viol bow strongly across the ends of its prongs. In about ten seconds stop the vibrations of A with the fingers, and you will see and hear the piece of glass rattling against the prong of the fork B; remove the glass, and place the ear near the fork B, or better, the open end of the box, and you may hear a distinct sound, showing that the fork B has been thrown into a state of vibration by the fork A.

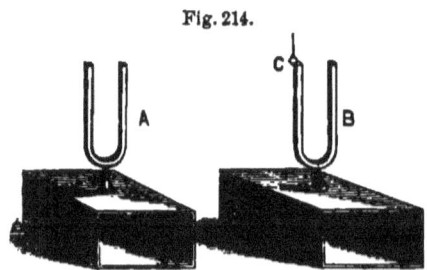

Fig. 214.

So the pulses that traverse the air between the forks, so gentle that only the sensitive organ of the ear can perceive them, become great enough to move the rigid steel when their blows, dealt at the rate of perhaps 512 in a second, add themselves together. The large number of blows make up for the feebleness of each by itself.

These experiments show that a vibrating body tends to make other bodies near it vibrate in its own period. The vibrations thus caused are called *forced vibrations*. These occur, in Exp. 2, with B, C, and E; in the vibrations of sounding-boards, and of the membrane and fluids of the ear (page 315), and in the air when transmitting a sound-wave, etc. But as the period of the incident waves coincides more and more nearly with the period of the second body, the amplitude of the vibrations of the latter becomes greater and greater, until finally its vibration is uniform, like D (Fig. 213), not irregular, like B, C, and E. Such are called *sympathetic vibrations*, as in Exps. 4 and 5.

§ **276. Distinction between noise and musical sound.** If the body that strikes the air deals it but a single blow, like the discharge of a fire-cracker, the ear receives but a single shock, and the result is called a *noise*. If several shocks are slowly received by the ear in succession, the ear distinguishes them as so many separate noises. If, however, the body that strikes the air is in vibration, and deals it a great number of little blows in a second, or if a large number of fire-crackers are discharged one after another very rapidly, so that the ear is unable to distinguish the individual shocks, the effect produced is that of one continuous sound, which may be pleasing to the ear; and, if so, it is called a *musical sound*. But continuity of sound does not necessarily render it musical. The sound produced by a hundred children beating various articles in a room with clubs might not be lacking in continuity, but it would be an intolerable noise. There would be wanting those elements that please the ear; viz., regularity both in periodicity and intensity of the shocks which it receives. The distinction between music and noise is, generally speaking, a distinction between the agreeable and the disagreeable, between regularity and confusion. *The characteristics of a musical sound are regularity and simplicity.*

XLIII. PITCH OF SOUNDS.

§ 277. On what pitch depends. — Draw the finger-nail slowly, and then rapidly, across the teeth of a comb. The two musical sounds produced are commonly described as *low* or *grave*, and *high* or *acute*, and the hight of a musical sound is called *pitch*. What is the cause of a difference in hight or pitch of two sounds?

Experiment. Procure a circular sheet-iron or pasteboard disk A, Figure 215, 30^{cm} in diameter. From the center of the disk describe a circle with a radius of 12^{cm}. In the circumference of this circle, with a punch, cut holes 8^{mm} in diameter, leaving equal intervals of about 2^{cm} between the holes. Insert in a rubber tube a piece of glass tube B, of 1^{cm} bore, drawn out at one end so that its orifice is about 4^{mm} in diameter. Attach the disk to some rotating apparatus, hold the small orifice of the glass tube opposite the holes, and blow steadily through the tube, and rotate the disk at first very slowly and then with gradually increasing rapidity. The breath, as it makes its exit from the tube, cannot escape continuously through the holes, but is cut up by the passing obstructions into a series of puffs, which at first are heard as so many distinct sounds; as the speed increases, the number of puffs in a second increases, until the ear can no longer separate them, when they blend together in a deep sound of a definite pitch.

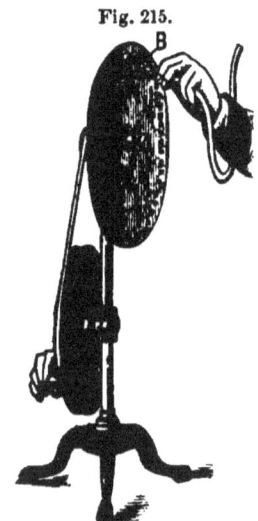

Fig. 215.

The peculiarity of this instrument is that it does not produce sound by its own vibrations. Every time the air is driven through a hole, it produces a pulse of condensation in the air beyond; and during the interval between the successive discharges, a pulse of rarefaction will be caused by the elasticity of the air, so that the result is the same, so far as the effect on the air medium is concerned, as if a body were vibrating in it. As the velocity increases, the pitch constantly rises,

SIREN. 299

until, at the greatest speed conveniently attainable, it becomes painfully shrill. Varying the force of the breath affects the loudness of the sound, but does not affect its pitch.

We learned on page 293 that on the number of vibrations in a second, called the *vibration-frequency*, depends the wavelength. So we have discovered the important fact that *pitch depends upon vibration-frequency or wave-length*, i.e., *the greater the number of vibrations per second, or the shorter the wavelength, the higher the pitch.*

QUESTIONS AND EXERCISES.

1. Why does the same bell always give a sound of the same pitch?
2. (*a*) What is the effect of striking a bell with different degrees of force? (*b*) What change in the vibrations is produced? (*c*) What property of sound remains the same?
3. (*a*) Strike a key of a piano and hold it down; what is the only change you observe in the sound produced while it remains audible? (*b*) What is the cause of this change?
4. Rake the teeth of a comb with a finger-nail, at first slowly, then quickly, and account for the difference in the character of the sounds produced.
5. (*a*) On what does pitch depend? (*b*) On what, loudness?

§ 278. **How to find the vibration-frequency of a tone.— Siren.**— The perforated wheel described above is a cheap imitation of a portion of an important instrument called a *siren*. The instrument completed has an attachment called a *counter*, which shows the number of revolutions the wheel makes in a given time.

Suppose that it is required to ascertain the number of vibrations per second necessary to produce a given pitch. Take some instrument that gives the required pitch, *e.g.*, a tuning-fork, set it in vibration ; also rotate the siren, causing the pitch of its sound gradually to rise until it corresponds with the pitch of the fork ; then, sustaining that pitch, set the counter in operation, and at the end of a given time read off the number of revolutions made by the wheel ; this number, multiplied by the number of holes in the wheel, gives the number of sound-waves produced by the wheel during the given time, and the number of vibrations made by the fork in the same time; and this number, divided by the number of seconds employed, gives the number of vibrations that must be made in a second by any instrument in order to produce a sound of the same pitch. With the siren we may even determine the number of vibrations made by the wing of a fly which buzzes around our ears.

§ 279. **Musical scale.** — Long before any one had attempted to find the frequency of vibration of a sounding body, men had used a succession of sounds, differing in pitch, that formed the so-called *musical scale*, or *gamut*, and were familiar with its intervals. Very different scales have satisfied musicians of different ages and nations. We can find a scale that will nearly or exactly satisfy modern musical ears among Europeans and Americans on a well-tuned piano or organ. On such a piano, by the siren or otherwise, it is found that the note called middle C (C') has, on the best American instruments, about 270 double vibrations per second; on German, 264; while the French legal standard is 261. Physical apparatus is usually based on $C' = 256$ vibrations, 256 being a power of 2. Assuming $C' = 264$ vibrations, if we extend our measures up and down the scale, or get a violinist or singer to perform near an instrument that counts the vibrations, *e.g.*, the siren, numbers agreeing very closely with those given in Figure 216 are obtained; no one's ear is accurate enough to play or sing *precisely* the same on two trials. Since the ear is wholly incapable of determining the number of vibrations corresponding to a given note, but is capable of determining with wondrous precision the ratio of the vibration numbers of two notes, it is

Fig. 216.

Notes.	Vibration numbers.	Vibration ratios.
C	132	1
D	148½	9/8
E	165	5/4
F	176	4/3
G	198	3/2
A	220	5/3
B	247½	15/8
C'	264	2
D'	297	9/4
E'	330	5/2
F'	352	8/3
G'	396	3
A'	440	10/3
B'	495	15/4
C''	528	4

LIMITS OF THE SCALE AND HEARING. 301

clear that all music must depend upon the recognition of such ratios. For this reason the numbers in the third column are of great importance. The eighth note, above or below a given note, counting the given note as one, is called an eighth, — more commonly, an *octave*, — above or below. Thus C' is the octave above C, and C_{-1} an octave below. In a similar manner D is called the *second*, and G the *fifth*, etc., in the scale in which C is the prime or first note.

PROBLEMS.

1. Find the vibration number for each note of the scale of which C'' is the first note.
2. What is the vibration number of C_{-1}, an octave below C?
3. Find the wave-length corresponding to each note of the scale of which C' is the first, when the temperature of the air is 16° C.?
4. Find the length of a resonance tube (disregarding its diameter), closed at one end, which will respond to C' when the temperature is 16° C.?

§ 280. **Limits of the scale and hearing.** — The lowest note of a $7\frac{1}{3}$ octave piano makes about $27\frac{1}{2}$, the highest, 4,224 vibrations per second; but these extreme notes have little musical value, and the lowest notes are only used for their harmonics (see page 395). The range of the human voice lies between 100 and 1,000 vibrations per second, or a little more than three octaves; an ordinary singer has about the compass of two octaves.

The ear is capable of hearing vibrations far exceeding in number the requirements of music. It can appreciate sounds arising from 32 to 38,000 vibrations[1] per second, *i.e.*, a range of about eleven octaves, and a corresponding range of wave-length between seventy feet and three or four-tenths of an inch. These numbers vary, however, considerably with the person. Exceptional ears can hear as many as 50,000 vibrations. Some ears can hear a bat's cry, or the creaking of a cricket; others cannot. Singing mice are sometimes placed on exhibition. Of those who go to hear them, some can hear nothing, others a little, and

[1] Preyer places the lowest limit for some ears at 16 vibrations per second.

others again can hear much. In the ability to hear sharp sounds, no animal is superior to the cat, which finds her prey in the dark by its squealing.

§ **281. Beats.**—**Experiment 1.** Strike simultaneously the lowest note of a piano and its sharp (black key next above), and listen to the resulting sound.

Experiment 2. Take two forks which make the same number of vibrations per second (page 296), load one of the prongs of one fork by sticking to it a small ball of wax, and thereby cause it to make a few less vibrations per second than the other. Set both forks in vibration, and note the result.

In both cases you will hear a peculiar wavy or throbbing sound, caused by an alternate rising and sinking in loudness. These alternations in loudness are called *beats*.

Fig. 217.

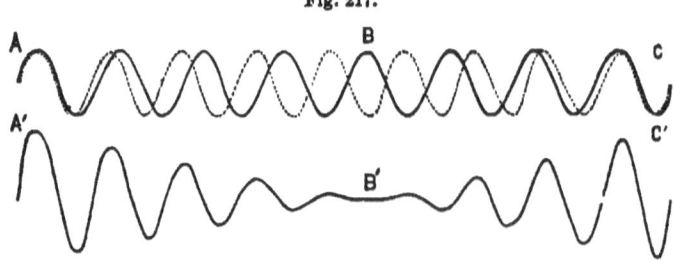

Let the continuous curve line A C (Fig. 217) represent a series of waves proceeding from the prongs of the loaded fork, and the dotted line a series of waves proceeding from the other fork. As explained on page 311, the elevations of these waves may represent the distance the air-particle has been moved in the condensed part of the wave; similarly with the depressions for the rarefied part. Now the waves from both forks may start together at A; but as the waves from the loaded fork are given less frequently, so are they correspondingly longer and lag behind; and at certain intervals, as at B, condensations will correspond with rarefactions, producing by their interference momentary silence, too short, however, to be perceived; but the sound as received by the ear is correctly represented in its varying loudness by the curved line in the lower part of the figure. This line represents the exact resultant of the two alternately concurring and opposing forces on the particles of the air between the forks and the ear.

If one of the forks makes 256 vibrations, and the other 255 vibrations in a second, it is apparent that once during the second the condensation of one series of waves will coincide with the condensation of the other, producing a sound of maximum intensity; and once during the same time the condensation of the one will coincide with the rarefaction of the other, producing a sound of minimum intensity; this will cause just one beat per second. If there is a difference of two vibrations per second between the two forks, then there will be two beats per second.

In every case *the number of beats per second due to two simple tones is equal to the difference of their respective vibration-numbers.* The sensation produced on the ear by such a throbbing sound, when the beats are sufficiently frequent, is unpleasant, for the similar reason that the sensation produced by flashes of light that enter the eye, when you walk on the shady side of a picket fence, is unpleasant. The unpleasant sensation, called by musicians *discord*, is found to be due to beats (see page 307).

XLIV. VIBRATION OF STRINGS.

§ 282. Sonometer. — Experiment. Take a piece of violin-string or piano-wire a little longer than your table. Fasten one end to a nail in one end of the table, and pass the other end over a pulley fastened to the other end of the table, and to this end of the string suspend a pail containing sand, the two weighing just a pound. Place under the string, near the ends of the table, two wedge-shaped bridges A and B (Fig. 218). An apparatus thus arranged is called a *sonometer*. Pluck the string with the fingers near the middle, causing it to vibrate, and note the pitch of the sound, and the length of the string between the bridges. Move the bridge A toward B; the pitch rises as the vibrating portion of the string is shortened. Vary the position of A until a pitch is obtained an octave above the pitch given at first,

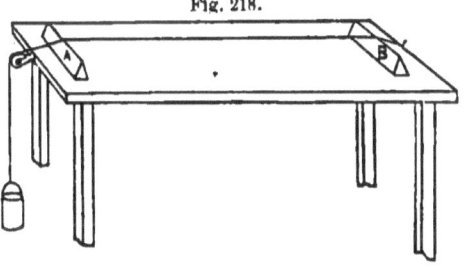

Fig. 218.

and it will be found that the string is just one-half its original length; i.e., *by halving the string its vibration-number is doubled.* At two-thirds its original length, it gives a note at an interval of a fifth above that given by its original length; and genera'.y *the reciprocals of the fractions* (page 300), *representing the relative vibration-numbers of the several notes of a scale, represent the relative lengths of the strings that produce these notes.*

Now, increasing the weight in the pail, the pitch rises, till, when the tension is four pounds, the pitch has risen an octave. Let the tension be the same; try another string, weighing, for the same length, four times as much; the pitch is an octave lower than that given by the lighter string. (These experiments will not give very accurate results.)

These conclusions may be summarized by saying: *The vibration-numbers of strings of the same material vary inversely as their lengths and square roots of their weights, and directly as the square roots of their tension.*

QUESTIONS AND PROBLEMS.

1. Why does a violinist finger the strings of the violin when playing?

2. Examine the strings of a piano, and ascertain the different methods by which a wide range of pitch is effected.

3. How does the length of the string that gives the note F compare with the length of the C-string below it, other things being equal?

XLV. OVERTONES AND HARMONICS.

§ 283. **Vibration in parts.** — Experiment 1. Hang up a rubber tube A C (Fig. 219), 3^m long, filled with sand, fastening both ends. Pluck it near the middle, and it will swing to and fro as a whole (2), at a rate dependent on its length, tension, etc. Hold it fast at B (3), and pluck it at a point half-way between A and B. Both halves are thrown into independent vibrations, and continue so to vibrate for a brief time after the hand is withdrawn from B. Again hold it fast at B, one-third its length above A (4), and pluck it half-way between A and B; the length BC instantly divides itself at B' into two equal parts, and on withdrawing the hand from B, the whole tube is seen to vibrate in three distinct and equal sections. In a similar manner it may be made to vibrate in four, five, etc., sections.

All of the above experiments may be repeated with the same results on the string of the sonometer. By placing paper riders[1] along the string, the ventral segments and the nodes can be easily discovered, as those placed near the center of the segments will be thrown off, while those at the nodes will remain comparatively at rest.

Fig. 219.

The sounds coming from a string or other body that vibrates in parts are called *overtones*. If, as is the case with a string or a column of air in an organ pipe (page 321), the vibration-number of the overtone is just two, three, four, etc., times that of the fundamental or lowest tone, the sound is called a *harmonic*. Many overtones can be produced from a steel bar or a metallic plate, but no harmonics. This distinction is of great importance, for, practically, no musical instruments are of much use unless their vibrating parts furnish harmonics.

§ **284. Complex vibrations.** — **Experiment 1.** Strike one of the lowest notes of a piano, hold the key down, and immediately apply the tip of the finger to some point of the wire struck, and notice any changes in tone that may occur after applying the finger. Repeat this at many points along the string. If, after touching the string, the fundamental tone coutinues, it shows that you have touched a node, and consequently have not stopped the vibrations by which this tone is produced; still you will notice that the sound, though not changed in pitch, is changed somewhat in quality (see page 309). If the funda-

[1] Made by folding narrow strips of paper in the middle, so that they may be hung on the string.

mental sound disappears, there will most probably be a sound of a higher pitch that will continue, showing that although you have stopped one set of vibrations, there were still other vibrations in the string of a higher vibration-period which you did not stop, and which now become audible since the louder fundamental is silenced.

Experiment 2. Press down the C'-key (middle C) gently, so that it will not sound; and while holding it down, strike the C-wire strongly. In a few seconds release the key, so that its damper will stop the vibrations of the string that was struck, and you will hear a sound which you will recognize by its pitch as coming from the C'-wire. Place your finger lightly on the C'-wire, and you will find that it is indeed vibrating. Press down the right pedal with the foot, so as to lift the dampers from all the wires, strike the C-key, and touch with the finger the C'-wire; it vibrates. Touch the keys next to C', viz., B and D'; they have only a slight forced vibration. Touch G'; it vibrates.

Now it is evident that the vibrations of the C' and G'-wires are sympathetic. But a C-wire vibrating as a whole cannot cause sympathetic vibrations in a C'-wire; but, if it vibrates in halves, it may. Hence, we conclude that when the C-wire was struck it vibrated, not only as a whole, giving a sound of its own pitch, but also in halves; and the result of this latter set of vibrations was, that an additional sound was produced by this wire, just an octave higher than the first-mentioned sound.

Again, the G'-wire makes 396 vibrations in a second, or three times as many (132) as are made by the C-wire; hence the latter wire, in addition to its vibrations as a whole and in halves, must have vibrated in thirds, inasmuch as it caused the G'-wire to vibrate. It thus appears that a string may vibrate at the same time as a whole, in halves, thirds, etc., and the result is that *a sensation is produced that is compounded of the sensations of several sounds of different pitch.*

Not only do stringed instruments produce compound tones, but no ordinary musical instrument is capable of producing a *simple tone*, i.e., a sound generated by vibrations of a single period. In other words, *when any note of any musical instrument is sounded, there is produced, in addition to the primary*

CAUSE OF HARMONY AND DISCORD. 307

tone, *a number of other tones in a progressive series, each tone of the series being usually of less intensity than the preceding.* The primary or lowest tone of a note is usually sufficiently intense to be the most prominent, and hence is called *the fundamental tone*.

§ 285. Cause of harmony and discord. — The harmonics in any note are produced successively by two, three, etc., times the number of vibrations made by its fundamental. Hence, if any two notes an octave apart, — for instance, C and C', — are sounded simultaneously, there will result for

$$\left. \begin{array}{l} C, \ 1, \ 2, \ 3, \ 4, \ 5, \ 6, \text{ etc.,} \\ C', \quad 2, \quad 4, \quad 6, \text{ etc.,} \end{array} \right\}$$ times the number of vibrations made

by the fundamental of C. So that the fundamental of C', and each of its overtones (with the exception of the highest, which are too feeble to affect the general result) coincides with one of the overtones of C. Not only is there perfect agreement among the overtones of two notes an octave apart when sounded together, as when male and female voices unite in singing the same part of a melody, but the richness and vivacity of the sound is much increased thereby. *That two notes sounded together may harmonize, it is essential not only that the pitch of their fundamental tones be so widely different that they cannot produce audible beats, but that no beats shall be formed by their overtones, or by an overtone and a fundamental.*

For example, the vibration-numbers of the fundamentals of C' and its octave C'' are respectively 264 and 528, and the number of beats that they give is 264 in a second. If, instead of C'', a note, the vibration-number of whose fundamental is 527, is sounded with C, the number of beats produced by their fundamentals would be 263, and no discord would result therefrom (why?); but there would be one beat per second between the first overtone of C' and the fundamental of C'', and this would introduce a discord.

Observe that the relation between the vibration-numbers of the fundamentals of C and C', C and G, C and F, and C, of any diatonic scale and any note in the same scale, can be expressed in terms of small numbers, *e.g.*, 1 : 2, 2 : 3, 3 : 4, etc. (see p. 300). Generally, *those notes and only those harmonize whose fundamental tones bear to one another ratios expressed by small numbers; and the smaller the numbers which express the ratios of the rates of vibration, the more perfect is the harmony of two sounds.*

It follows, from what has been said, that only a limited number of notes can be sounded with any given note assumed as a prime without generating discord. Hence, the musical scale is limited to certain determinate degrees, represented by the eight notes of the so-called *musical* or *diatonic scale*. This scale is not the result of any arbitrary or fanciful arrangement, but is determined by the possibility of its notes harmonizing with the prime of the scale, both as regards their fundamental tones and their overtones.

EXERCISES.

1. Prepare a table of the series of overtones of C and G respectively, as on page 307, and ascertain what overtones of the two series harmonize.

2. Arrange the notes of the diatonic scale in a single octave in the order of their rank with reference to their harmonizing with the prime of the scale, on the principle that "the smaller the numbers which express the ratio," etc.

3. Verify your conclusions as follows: Strike the C-key of a piano, together with each of the seven white keys above it, consecutively, and compare the results of the different pairs with reference to harmony.

ANALYSIS OF SOUNDS. 309

XLVI. QUALITY OF SOUND.

Let the same note be sounded with the same intensity, successively, on a variety of musical instruments, *e.g.*, a violin, cornet, clarinet, accordion, jews-harp, etc.; each instrument will send to your ear the same number of waves, and the waves from each will strike the ear with the same force, yet the ear is able to distinguish a decided difference between the sounds, — a difference that enables us instantly to identify the instruments from which they come. Sounds from instruments of the same kind, but by different makers, usually exhibit decided differences of character. For instance, of two pianos, the sound of one will be described as richer and fuller, or more ringing, or more "wiry," etc., than the other. No two human voices sound exactly alike. That difference in the character of sounds, not due to pitch or intensity, that enables us to distinguish one from another, is called *quality*. Two sounds may differ from one another *in loudness, pitch, or quality; they can differ in no other respect*.

Pitch depends on frequency of vibrations, loudness on their amplitude; *on what does quality depend?*

§ 286. Analysis of sounds. — The unaided ear is unable, except to a very limited extent, to distinguish the individual tones that compose a note. Helmholtz arranged a series of resonators consisting of hollow spheres of brass, each having two openings: one (A, Fig. 220) large, for the reception of the sound-waves, and the other (B) small and funnel-shaped, and adapted for insertion into the ear. Each resonator of the series was adapted by its size to resound powerfully to only a single tone of a definite pitch. When any musical sound is produced in front of these resonators, the ear, placed at the orifice of any one,

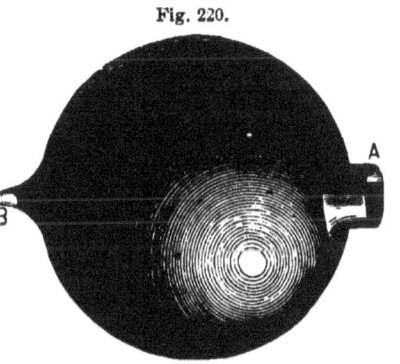

Fig. 220.

is able to single out from a collection that overtone, if present, to which alone this resonator is capable of responding. It is found that, when a note is produced on a given instrument, not only is there a great variety of intensity represented by the overtones, but all the possible overtones of the series are by no means present. Which are wanting depends very much, in stringed instruments, upon the point of the string struck. For example, if a string is struck in its middle, no node can be formed at that point; consequently, the two important overtones produced by 2 and 4 times the number of vibrations of the fundamental will be wanting. Strings of pianos, violins, etc., are generally struck near one of their ends, and thus they are deprived of only some of their higher and feebler overtones.

§ **287.** Synthesis of sounds. — The sound of a tuning-fork, when its fundamental is reënforced by a suitable resonance-cavity, is very nearly a simple tone. By sounding simultaneously several forks of different but appropriate pitch, and with the requisite relative intensities, Helmholtz succeeded in reproducing sounds peculiar to various musical instruments, and even in imitating most of the vowel sounds of the human voice.

Thus it appears that he has been able to determine, both analytically and synthetically, that *the quality of a given sound depends upon what overtones combine with its fundamental, and on their relative intensities;* or, we may say more briefly, *upon the form of vibration*, since the form must be determined by the character of its components.

XLVII. COMPOSITION OF SONOROUS VIBRATIONS, AND THE RESULTANT WAVE-FORMS.

§ 288. Method of representing sound vibrations graphically. — It is evident that there must be a particular aërial wave-form corresponding to each compound vibration, otherwise the ear would not be able to appreciate a difference in quality of sounds to which these combination-forms give rise. Every particle of air engaged in transmitting a compound sound is simultaneously acted upon by several sets of vibratory movements, and it remains to investigate what its motion will be under their joint influence.

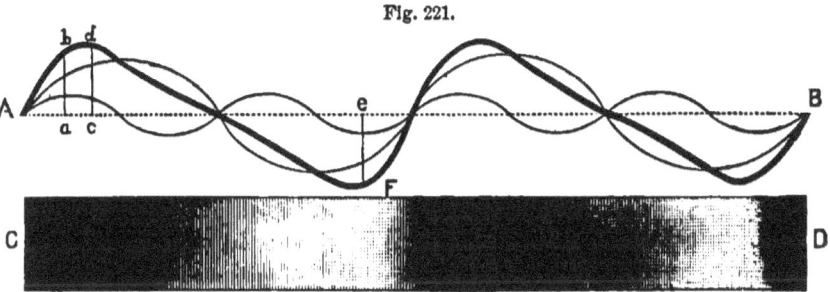

Fig. 221.

The light wave-lines A B (Fig. 221) represent typically two series of aerial sound-waves, corresponding respectively to a fundamental and its first overtone. The heavy line represents the form of the joint wave which results from the combination of the two constituents. If we suppose lines perpendicular to the axis, that is, to the dotted line, or line of repose, to be drawn to each point in this line, as ab, cd, eF, etc., they will represent by their varying lengths the displacement of any particle in a vibrating body, or any particle of air traversed by sound-waves, from its normal position.

The rectangular diagram C D is intended to represent a portion of a tranverse section of a body of air traversed by the joint wave represented by the heavy wave-line above. The

312 SOUND.

depth of shading in different parts indicates the degree of condensation at those parts.

Figure 222 represents wave-lines drawn by an instrument called a *vibrograph*. The second line represents a sound two octaves above that which the first line represents, and the third line shows the result of the combination of the two sets of vibrations.

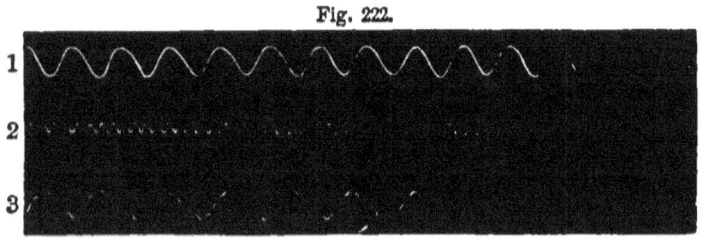
Fig. 222.

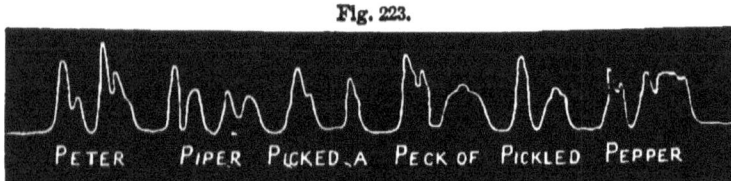
Fig. 223.

In an elaborate apparatus called the *logograph*, a thin membrane of gold-beater's skin carries a marker resembling the point of a stylographic pen. When a person sings or talks to this membrane, it traces upon paper a graphic representation of the varying air pressure. That is, all the changes in the density of the air, and all the movements of a given air-particle during the passage of the sound-waves, are faithfully depicted in a line traced by the marker on a passing paper; just as the heavy wave-line A B (Fig. 221) may be said to represent the condition of the air C D, or of the motion of any particle of it, supposing that a marker were attached to it and a paper drawn beneath it at right angles to the path of its motion. The diagram in Figure 223 shows the result produced by pronouncing the sentence there given at the rate of six syllables in a second.

§ **289.** **Manometric flames.** — Apparatus like that shown in Figure 224 may be very easily prepared, and will serve to illustrate in a pleasing manner many facts pertaining to sound. Procure a wooden pill-box or tooth-pick box A, having a capacity of 50 to 100^{ccm}. Across the top of the open box stretch tightly a circular piece of gold-beater's

MANOMETRIC FLAMES.

skin a, and glue it at its edges so that it may cover the box like the head of a drum. Crowd on the cover, and the box will have two compartments, b and c. Through the bottom of the box, and through the cover, pass glass tubes e and d, opening into the compartments. Also introduce another tube n through the side of the cover. Connect the last tube by means of a rubber tube with a gas burner. Attach a piece of large-sized rubber tube to the glass tube e, and into the other extremity of the rubber tube introduce the small end of a pasteboard cone B. The tube d should be drawn out so as to be able to give a small flame. Place two thin glass mirrors M, about 14^{cm} square, back to back, and secure them by light frames at the top and bottom, and in the center of each frame insert small rods C and D.

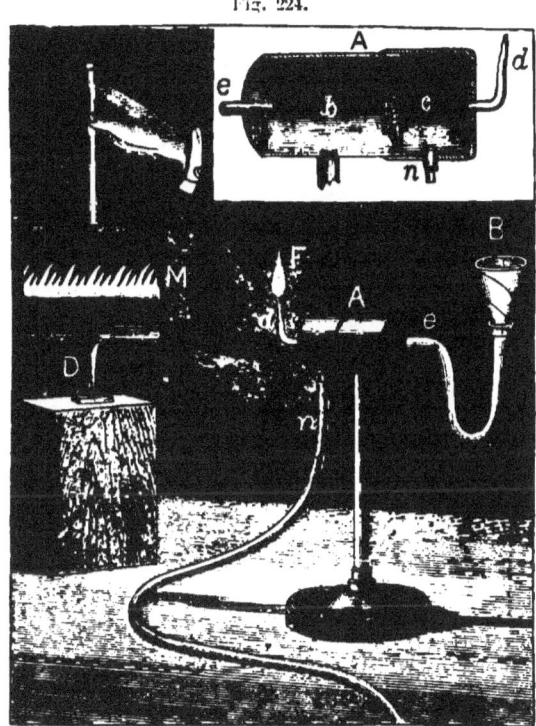

Fig. 224.

Light the gas[1] at the extremity of d, and hold the mirror vertically, and at a short distance from the flame F; an image of the flame will appear in the mirror, as represented by A (Fig. 225). Rotate the mirror, and the flame appears drawn out in a band of light, as shown in B of the same figure.

Now sing into the cone B (Fig. 224), the sound of oo in tool, and waves of air will run down the tube, beat against the membrane a, as against the drum-head of the ear (see § 290), causing it to vibrate, and the membrane in turn acts upon the gas in the compartment c, throwing it into vibration. The result is, that instead of a flame appearing in the rotating mirror as a continuous band of light, it is divided up into a

[1] If gas is not accessible, the end of the tube d may be inserted in a candle flame, and good results obtained.

314 SOUND.

series of tongues of light, as shown in C of Figure 225, each condensation being represented by a tongue, and each rarefaction by a dark interval between the tongues. If a note an octave higher than the last is sung, we obtain, as we should expect, twice as many tongues in the

Fig. 225.

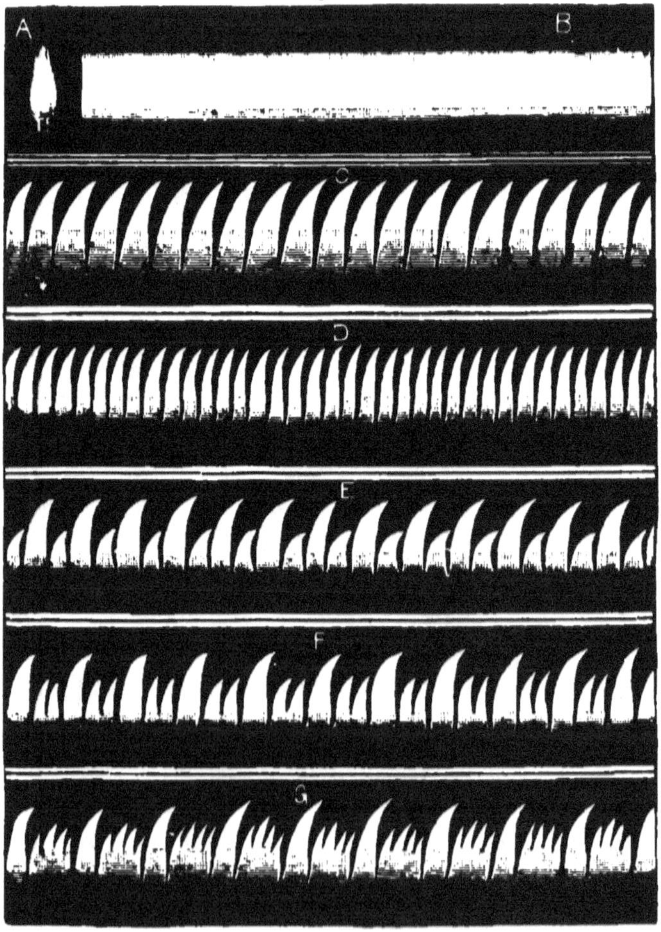

same space, as shown in D. E represents the result when the two tones are produced simultaneously, and illustrates in a striking manner the effect of interference. (Explain.) F represents the result when the vowel *e* is sung on the key of C'; and G, when the vowel *o* is sung on the same key. These are called *manometric flames*.

THE EAR. 315

XLVIII. SOME SOUND-RECEIVING INSTRUMENTS.

§ **290. The ear.** — In Figure 226, A represents the external ear-passage; *a* is a membrane, called the *tympanum*, a little thicker than gold-beater's skin, stretched across the bottom of the passage, and thus closing the orifice of a cavity *b* in the bones of the skull called the *drum;* *c* is a chain of small bones stretching across the drum, and connecting the tympanum with the thin membranous wall of the *vestibule* *e;* *ff* are a series of semicircular canals opening into the vestibule;

Fig. 226.

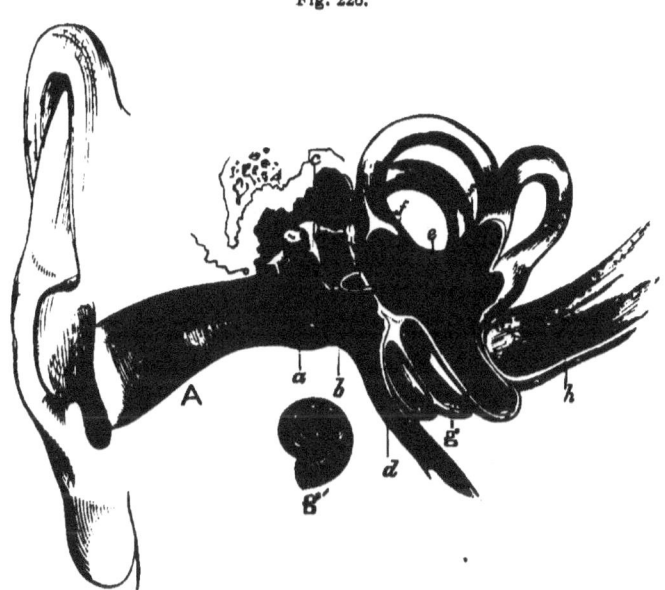

g is the opening into another canal in the form of a snail-shell *g'*, hence called the *cochlea* (this is drawn on a reduced scale); *d* is a tube (the *Eustachian tube*) connecting the drum with the throat; and *h* is the auditory nerve. The vestibule and all the canals opening into it are filled with a transparent liquid which is mainly water. The drum of the ear contains air, and the Eustachian tube forms a means of ingress and egress of air through the throat.

Now how does the ear hear? and how is it able to distinguish between the infinite variety of form, rapidity, and intensity of aerial sound-waves, so as to interpret correctly the corresponding quality, pitch, and loudness of sound? Sound-waves enter the external ear-passage A as ocean-waves enter the bays of the sea-coast, are reflected inward, and strike the tympanum. The air-particles, beating against this drum-head, impress upon it the precise wave-form that is transmitted to it through the air from the sounding body. The motion received by the drum-head is transmitted by the chain of bones to the membranous wall of the vestibule. From the walls of this cavity project into its liquid contents thousands of fine elastic threads or fibres, which we may, for convenience, call *bristles*. Especially in the spiral passage of the cochlea, as it becomes smaller and smaller, these vibratile bristles become of gradually diminishing length and size (such as the wires of a piano may roughly represent), and are therefore suited to respond sympathetically to a great variety of vibration-periods. This arrangement is sometimes likened to a harp of three thousand (this being about the number of bristles) strings. The auditory nerve at its extremity is divided into a large number of filaments, like a cord unravelled at its end, and one of these filaments is attached to each bristle. Now, as the sound-waves reach the membranous wall of the vestibule, they set it, and by means of it the liquid contents, into forced vibration, and so through the liquid all the fibres receive an impulse. Those bristles whose vibration-periods correspond with the periods of the constituents forming the compound wave are thrown into sympathetic vibration. The bristles stir the nerve filaments, and the nerve transmits to the brain the impressions received. Just as a piano, when its dampers are raised and a person sings into it, may be said to analyze each sound, and show by the vibrating strings of how many tones it is composed, as well as their respective pitch, and by the amplitude of their vibrations their respective intensities; so it is thought this wonderful harp of the ear analyzes every complex sound-wave into a series of simple vibrations. Tidings of the disturbances are communicated to the brain, and there, in some mysterious manner, these disturbances are interpreted as sound of *definite quality, pitch, and intensity.*

§ 291. Phonograph.—Figure 227 represents a vertical section of the Edison phonograph. A metallic cylinder A is rotated by means of a crank B in the direction indicated by the arrow. On the surface of the cylinder is cut a shallow spiral groove running around the cylinder from end to end, like the thread of a screw. A small metallic point, or style, projecting from the under side of a thin metallic disk *o*, which closes one orifice of the mouthpiece C, stands directly over the thread. By a simple device the cylinder, when the crank is turned, is made to advance just rapidly enough to allow the groove to keep constantly under the style. The cylinder is covered with tinfoil. The space E represents the space (greatly exaggerated) between the tinfoil and the bottom of the groove.

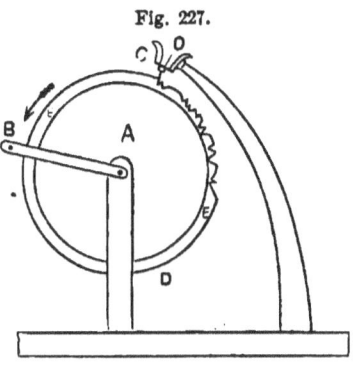

Fig. 227.

Now, when a person directs his voice toward the mouth-piece, the aerial waves cause the disk *o* to participate in every motion made by the particles of air as they beat against it, and the motion of the disk is communicated by the style to the tinfoil, producing thereon impressions or indentations as it passes on the rotating cylinder. The result is that there is left upon the foil an exact representation in relief of every movement made by the style. Some of the indentations are quite perceptible to the naked eye, while others are visible only with the aid of a microscope of high power. Figure 228 represents a piece of the foil as it would appear inverted after the indentations (here greatly exaggerated) have been imprinted upon it.

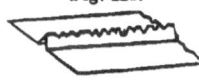

Fig. 228.

The words addressed to the phonograph having been thus impressed upon the foil, the mouth-piece and style are temporarily removed, while the cylinder is brought back to the position it had when the talking began, and then the mouth-piece is replaced. Now, evidently, if the crank is turned in the same direction as before, the style, resting upon the foil beneath, will be made to play up and down as it passes over ridges and sinks into depressions; this will cause the disk *o* to reproduce the same vibratory movements that caused the ridges and depressions in the foil. The vibrations of the disk are communicated to the air, and through the air to the ear; and thus the words spoken to the apparatus may be, as it were, shaken out into the air again at any subsequent time, even centuries after, accompanied by the exact accents, intonations, and quality of sound of the original.

§ 292. String telephone. — In the phonograph, the metallic disk serves, as it were, alternately, as an ear and a tongue. If, instead of the same disk being made to do double duty, two disks (or, better, two membranes of gold-beater's skin or bladder) connected by a thread are used, either one of which may serve as a tongue and the other simultaneously as an ear, conversation may be carried on by means of them through considerable distances. Figure 229 represents such an arrangement, which constitutes the well-known, instructive toy, called the *lover's telegraph*, though it is more properly a *telephone*.

Fig. 229.

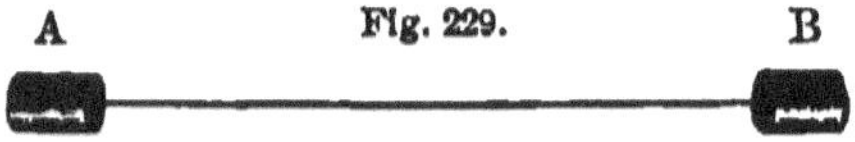

The thread is attached at each extremity to the centers of the membranes which cover one orifice of each of the tin speaking-tubes A and B, by passing the thread through the membranes, and tying the knots at the ends. A person speaking into one of the tubes throws its membrane into vibration; these impulses are communicated through the string to the other membrane, which is thus caused to vibrate in unison with the first. If now another person place his ear near the latter membrane, he can hear distinctly the words spoken by the first person, though a quarter of a mile distant, while other persons stationed midway between these two hear nothing.

It seems fair to presume, that if the movements of the hand or of machinery could be rendered sufficiently delicate to imitate these minute movements of the membrane, talking might be accomplished with the hand or machinery; for *talking*, after all, *is only mechanical motion*.

§ 293. Electric telephone. — In this telephone the vibrations of one disk are reproduced in another through the agency of electricity, as explained on page 270.

XLIX. MUSICAL INSTRUMENTS.

§ 294. Musical instruments may be grouped in three classes: (1) Stringed instruments; (2) wind instruments, in which the sound is due to the vibration of columns of air confined in tubes; (3) instruments in which the vibrator is a membrane or plate. The first class has received its share of attention; the other two merit a little further consideration.

§ 295. **Sounding air-columns.**— Experiment 1. Take four glass tubes, A, B, C, and D, respectively 48, 48, 24, and 12^{cm} long, and about 2.5^{cm} diameter. Blow gently across one of the ends of each; C gives a sound an octave higher than A or B, and D an octave higher than C. Close one of the ends of B, C, and D, and repeat the experiment, and you will find that the notes obtained from these three have still the same relation to one another. Blow across one end of A, which is open at both ends, and across the open end of B; A gives a note about an octave higher than B.

These experiments show (1) that the pitch of vibrating air-columns, as well as of strings, varies with the length, and *in both stopped*[1] *and open pipes the number of vibrations is inversely proportional to the length of the pipe*[2]; (2) that *an open pipe gives a note an octave higher than a closed pipe of the same length.*

Experiment 2.— Blow across the orifice of B as before, gradually increasing the force of the current. It will be found that only the gentle current will give the full musical fundamental tone of the tube, — a little stronger current produces a mere rustling sound; but when the force of the current reaches a certain limit, an overtone will break forth; and, on increasing still further the power of the current, a still higher overtone may be reached.

Figure 230 represents an open organ-pipe provided with a glass window A in one of its sides. A wire hoop B has, stretched over it, a membrane, and the whole is suspended by a thread within the pipe. If the membrane is placed near the upper end, a buzzing sound proceeds

---

[1] A *stopped* pipe is one which is closed at one end.
[2] The diameter has the same influence here as in the resonance-jar (p. 293), but we shall neglect it.

from the membrane when the fundamental of the pipe is sounded; and sand placed on the membrane will dance up and down in a lively manner. On lowering the membrane, the buzzing sound becomes fainter, till, at the middle of the tube, it ceases entirely, and the sand becomes quiet. Lowering the membrane still further, the sound and dancing recommence, and increase as the lower end is approached.

Fig. 230.

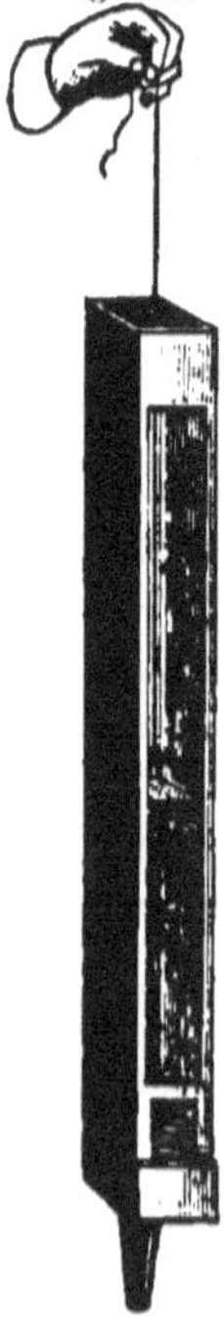

It is thus found, that (3) *when the fundamental of an open pipe is sounded, its air-column divides itself into two equal vibrating sections, with the antinodes at the extremities of the tube, and a node in the center.*

If the pipe is stopped, there is a node at the stopped end; if it is open, there is an antinode at the open end; and in both cases there is an antinode at the end where the wind enters, which is always to a certain extent open.

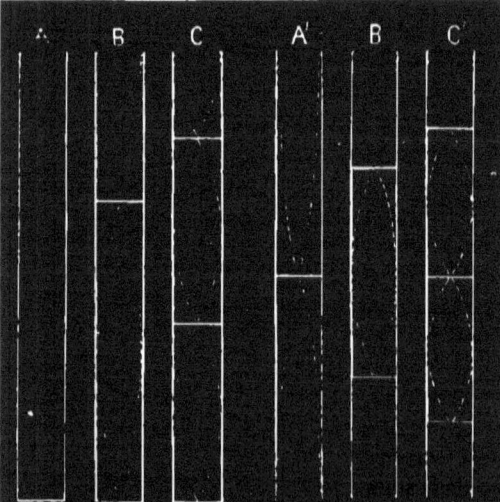

Fig. 231.

A, B, and C of Figure 231 show respectively the positions of the nodes and antinodes for the fundamental and first and second overtones of a closed pipe; and A', B', and C' show the positions of the same in an open pipe of the same length. The distance between the dotted lines shows the relative amplitudes of the vibrations of the air-particles at various points along the tube.

Now the distance between a node and its nearest antinode is a quarter of a wave-length. Comparing then A and A', it will be seen that the

SOUNDING PLATES. 321

wave-length of the fundamental of the closed pipe must be twice the wave-length of the fundamental of the open pipe; hence the vibration-period of the latter is half that of the former; consequently the fundamental of the open pipe must be an octave higher than that of the closed pipe.

The number of segments into which the length of the air-column is divided, in the three cases of the closed tube, are respectively $\frac{1}{2}$, $\frac{3}{2}$, and $\frac{5}{2}$; hence the corresponding vibration-numbers are as $1:3:5$, etc. Hence, (4) *in closed tubes, only those overtones whose vibration-numbers correspond to the odd multiples of the fundamental are present.*

The number of segments into which the length of the air-column is divided, in the three cases of the open tube, are respectively $\frac{2}{2}$, $\frac{4}{2}$, and $\frac{6}{2}$; their vibration-numbers are therefore as $1:2:3$, etc. Hence, (5) *in open tubes, the complete series of overtones corresponding to its fundamental are present.*

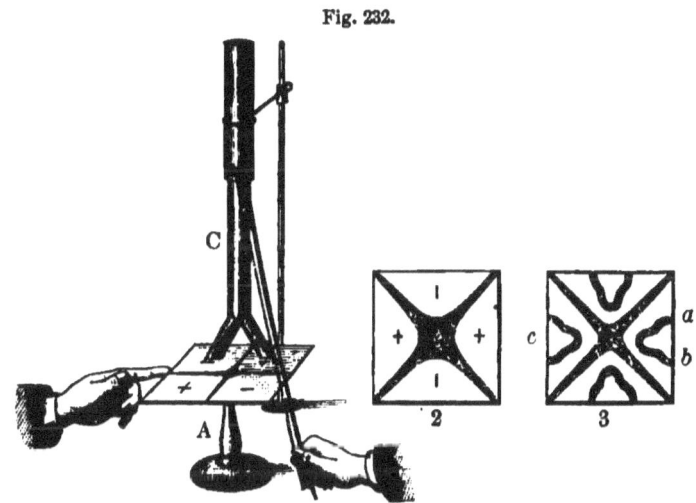

Fig. 232.

§ 296. Sounding plates. — Experiment. Procure at a hardware store a perfectly flat piece of sheet brass 2^{mm} thick and 20^{cm} square. Fasten it at its center to a supporting rod A, Figure 232. Scatter on the plate some fine sand, and draw a resined bow steadily and firmly over one of its edges near a corner; and at the same time touch the middle of one of its edges with the tip of the finger; a musical sound will be produced, and the sand will dance up and down, and quickly collect in two rows, extending across the plate at right angles to one another. Draw the bow across the middle of an edge, and touch with a finger one of its corners, and the sand will arrange itself in two diagonal rows (2) across the plate, and the pitch of the note will be a fifth higher. Touch, with the nails of the thumb and forefinger, two points a and b (3) on one edge, and draw the bow across the middle c of the opposite edge, and you will obtain additional rows and a shriller note.

By varying the position of the points touched and bowed, a great variety of patterns can be obtained, some of them exceedingly complicated and beautiful. It will be seen that the effect of touching the plate with a finger is to prevent vibration at that point, and consequently a node is there produced. The whole plate then divides itself up into segments with nodal division lines in conformity with the node just formed. The sand rolls away from those parts which are alternately thrown into crests and troughs, to the parts that are at rest.

§ 297. Interference. — Experiment. Provide a tin tube C, Figure 232, 1^m long and 5^{cm} in diameter, made in two parts so as to telescope one within the other. The extremity of one of the parts terminates in two slightly smaller branches. Bow the plate, as in the first experiment (1), place the two orifices of the branches over the segments marked with the + signs, and regulate the length of the tube so as to reënforce the note given by the plate, and set the plate in vibration. Now turn the tube around, so that one orifice may be over a +segment, and the other over a −segment; the sound due to resonance entirely ceases. It thus appears that the two segments marked + pass through the same phases together; likewise the phases of −segments correspond with one another; i.e., when one +segment is bent upward, the other is bent upward, and at the same time the two −segments are bent downward; for, when the two orifices of the tube are placed over two +segments or two −segments, two condensations followed by two rarefactions pass up these branches and unite at their junction to produce a loud sound; but when one of the orifices is over a +segment and the other over a −segment, a condensation passes up one branch at the same time that a rarefaction passes up the other, and the two destroy one another when they come together; i.e., the two sound-waves combine to produce silence.

§ 298. Bells. — A bell or goblet is subject to the same laws of vibration as a plate.

Experiment. — Nearly fill a goblet with water, strew upon the surface lycopodium powder, and draw a resined bow gently across the edge of the glass. The surface of the water will become rippled with wavelets radiating from four points 90° apart, corresponding to the centers of four ventral segments into which the bell is divided, and the powder will collect in lines proceeding from the nodal points of the bell. By touching the proper points of a bell or glass with a fingernail, it may be made to divide itself, like a plate, into 6, 8, 10, etc., (always an even number) vibrating parts.

§ 299. Vocal organs. — It is difficult to say which is more to be admired, — the wonderful capabilities of the human voice or the extreme simplicity of the means by which it is produced. The organ of the voice is a reed instrument situated at the top of the windpipe or trachea. A pair of elastic bands aa, Figure 233, called the *vocal chords*, is stretched across the top of the windpipe. The air-passage b, between these chords, is open while a person is breathing; but when he speaks or sings they are brought together so as to form a narrow, slit-like opening, thus forming a sort of double reed, which is made to vibrate, when air is forced from the lungs through the narrow passage, somewhat like the little

Fig. 233.

tongue of a toy trumpet. The sounds are grave or high according to the tension of the chords, which is regulated by muscular action. The cavities of the mouth and the nasal passages form a compound resonance-tube. This tube adapts itself, by its varying width and length, to the pitch of the note produced by the vocal chords. Place a finger on the protuberance of the throat called the Adam's apple, and sing a low note; then sing a high note, and you will observe that the protuberance rises in the latter case, thus shortening the distance between the vocal chords and the lips. Set a tuning-fork in vibration, open the mouth as if about to sing the corresponding note, place the fork in front of it, and the cavity of the mouth will resound to the note of the fork, but will cease to do so when the mouth adapts itself to the production of some other note. The different qualities of the different vowel sounds are produced by the varying form of the resonating mouth-cavity, the pitch of the fundamentals given by the vocal chords remaining the same. This constitutes *articulation*.

CHAPTER VI.

RADIANT ENERGY.—LIGHT.

L. INTRODUCTION.

§ 300. Light a form of energy. — Exposed to the sun, the skin is warmed, and thus the sense of touch is affected; it is illuminated, and thereby the sense of sight is affected; it is tanned, and thereby its chemical condition is changed. It is evident that we receive something which must come to us from the sun. To the sense of touch it appears to be heat; to the eye it is light; to certain substances it is a power to produce chemical changes. But *what is it that we receive from the sun?*

Fig. 234.

Experiment. — Blacken one-half of one side of a slip of glass with candle-smoke. With a convex lens, sometimes called a "burning-glass," converge the sun's light upon the blackened portion so as to produce a small luminous spot on the black surface. This spot quickly becomes very hot, but the lens meantime remains comparatively cold. Move the luminous spot to the unblackened portion of the glass. The spot becomes only slightly heated. Place a piece of paper behind and in contact with the glass, and it quickly burns.

Whether we receive heat from the sun or not, it is evident that we receive something that can be converted into heat.

Figure 234 represents an instrument called a *radiometer*. The moving part is a small vane resting on the point of a needle. It is so nicely poised on this pivot that it rotates with the greatest freedom. To the extremities of each of the four arms of the vane are attached disks of aluminum which are white on one side and black on the other. The whole is enclosed in a glass bulb from which the air is exhausted till less than $\frac{1}{1000}$ of the original quantity is left. If the instrument is exposed to the sun's light, or even to the light of a candle, the wheel will rotate with the unblackened faces in advance.

In just what manner it is caused to rotate does not concern us; but the fact that it does rotate, and that it is caused to rotate directly or indirectly by something that comes from the sun or the candle, is pertinent to the question before us. Whenever a body is caused to move or increase its rate of motion, energy must be imparted to it; hence energy must be imparted to the radiometer-vane by the sun or candle.

Bell, the inventor of the telephone, has succeeded in producing musical sounds by the action of sun-light and other intense lights. But sound always originates in motion, and motion springs only from some form of energy. *So, then, that which we receive from the sun, whether it affects the sense of touch and is called heat, or the eye and is called light, or produces chemical changes and is called chemism, is in reality some form of energy.*

§ 301. **Ether the medium of motion.** — If light is motion, what moves? Our atmosphere is but a thin investment of the earth, while the great space that separates us from the sun contains no air or other known substance. But *empty space can neither receive nor communicate motion.* It is assumed — it is necessary to assume — that there is some medium filling the interplanetary space, in fact, filling all otherwise unoccupied space (*i.e.*, where matter is not, ether is), by which motion can be communicated from one point in the otherwise empty space to another. This medium has received the name of *ether.* Ether is supposed to penetrate even among the molecules of liquid and solid matter, and thus surrounds every molecule of matter in the universe, as the atmosphere surrounds the earth.

No vacuum of this medium can be obtained; an attempt to pump it out of a space would be like trying to pump water with a sieve for a piston. We cannot see, hear, feel, taste, smell, weigh, nor measure it. What evidence, then, have we that it exists? You believe that a horse can see; you have no absolute knowledge of the fact. But you reason thus: he behaves *as if* he could see; in other words, you are able to account for his actions on the hypothesis that he can see, and on no other. Phenomena occur just *as* they would occur *if* all space were filled with an ethereal medium capable of transmitting motion, and we can account for these phenomena on no other hypothesis; hence our belief in the existence of the medium.

The transmission of energy through the medium of ether is called *radiation;* energy so transmitted is called *radiant energy*, and the body emitting energy in this manner is called a *radiator*. Sound is another form of radiant energy transmitted through solid, liquid, or gaseous media.

§ 302. **Undulatory theory of light.** — Is motion communicated by a transfer of a medium or by a transfer of vibrations, *i.e.*, by undulations? All evidence points to one conclusion: that we receive energy from the sun in the form of vibrations or wave-action; that these vibrations, inaudible to our ears, cause through the eye the sensation of sight, and through the hand the sensation of warmth. This is known as the *undulatory theory of light*. To learn what the special evidences of the correctness of this theory are, the pupil must wait for further development of our subject; but it should be borne in mind that *the strongest proof of the correctness of any theory is its exclusive competence to explain phenomena. Light is vibration that may be appreciated by the organ of sight.*

§ 303. **Light itself invisible.** — Darken a room, and admit a sunbeam through a small nail- or key-hole. You can trace its path through the room only by particles of dust floating in the room. But if the air in a certain space is cleansed of dust, the path of a sunbeam through this space will be totally dark. If the eye is placed in its path, or any object upon which it may strike, you become aware of its presence, not by seeing the light, but by seeing the object which sends you the light.

§ 304. **Light travels in straight lines.** — The path of the light admitted into a darkened room through a small aperture, as indicated by the illuminated dust, is perfectly straight. *An object is seen by means of light it sends to the eye.* A small object placed in a straight line between the eye and a luminous point may intercept the light in that path, and the point become invisible. Hence, we cannot see around a corner, or through a tube bent so that a straight string cannot be drawn through its bore.

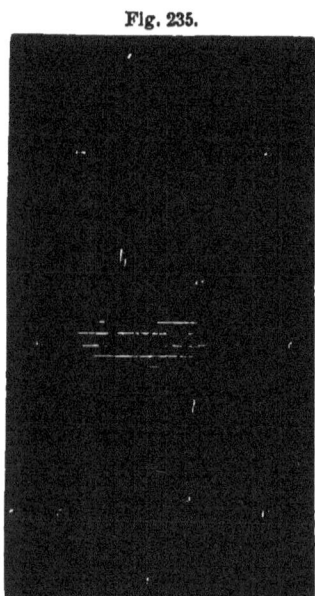

Fig. 235.

§ 305. **Ray, beam, pencil.** — Any line RR, Figure 235, which pierces the surface of a wave of light ab perpendicularly is called a *ray* of light. It is an expression for the direction in which motion is propagated, and along which the successive effects of light occur. If the wave-surface $a'b'$ is a plane, the rays $R'R'$ are parallel, and a collection of such rays is called a *beam* of light. If the wave-surface $a''b''$ is spherical or concave, the rays $R''R''$ have a common point at the center of curvature, and a collection of such rays is called a *pencil* of light.

§ 306. **Transparent, translucent, and opaque bodies.** — Bodies are *transparent*, *translucent*, or *opaque*, according to the manner in which they act upon the luminiferous waves which pass through them. Generally speaking, those objects are *transparent* that allow other objects to be seen through them distinctly; *e.g.*, air, glass, and water. Those objects are *translucent* that allow light to pass, but in such a scattered condition that objects are not seen distinctly through them; *e.g.*, fog, ground glass, and oiled paper. Those objects are *opaque* that apparently cut off all the light and prevent objects from being seen through them.

§ 307. **Luminous and illuminated objects.** — Some bodies are seen by means of light, which they generate and emit; *e.g.*, the sun, a candle flame, and a "live coal"; they are called *luminous bodies*. Other bodies are seen only by means of light which they receive from luminous ones, and when thus rendered visible, are said to be *illuminated; e.g.*, the moon, a man, a cloud, and a "dead" coal.

§ 308. **Every point of a luminous body an independent source of light.** — Place a candle flame in the center of a darkened room; every wall and every point of each wall becomes illuminated. Place your eye in any part of the room, *i.e.*, in any direction from the flame; it is able to see not only the flame, but every point of the flame; hence every point of the flame must emit light in every direction. *Every point of a luminous body is an independent source of light and emits light in every direction.* Such a point is called a *luminous point*. In Figure 236 there are represented a few of the infinite number of pencils of light emitted by three luminous points of a candle flame. Every point of an illuminated object *ab* receives light from every luminous point.

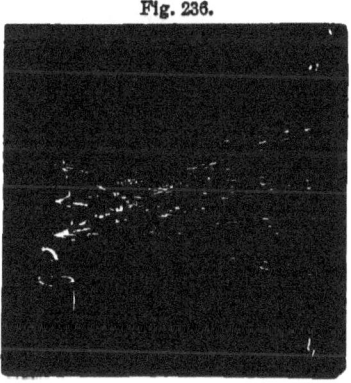

Fig. 236.

§ **309. Images formed through small apertures.**—**Experiment.**—Cut a hole about 8^{cm} square in one side of a box; cover the hole with tin-foil, and prick a hole in the foil with a pin. Place the box in a darkened room, and a candle flame in the box near to the pinhole. Hold an oiled-paper screen before the hole in the foil; an inverted image of the candle flame will appear upon the translucent paper. An *image* is a kind of picture of an object.

If light from objects illuminated by the sun — *e.g.*, trees, houses, clouds, or even an entire landscape — is allowed to pass through a small aperture in a window shutter and strike a white screen, or a white wall in a dark room, rays carrying with them the color of the points from which they issue will imprint their own color on the screen, and inverted images of the objects in their true colors will appear upon it. The cause of these phenomena is easily understood. When no screen intervenes between the candle and the screen A, Figure 237, every point of the screen receives light from every point of the candle; consequently, on every point on A, images of the infinite number of points of the candle are formed. The result of the confusion of images is equivalent to no image. But let the screen B, containing a small hole, be interposed; then, since light travels only in straight lines, the point Y′ can only receive an image of the point Y, the point Z′ only of the point Z, and so for intermediate points; hence a distinct image of the object must be formed on the screen A. *That an image may be distinct, the rays from different points of the object must not mix on the image, but all rays from each point on the object must be carried to its own point on the image.*

Fig. 237.

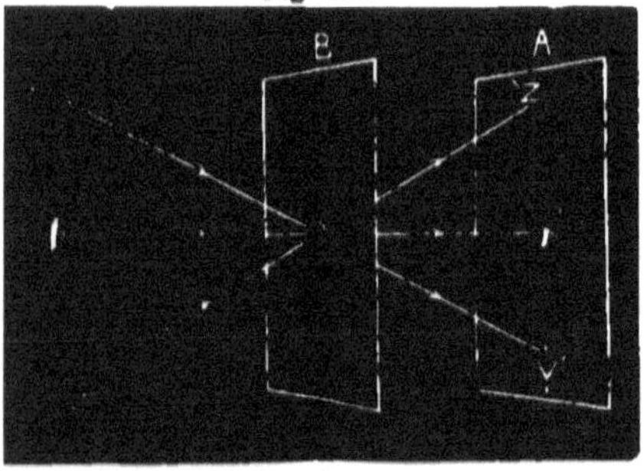

QUESTIONS.

1. Why are images, formed through apertures, inverted?
2. Why is the size of the image dependent on the distance of the screen from the aperture?
3. Obtain the dimensions, respectively, of an object and its image, and their respective distances from the intervening screen, and ascertain the law that determines in all cases the size of an image.
4. Why does an image become dimmer as it becomes larger?
5. Why do we not imprint an image of our person on every object in front of which we stand?
6. Can rays of light cross one another without interfering?
7. What fact does a gunner recognize in taking sight?

§ 310. **Shadows.** — **Experiment 1.** Procure two pieces of tin or card-board, one 18^{cm} square, the other 3^{cm} square. Place the first between a white wall and a candle flame in a darkened room. The opaque tin intercepts the light that strikes it, and thereby excludes light from a space behind it.

This space is called a *shadow*. That portion of the surface of the wall that is darkened is a *section of the shadow*, and represents the form of a section of the body that intercepts the light. A section of a shadow is frequently for convenience called a shadow. Notice that the shadow is made up of two distinct parts, — a dark center bordered on all sides by a much lighter fringe. The dark center is called the *umbra*, and the lighter envelope is called the *penumbra*. .

Experiment 2. Carry the tin nearer the wall, and notice that the penumbra gradually disappears and the outline of the umbra becomes more distinct. Employ two candle flames, a little distance apart, and notice that two shadows are produced. Move the tin toward the wall, and the two shadows approach one another, then touch, and finally overlap. Notice that where they overlap the shadow is deepest. This part gets no light from either flame and is the umbra; while the remaining portion gets light from one or the other and is the penumbra.

Just so *the umbra of every shadow is the part that gets no light from a luminous body, while the penumbra is the part that gets light from some portion of the body, but not from the whole.*

Experiment 3. Repeat the above experiments, employing the smaller piece of tin, and note all differences in phenomena that occur.

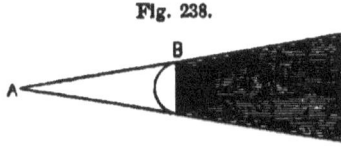

Fig. 238.

Hold a hair in the sunlight, about a centimeter in front of a fly-leaf of this book, and observe the shadow cast by the hair. Then gradually increase the distance between the hair and the leaf, and note the change of phenomena. If the source of light were a single luminous point, as A, Figure 238, the shadow of an opaque body B would be of infinite length, and would consist only of an umbra. But, if the source of light has a sensible size, the opaque body will intercept just as many separate pencils of light as there are luminous points, and consequently will cast an equal number of independent shadows.

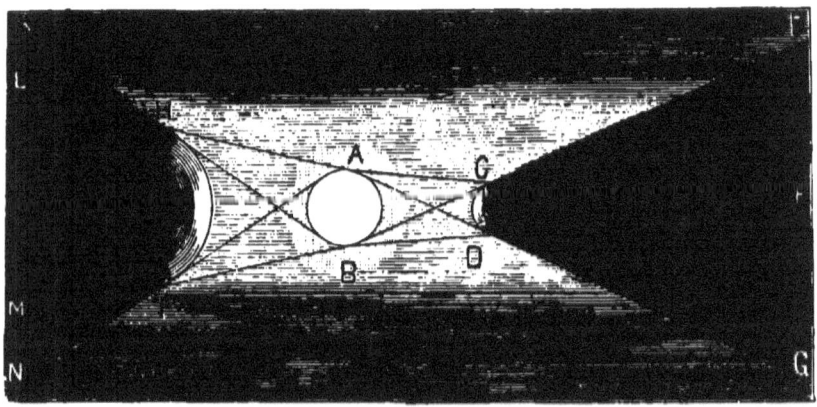

Fig. 239.

Let A B, Figure 239, represent a luminous body, and C D an opaque body. The pencil from the luminous point A will be intercepted between the lines C F and D G, and the pencil from B will be intercepted between the lines C E and D F. Hence, the light will be wholly excluded only from the space between the lines C F and D F, which enclose the umbra. The enveloping penumbra, a section of which is included between the lines C E and C F, and between D F and D G, receives light from certain points of the luminous body, but not from all.

LAW OF INVERSE SQUARES.

QUESTIONS.

1. Explain the umbra and penumbra cast by the opaque body H I, Figure 239.
2. When will a transverse section of an umbra of an opaque body be larger than the object itself?
3. When has an umbra a limited length?
4. What is the shape of the umbra cast by the sphere C D, Figure 239?
5. If C D should become the luminous body, and A B a non-luminous opaque body, what changes would occur in the umbra and the shadow cast?
6. Why is it difficult to determine the exact point where the umbra of a church-steeple terminates on the ground?
7. What is the shape of a section of a shadow cast by a circular disk placed obliquely between a luminous body and a screen? What is its shape when the disk is placed edgewise?
8. The section of the earth's umbra on the moon in an eclipse always has a circular outline. What does this show respecting the shape of the earth?

LI. PHOTOMETRY.

§ 311. **Law of inverse squares.** — *Experiment 1.* Arrange apparatus as follows: Lay a silver half-dollar on the center of a circular piece of stiff, white, unglazed paper of 15^{cm} diameter, and rub the entire surface, except the portion covered by the coin, with a sperm or a tallow candle. Hold the paper in a warm oven for a minute. When the paper is placed between two lights in a darkened room, the ungreased spot will appear light on a dark background on the side which receives the more light, and dark on a light background on the side which receives less light; but the spot becomes nearly invisible when both sides are equally illuminated. Draw a straight chalk line across a table, and place at right angles to this line a row of four lighted candles, and on the same line, at a distance, a single lighted candle. Half-way between this candle and the row of candles place the prepared paper, as in Figure 240. It is evident that one side of the paper receives four times the

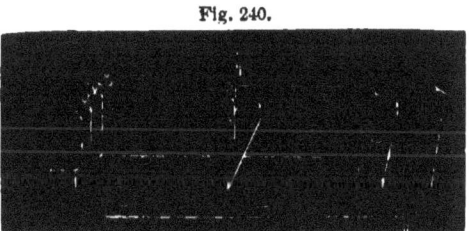

Fig. 240.

light that the other does. Move the row of lights slowly away from the paper, or move the single light toward the paper, and a point will be found in either case where the spot will nearly disappear. When this occurs it will be found that the row of lights is twice as far from the paper as the single light. The paper now receives the same amount of light from the single light as from the four lights.

Thus, by doubling the distance, the intensity of illumination is diminished four-fold. In a similar manner it may be shown that at three times the distance it takes nine lights to be equivalent to one light. Hence, *the intensity of light diminishes as the square of the distance increases.* This is called the *law of inverse squares.*

Experiment 2. Introduce the paper disk, as above, between a candle light and a kerosene light or a gas flame, and so regulate the distance that the central spot will disappear, and calculate the relative intensities of the two lights in accordance with the law of inverse squares.

Apparatus arranged for this purpose is called a *photometer.* "The *candle power*, which is the unit of light generally employed in photometry, is the amount of light given by a sperm candle weighing one-sixth of a pound, and burning one hundred and twenty grains an hour." The relative brightness of the common sources of light are approximately as follows [1] : —

Sunlight at the sun's surface.........	190,000 candle power.
Most powerful electric arc..........	55,900 " "
Most powerful calcium light.........	1,300 " "
Light of ordinary gas-burner.........	12 to 16 " "
Standard candle....................	1 " "

"The total quantity of light emitted by the sun is equivalent to the light of 6,300,000,000,000,000,000,000,000,000 (six thousand three hundred billions of billions) candles." Of this enormous quantity of light the earth intercepts an extremely small fraction.

[1] C. A. Young.

VISUAL ANGLE. 335

QUESTIONS.

1. Suppose that a lighted candle is placed in the center of each of three cubical rooms respectively 10, 20, and 30 feet on a side; would a single wall of the first room receive more or less light than a single wall of either of the other rooms?

2. Would one square foot of a wall of the third room receive as much light as would be received by one square foot of a wall of the first room? If not, what difference would there be, and why the difference?

3. If a board 10^{cm} square is placed 25^{cm} from a candle flame, the area of the shadow of the board cast on a screen 75^{cm} distant from the candle will be how many times the area of the board? Then the light intercepted by the board will illuminate how much of the surface of the screen if the board is withdrawn?

4. Give a reason for the law of Inverse Squares.

5. To what besides light has this law been found applicable?

6. The two sides of a paper disk are illuminated equally by a candle flame 50^{cm} distant on one side and a gas flame 200^{cm} distant on the other side; compare the intensities of the two lights at equal distances from their sources.

Fig. 241.

LII. VISUAL ANGLE, ETC.

§ 312. Visual angle. — **Experiment.** Prick a pin-hole in a card, place an eye near the hole, and look at a pin about 20^{cm} distant. Then bring the pin slowly toward the eye, and the dimensions of the pin will appear to increase as the distance diminishes.

Why is this? We see an object by means of its image formed on the retina of the eye, and its apparent magnitude is determined by the extent of the retina covered by its image. Rays proceeding from opposite extremities of an object, as AB, Figure 241, meet and cross one another in the window of the eye, usually called the *pupil*. Now, as the distance between the points of the blades of a pair of scissors depends upon the angle that the handles form with one another, so the size of the image formed on the retina depends upon the size of the angle, called the *visual angle*, formed by these rays as they enter the eye. But the size of the visual angle diminishes as the distance of the object from the eye increases, as shown in the diagram; *e.g.*, at twice the distance the angle is one-half as great, at three times the distance the angle is one-third as great, and so on. Hence, *the apparent size of an object diminishes as its distance from the eye increases.*

QUESTIONS.

1. Why do the rails of a railroad track appear to converge as their distance from the observer increases?
2. Why, in looking through a long hall or tunnel, do the floor and the ceiling appear to approach one another?
3. Why do parallel lines, retreating from the eye, appear to converge?
4. Why can a book, held in front of the face, entirely conceal from view a house?

§ 313. **Methods of estimating size.** — Let a man stand beside a boy of half his hight, and to an observer, twenty feet distant, the former will subtend a visual angle twice as great as the latter, and will appear twice as tall. Then, let the man move back twenty feet farther from the observer, and he and the boy will then subtend equal angles, but they will not appear to be of equal hight, nor will the man's hight appear diminished in a very perceptible degree. The sun and the moon are about 4,000 miles nearer to us when they are in the zenith than when near the horizon, but in the latter case they appear much larger. It makes a great difference in the variation of the apparent size of a pin, as it moved to and from the eye, whether it is seen through a pin-hole in a card or whether the card is removed; and, again, whether it is seen with one eye or both eyes. The fact is, that in estimating the size of objects, our judgment is influenced by many other things besides the visual angles which they subtend. Our knowledge of the real size of an object, also of the fact that the tendency of an increase in distance is to diminish the apparent size of a body, and that an object *does not become shorter* as it moves away from us, does much toward correcting an estimate based on the size of the visual angle. Our estimate of the size of objects whose size is unknown is influenced much by comparison with objects in their vicinity whose size is known, as in the case of the sun and the moon when they are in range with other objects in the horizon, and in the case of the pin, whether it is seen alone through a hole or in conjunction with other objects. Again, when we look at an object with both eyes we are obliged to turn the eyes inward or outward, according as an object approaches or recedes, in order that light from the object may continue to enter the eye. The effort necessary to adapt the position of the eyes, so as to see objects at different distances, helps in forming a correct estimate of their size. Hence, the pin seen by both eyes does not appear to undergo so great a change in size, as it moves to and from the observer, as when seen by one eye. We are not at the time conscious of going through the processes of reasoning indicated above, because it has become a matter of habit with us. If a man born blind suddenly acquires the power of seeing, he at first makes ludicrous mistakes in judging of size and distance of objects, because he has not acquired these methods of reasoning. An infant will reach out its hands to seize a bird that may be flying many yards above.

§ 314. **Velocity of Light.** — We must believe that light-waves require time to traverse space, although their speed is so great that no ordinary means can measure the time, it is so short. But the distances of the heavenly bodies are so great that the time that their light requires to reach us may be easily measured.

To illustrate one method, let J, in Figure 242, represent a clock striking a single stroke every hour, and the circle EE' a road around which a person W travels; the length of the straight line EE' is four miles. So long as W remains at E, the strokes come exactly once an hour by his watch; but, as he moves away, the intervals become slightly longer, so that, however long he is on the road, if the watch and clock run accurately, when he has reached E' the sound of the bell reaches him about twenty seconds after the hour. As he continues back to E, the sounds come more and more nearly on time, so that at E they are just at the proper time. Similarly, at regular intervals in the heavens an eclipse of one of Jupiter's moons takes place; the average interval being known, add it to the time at which an eclipse is observed when the earth is near E, and thus we may predict the times of an eclipse for years ahead. All the eclipses, except when the earth is at E, are observed to be a little behind the predicted times; at E' as much as 16¼ minutes. But at E' the light has had to travel 184,000,000 miles farther to reach the eye than at E.

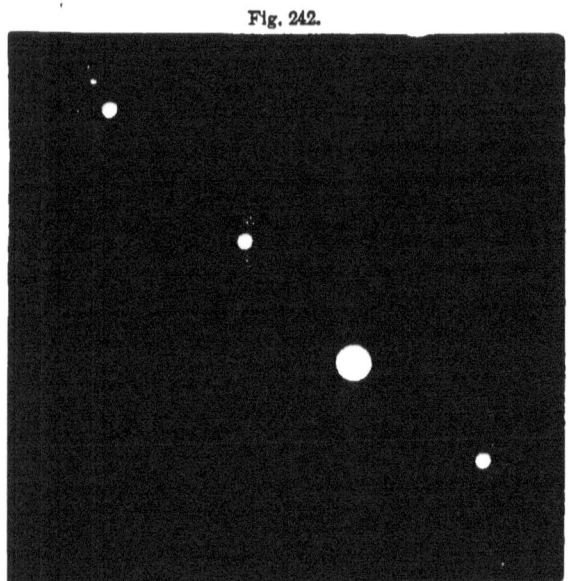

Fig. 242.

Hence, light must travel at the rate of $184{,}000{,}000 \div (16\tfrac{1}{4} \times 60)$ = about 186,000 miles (about $300{,}000^{km}$) in a second.

Sound creeps along at the comparatively slow pace of about one-fifth of a mile (or $\tfrac{1}{3}^{km}$) per second. The former is the velocity with which waves in ether are transmitted; the latter, the velocity with which waves in air move forward. This great difference can be accounted for only on the supposition that *the rarity and elasticity of ether are enormously greater than that of air* (see page 284).

LIII. REFLECTION OF LIGHT.

§ 315. Law of reflection. — Arrange apparatus as follows: A B, Figure 243, is a board 12cm square, having a mirror 8cm square fastened to one of its sides. E is a rod 24cm long inserted in the board close to the middle of one of the edges of the mirror, and perpendicular to the surface of the board. D F is an arc of pasteboard supported by the rod. The outer edge of the arc is described by a radius equal to the length of the rod, and is divided into degrees. Cover the opening orifice of the tube C of the *porte lumière* [1] with a circular tin pierced in its center by a circular hole m, 7mm in diameter, and admit a slender beam of sunlight mc.

Experiment. Place the mirror so that the beam of light may strike it obliquely, and just graze the arc so as to illuminate it at one point. A beam of light as it approaches an object is termed an *incident beam*. The beam, unable to pass through the opaque silvered surface of the mirror, is reflected by this surface obliquely, but on the opposite side of the perpendicular oc. A beam of light after reflection is termed a *reflected beam*. The spot of light on the arc produced by the reflected beam will be found to be the same number of degrees distant from the perpendicular as the spot produced by the incident beam. Hence, the angle nco, called the *angle of reflection*, is equal to the angle mco, called the *angle of incidence*. Incline the mirror so that the incident beam may strike the mirror more or less obliquely, and the reflected beam will leave it always at an equal angle. Render the path of the incident and reflected beam luminous by introducing a cloud of smoke from touch

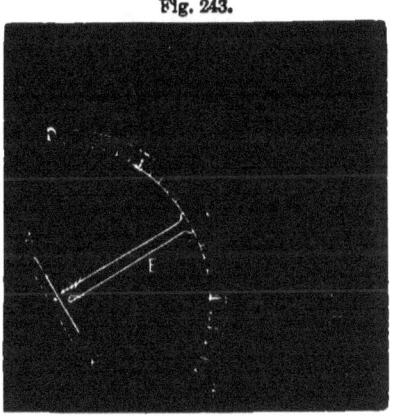

Fig. 243.

---

[1] Some means of introducing a beam of sunlight into a darkened room is indispensable in experimenting with light. The experiments on this subject will be given on the supposition that the pupil is provided with means of accomplishing this. Directions for constructing apparatus suited to this purpose, usually called a *porte lumière*, may be found in Mayer and Barnard's little book on "Light," published by D. Appleton & Co., New York, and in Dolbear's "Art of Projection," published by Lee & Shepherd, Boston. A description of an inexpensive apparatus devised by the author may be found in Section H of the Appendix.

340 RADIANT ENERGY. — LIGHT.

paper, and the angles formed with the perpendicular will be quite apparent. *Light, as well as sound, conforms to the general law of reflection.* (See page 118.)

§ 316. Diffused light. — **Experiment 1.** Introduce a small beam of light into a darkened room, by means of a *porte lumière*, and place in its path a mirror. The light is reflected in a definite direction. If the eye is placed so as to receive the reflected light, it will see, not the mirror, but the image of the sun, and the light will be painfully intense. Substitute for the mirror a piece of unglazed paper. The light is not reflected by the paper in any definite direction, but is scattered in every direction, illuminating objects in the vicinity and rendering them visible. Looking at the paper, you see, not an image of the sun, but the paper, and you may see it equally well in all directions.

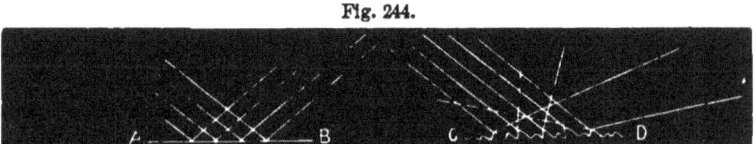

Fig. 244.

The *dull* surface of the paper receives light in a definite direction, but reflects it in every direction; in other words, it scatters or *diffuses* the light. The difference in the phenomena in the two cases is caused by the difference in the smoothness of the two reflecting surfaces. AB, Figure 244, represents a smooth surface, like that of glass, which reflects nearly all the rays of light in the same direction, because nearly all the points of reflection are in the same plane. CD represents a surface of paper having the roughness of its surface greatly exaggerated. The various points of reflection are turned in every possible direction; consequently, light is reflected in every direction. Thus, the dull surfaces of various objects around us reflect light in all directions, and are consequently visible from every side. Objects rendered visible by reflected light are said to be *illuminated*.

By means of regularly reflected light we see images of objects in mirrors, but only in definite directions; by means of diffused light we see the mirror itself in every direction. Whether we see the image of the source of the light (the eye being situated so as to receive the

REFLECTION FROM PLANE MIRRORS. 341

regularly reflected light), or the object on which the light falls, or both at the same time, depends largely upon the degree of smoothness possessed by the object that reflects the light. Smooth surfaces are called *mirrors*. Polished metals are the best mirrors. Surfaces of liquids at rest are excellent mirrors. It is sometimes difficult to see a smooth surface of a pond surrounded by trees and overhung by clouds, as the eye is occupied by the reflected images of these objects: but a faint breath of wind, slightly rippling the surface, will reveal the water.

Experiment 2. Place a basin of water on a table, and hold a candle flame so that its rays may form a large angle with the liquid surface, and notice the brightness of its image. Lower the candle and the eye so that the incident and reflected rays, as nearly as possible, graze the surface of the liquid, and notice how much brighter the image becomes. Notice how much brighter the varnished surfaces of furniture appear when viewed very obliquely, than when seen by light reflected less obliquely. Also notice how much more dazzling is the light reflected from the surface of a pond just before the sun sets, than at noon when the sun is overhead. This is due in part to our being at a suitable position to observe it.

The amount of light reflected from a smooth surface increases rapidly as the angle of incidence increases. Thus, at a perpendicular incidence, out of 1,000 parts of light that strike a surface of water, only 18 parts are reflected; at 40°, 22 parts are reflected; at 80°, 333 parts; and at $89\frac{1}{4}$°, 721 parts. The above is not even approximately true of metals or substances having metallic reflection, such as galena, etc.

§ 317. Reflection from plane mirrors; virtual images.— M M (Fig. 245) represents a plane mirror, and A B a pencil of divergent rays proceeding from the point A of an object A H. Erecting perpendiculars at the points of incidence, or the points where these rays strike the mirror, and making the angles of reflection equal to the angles of incidence, the paths B C and E C of the reflected rays are found.

It appears that *divergent incident rays remain divergent after reflection from a plane mirror*. In like manner construct a diagram, and show that *parallel incident rays are parallel after reflection*. Construct another diagram, and show that *convergent incident rays are convergent after reflection.* To an eye placed at C, the points from which the rays appear to come are of course in the direction of the rays as they enter the eye. These points may be found by *continuing* the rays CB and CE behind the mirror, till they meet at the points D and N. Every point of the object AH sends out its pencils of rays, and those that strike the mirror at a suitable angle to be reflected to the eye, produce on the retina of the eye an image of that point, and the point from which the light appears to emanate is found, as previously described. Thus, the pencils EC and BC appear to emanate from the points N and D, and the whole body of light received by the eye seems to come from an *apparent object ND*, behind the mirror. This apparent object is called an *image*, but as of course there can be no real image formed there, it is called a *virtual* or an *imaginary* image. It will be seen, by construction, that *an image in a plane mirror appears as far behind the mirror as the object is in front of it, and is of the same size and shape as the object.*

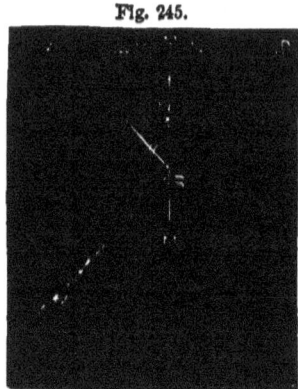

Fig. 245.

If the mirror is vertical, objects appear in their proper relations to the horizon; but, if the mirror has any other position, objects assume unnatural postures. Thus, turn this book so that the mirror MM (Fig. 245) may represent a horizontal mirror, and AH a vertical object above it, and it will be seen that the image appears inverted. To verify this, place a mirror in a horizontal position, and set on it a goblet of water. Also show by construction that, in a mirror making an angle of 45° with the horzon, vertical objects appear horizontal and vice versa. Verify this by experiment. Pupils may amuse themselves at their leisure, and at the same time be instructed, by performing the following experiments: —

Experiment 1. Place a printed page in front of a mirror, and attempt to read the print from the mirror. It will be seen that there is always a *lateral inversion;* for the same reason that when two persons stand facing one another, the right hand of one is opposite the left hand of the other.

Experiment 2. Place two mirrors facing one another and about 15^{cm} apart. Hold a pencil half-way between the mirrors, and look obliquely into one mirror just over the edge of the other, and you will see a large number of images of the pencil arranged at equal distances behind one another. Account for these images.

Experiment 3. Place two mirrors edge to edge so as to form an angle of 45° with one another. Place the face in the opening, and gradually close the mirrors till they touch the head.

§ 318. **Multiple reflection.** — **Experiment 1.** Allow the beam of light in the last experiment to strike a wall of the room. There will be projected upon the wall two, and perhaps more, circular images of the sun overlapping one another. It appears as though the beam of light is somehow split, by reflection from the mirror, into two or more parts, and that these parts travel thereafter in slightly different paths.

Fig. 246.

Experiment 2. Hold a candle flame in such a position (Fig. 246) that its light may strike a mirror (one having very thick glass is best) very obliquely, and place the eye so that it may receive the reflected light, and you may see many images of the flame.

Experiment 3. Place a pencil perpendicular to a mirror, with the point touching the glass, and you will see two images of the pencil, — one touching the point of the pencil, and the other at a distance equal to twice the thickness of the glass.

How are these phenomena produced? As you travel the sidewalk and pass windows, you frequently see your own image and images of other outdoor objects reflected by the glass, showing that even so transparent a substance as glass does not allow all the light that strikes it to pass through it, but reflects a portion. Let a beam of light Aa, Figure 247, strike a mirror BC obliquely; a portion of the light is reflected from the point of incidence a, and strikes the screen DE at b. Another portion of the light enters the glass, and a portion of it is reflected from the point c, and a portion of this last reflected light strikes the screen at d, while the remainder is reflected from e to f, and again from f, and a portion of it reaches the screen at g,

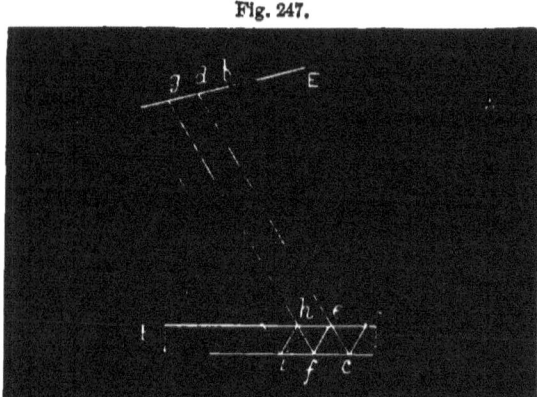

Fig. 247.

while the remainder is reflected from h to i, and undergoes further reflections and splittings, until the light, in consequence of the loss occasioned by successive divisions, becomes too feeble to produce distinct effects. If the eye take the place of the screen, since an object is seen in the direction in which the light comes to the eye, the point A will appear to lie somewhere on the line ba, extended; for the same reason it will appear to lie on the lines de, gh, etc.; but as these lines have no point in common, it is clear that the effect would be that of multiple images. (Show the application of this explanation in accounting for the phenomena obtained in the above experiments.)

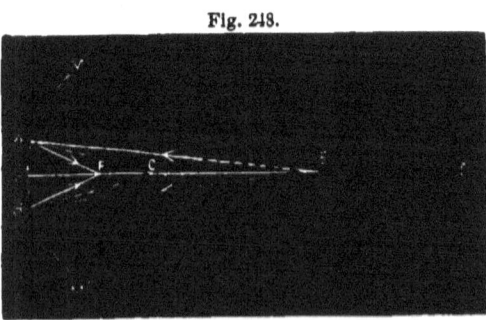

Fig. 248.

§ **319. Reflection from concave mirrors.** — Let MM', Figure 248, represent a section of a concave mirror, which may be regarded as a small part of a hollow spherical shell having a polished interior surface. The distance MM' is called the *aper-*

ture of the mirror. C is the center of the sphere, and is called the *center of curvature.* G is the *vertex* of the mirror. A straight line DG, drawn through the center of curvature and the vertex is called the *principal axis* of the mirror. A concave mirror may be considered as made up of an infinite number of small plane surfaces. All radii of the mirror, as CA, CG, and CB, are perpendicular to the small planes which they strike. If C be a luminous point, it is evident that all light emanating from this point, and striking the mirror, will be reflected back to its source at C.

Let E be any luminous point in front of a concave mirror. To find the direction that rays emanating from this point take after reflection, draw any two lines from this point, as EA and EB, representing two of the infinite number of rays composing the divergent pencil of light that strikes the mirror. Next draw radii to the points of incidence A and B, and draw the lines AF and BF, making the angles of reflection equal to the angles of incidence. Place arrow-heads on the lines representing rays of light to indicate the direction of the motion. The lines AF and BF represent the direction of the rays after reflection.

It will be seen that the rays after reflection are convergent, and meet at the point F, called the *focus.* This point is the focus of all reflected rays that emanate from the point E. It is obvious that if F were the luminous point, the lines AE and BE would represent the reflected rays, and E would be the focus of these rays. Since the relation between two such points is such that light emanating from either one is brought by reflection to a focus at the other, they are called *conjugate foci.* *Conjugate foci are two points so related that the image of one is formed at the other.* The rays EA and FB emanating from E are less divergent than rays FA and FB, emanating from a point F less distant from the mirror, and striking the same points. Rays emanating from D, and striking the same points A and B, will be still less divergent; and if the point D were removed to a distance of many miles, the rays incident at these points would be very nearly parallel. Hence rays may be regarded as practically parallel when their source is at a very great distance, e.g., the sun's rays. If a sunbeam, consisting of a bundle of parallel rays, as E A, D G, and H B (Fig. 249), strike a concave mirror parallel with its principal axis, they become convergent by reflection, and meet at a point (F) in the principal axis. This point, called *the principal focus, is just half-way between the center of curvature and the vertex of the mirror.*

Fig. 249.

On the other hand, it is obvious that *divergent rays emanating from the principal focus of a concave mirror become parallel by reflection.*

If a small piece of paper is placed at the principal focus of a concave mirror, and the mirror is exposed to the parallel rays of the sun, the paper will quickly burn, showing that *the focus of light is also a focus of heat;* or, in other words, that *all forms of radiant energy follow the same laws of reflection as light.*

Construct a diagram, and show that rays of light proceeding from a point between the principal focus and the mirror are divergent after reflection, but less divergent than the incident rays. Reversing the direction of the light, the same diagram will show that convergent rays of light are rendered more convergent by reflection from concave mirrors. *The general effect of a concave mirror is to increase the convergence or to decrease the divergence of incident rays.*

The statement, that parallel rays after reflection from a concave mirror meet at the principal focus, is only approximately true. The smaller the aperture of the mirror, the more nearly true is the statement. It is strictly true only of parabolic mirrors, such as are used with the head-lights of locomotives. Construct a diagram representing a mirror of large aperture, and it will be found that those rays that strike the mirror at considerable distance from its center, intersect the principal axis after reflection at points nearer to the mirror than the principal focus.

§ **320. Formation of images.** — **Experiment 1.** In a dark room hold the concave side of a bright silver dessert spoon a little distance

FORMATION OF IMAGES. 347

in front of the face, and introduce a candle flame between the spoon and your eyes; you will see a small *inverted* image of the flame about a centimeter *in front* of the spoon.

Experiment 2. Turn the convex side of the spoon toward you, and you will see a small *erect* image of the flame a little back of the spoon.

Experiment 3. Repeat the two preceding experiments, holding the spoon between the flame and the eyes, but not so as to screen the face from the light, and you will see similar images of yourself.

To determine the position and kind of images formed of objects placed in front of concave mirrors, proceed as follows: Locate the object, as D E, Figure 250. Draw lines, E A and D B, from the extremities of the object through the center of curvature of the mirror, to meet the mirror. These lines are called the *secondary axes*. Incident rays along these lines will return by the same paths after reflection. (Why?) Draw another line from D to any point in the mirror, *e.g.*, to F, to represent any other of the infinite number of rays emanating from

Fig. 250.

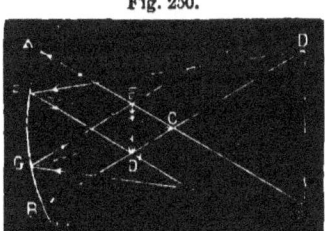

D. Make the angle of reflection C F D' equal to the angle of incidence C F D, and the reflected ray will intersect the secondary axis D B at the point D'. This point is the conjugate focus of all rays proceeding from D. Consequently, an image of the point D is formed at D'. This image is called a *real image*, because rays actually meet at this point. In a similar manner, find the point E', the conjugate focus of the point E. The images of intermediate points between D and E lie between the points D' and E'; and, consequently, the image of the object lies between those points as extremities.

Fig. 251.

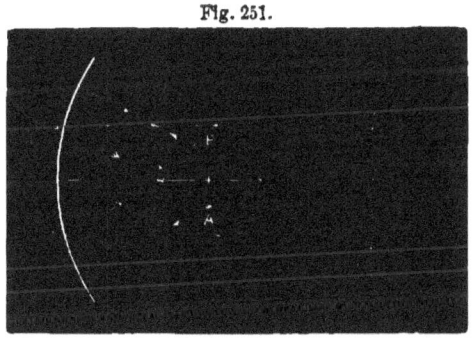

If, for the second ray to be drawn from any point, we select that ray which is parallel with the principal axis, as A G, Figure 251, it will not be necessary to measure angles. For this ray, after reflection, must pass through the principal focus F; and consequently the conjugate focus A' is easily found, and so for the point B' and intermediate points. Both methods of constructing images should be practised by the pupil.

It thus appears that *an image of an object placed beyond the center of curvature of a concave mirror is real, inverted, smaller than the object, and located between the center of curvature and the principal focus of the mirror.* An eye placed in a suitable position to receive the light, as at H (Fig. 252), will receive the same impression from the reflected rays as if the image E' D' were a real object. For a cone of rays originally emanates from (say) the point D of the object, but it enters the eye as if emanating from D', and consequently appears to originate from the latter point. A person standing in front of such a mirror, at a distance greater than its radius of curvature, will see an image of himself suspended, as it were, in mid-air. Or, if in a darkened room an illuminated object is placed in front of the mirror, and a small oiled-paper screen is placed where the image is formed, a large audience may see the image projected upon the screen.

Fig. 252.

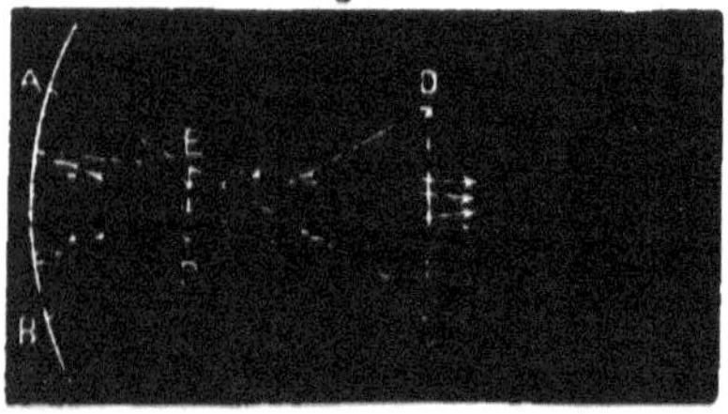

If E' D' (Fig. 250) is taken as the object, then the direction of the light in the diagram will be reversed, and E D will represent the image. Hence, *the image of an object placed between the principal focus and the center of curvature is also real and inverted, but larger than the object, and located beyond the center of curvature.* The image in this case may be projected upon a screen, but it will not be so bright as in the former case, because the light is spread over a larger surface.

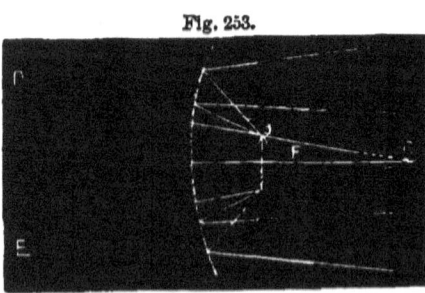

Fig. 253.

FORMATION OF IMAGES.

Construct the image of an object placed between the principal focus and the mirror, as in Figure 253. It will be seen in this case that a pencil of rays proceeding from any point of an object, e.g., D, has no actual focus, but appears to proceed from a *virtual* focus D', back of the mirror, and so with other points, as E. *The image of an object placed between the principal focus and the mirror is virtual, erect, larger than the object, and is back of the mirror.*

Fig. 254.

QUESTIONS.

Ascertain the answers to the following questions by constructing suitable diagrams, and afterwards verify your conclusions by experiment, if convenient.

1. When an object is located at a distance from a concave mirror equal to its radius, will any image be formed? Why?
2. What is the effect of placing the object at the principal focus? Why?
3. (a) When is the real image formed by a concave mirror smaller than the object? (b) When is it larger?
4. (a) When is the image formed by a concave mirror real? (b) When is it virtual?
5. (a) Is the image of an object formed by a convex mirror real or virtual? (b) Is it larger or smaller than the object? (c) Is it erect or inverted?

NOTE. — The diagram in Figure 254 will be found sufficiently suggestive as to the method of finding the disposition of a pencil of rays emanating from any point, e.g., A, after reflection from a convex mirror.

6. Is the general effect of a convex mirror to collect or to scatter rays?

LIV. REFRACTION.

Experiment 1. Across the bottom of a rectangular tin basin A B C D, Figure 255, mark a scale of millimeters. Into a darkened room admit a beam of sunlight, so that its rays may fall obliquely on the bottom of the basin, and note the place on the scale where the edge of the shadow D E cast by the side of the basin D C meets the bottom at E. Then, without moving the basin, fill it even full with water slightly clouded with milk, or with a few drops of a solution of mastic in alcohol. It will be found that the edge of the shadow has moved from D E to D F, and meets the bottom at F. Beat a blackboard rubber, and create a cloud of dust in the path of the beam in the air, and you will discover that the rays G D that graze the edge of the disk at D become bent at the point where they enter the water, and now move in the bent line G D F, instead of, as formerly, in the straight line G E. The path of the light in the water is now nearer to the vertical side D C; in other words, this part of the beam *is more nearly vertical than before.*

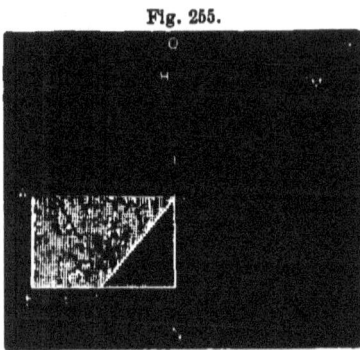

Fig. 255.

Experiment 2. Place a coin (A, Fig. 256) on the bottom of an empty basin, so that, as you look through a small hole in a card B C over the edge of the vessel, the coin is just out of sight. Then, without moving the card or basin, fill the latter with water. Now, on looking through the aperture in the card, the coin is visible. The beam of light A E, which formerly moved in the straight line A D, is now bent at E, where it leaves the water, and, passing through the aperture in the card, enters the eye. Observe that, as the light passes from the water into the air, it is turned farther from a vertical line E F; in other words, *the beam is farther from the vertical than before.*

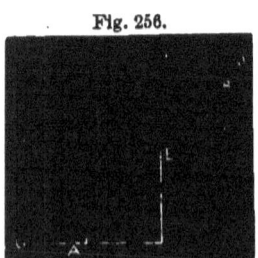

Fig. 256.

Experiment 3. From the same position as in the last experiment, direct the eye to the point G in the basin filled with water. Reach your hand around the basin, and place your finger where that point appears to be. On examination, it will be found that your finger is considerably above the bottom. Hence, *the effect of the bending of rays of light, as they pass obliquely out of water, is to cause the bottom to appear more elevated than it really is;* in other words, *to cause the water to appear shallower than it is.*

Experiment 4. Thrust a pencil obliquely into water, and it will appear shortened, bent at the surface of the water, and the immersed portion elevated.

Fig. 257.

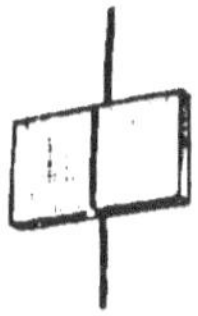

Experiment 5. Place a piece of wire (Fig. 257) vertically in front of the eye, and hold a narrow strip of thick plate glass horizontally across the wire, so that the light from the wire may pass obliquely through the glass to the eye. The wire will appear to be broken at the two edges of the glass, and the intervening section will appear to be moved to the right or left according to the inclination of the glass; but, if the glass is not inclined to the one side or the other, the wire does not appear broken.

When a beam of light passes from one medium into another of different density, it is bent or *refracted* at the boundary plane between the two media, unless it falls exactly perpendicularly on this plane. *If it passes into a denser medium, it is refracted toward a perpendicular to this plane; if into a rarer medium, it is refracted from the perpendicular.* The angle GDO (Fig. 255) is called the *angle of incidence;* FDN, the *angle of refraction;* and EDF, the *angle of deviation.*

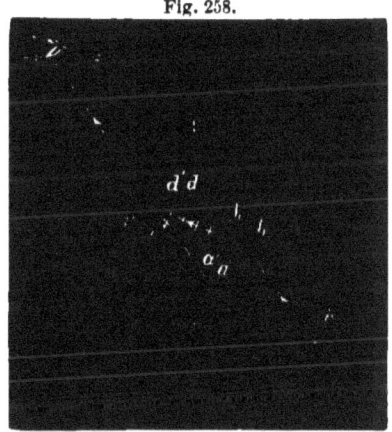

Fig. 258.

§ **321. Cause of refraction.** — *Careful experiments have proved that the velocity of light is less in a dense than in a rare medium.* Let the series of parallel lines A B (Fig. 258) represent a series of wave-fronts leaving an object C, and passing through a rectangular piece of glass D E, and constituting a beam of light. Every point in a wave-front moves with equal velocity as long as it traverses the same medium; but the point

a of a given wave ab enters the glass first, and its velocity is impeded, while the point b retains its original velocity; so that, while the point a moves to a', b moves to b', and the result is that the wave-front assumes a new direction (very much in the same manner as a line of soldiers execute a wheel), and a ray or a line drawn perpendicularly through the series of waves is turned out of its original direction on entering the glass. Again, the extremity c of a given wave-front cd first emerges from the glass, when its velocity is immediately quickened; so that, while d advances to d', c advances to c', and the direction of the ray is again changed. The direction of the ray, after emerging from the glass, is parallel to its direction before entering it, but it has suffered a lateral displacement. Let C represent a section of the wire used in Exp. 5, and the cause of the phenomenon observed will be apparent. If the beam of light strikes the glass perpendicularly, all points of the wave will be checked at the same instant on entering the glass; consequently it will suffer no refraction.

§ 322. **Index of refraction.** — The deviation of light, in passing from one medium to another, varies with the medium and with the angle of incidence. It diminishes as the angle of incidence diminishes, and is zero when the incident ray is normal (*i.e.*, perpendicular to the surface of the medium). It is highly important, knowing the angle of incidence, to be able to determine the direction which a ray of light will take on entering a new medium. Describe a circle around the point of incidence A (Fig. 259) as a center, with a radius of (say)

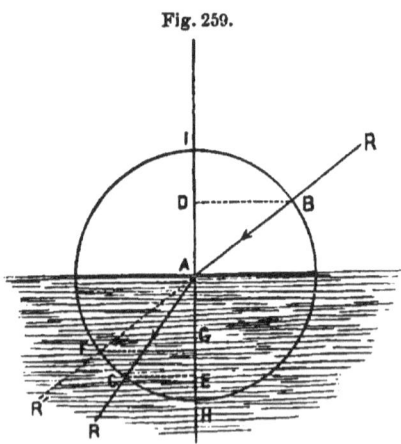

Fig. 259.

INDICES OF REFRACTION. 353

10^{cm}; through the same point draw I H perpendicular to the surfaces of the two media, and to this line drop perpendiculars B D and C E from the points where the circle cuts the ray in the two media. Then suppose that the perpendicular B D is $\frac{8}{10}$ of the radius A B; now this fraction $\frac{8}{10}$ is called (in Trigonometry) the *sine* of the angle D A B. Hence, $\frac{8}{10}$ is the *sine of the angle of incidence*. Again, if we suppose that the perpendicular C E is $\frac{6}{10}$ of the radius, then the fraction $\frac{6}{10}$ is the *sine of the angle of refraction*. The sines of the two angles are to one another as $\frac{8}{10} : \frac{6}{10}$, or as 4 : 3. The quotient (in this case $\frac{4}{3}$) obtained by dividing the sine of the angle of incidence by the sine of the angle of refraction is called the *index of refraction*. It can be proved to be the *ratio of the velocity of the incident to that of the refracted light*. It is found that, *for the same media the index of refraction is a constant quantity;* i.e., the incident ray might be more or less oblique, still the quotient would be the same.

§ 323. **Indices of refraction.** — The index of refraction for light in passing from air into water is approximately $\frac{4}{3}$, and from air into glass $\frac{3}{2}$; and, of course, if the order is reversed, the reciprocal of these fractions must be taken as the indices; *e.g.*, from water into air the index is $\frac{3}{4}$, from glass into air $\frac{2}{3}$. When a ray passes from a vacuum into a medium, the refractive index is greater than unity, and is called the *absolute index of refraction*. The *relative index of refraction, from any medium A into another B, is found by dividing the absolute index of B by the absolute index of A*.

The refractive index varies with the color of the light. (See page 365.) The following table is intended to represent *mean indices:* —

TABLE OF ABSOLUTE INDICES.

Air at 0° C. and 760mm pressure	1.000294	Carbon bisulphide		1.641
Pure water	1.33	Crown glass (about)		1.53
Alcohol	1.37	Flint glass (about)		1.61
Spirits of turpentine	1.48	Diamond (about)		2.5
Humors of the eye (about)	1.35	Lead chromate		2.97

354 RADIANT ENERGY. — LIGHT.

EXERCISES.

1. Draw a straight line to represent a surface of flint glass, and draw another line meeting this obliquely to represent a ray of light passing from a vacuum into this medium. Find the direction of the ray after it enters the medium, employing the index as given in the above table.

2. (a) Determine the index of refraction for light in passing from water into diamond. (b) In passing from water into air.

3. Ascertain the index of refraction for water in Exp. 1, p. 350, in which sine I (sine of angle of incidence) = $\frac{EC}{ED}$ (Fig. 255), and sine R (sine of angle of refraction) = $\frac{FC}{FD}$. Hence, the index of refraction
$$= \frac{\text{sine I}}{\text{sine R}} = \frac{EC}{ED} \div \frac{FC}{FD}$$

Fig. 260.

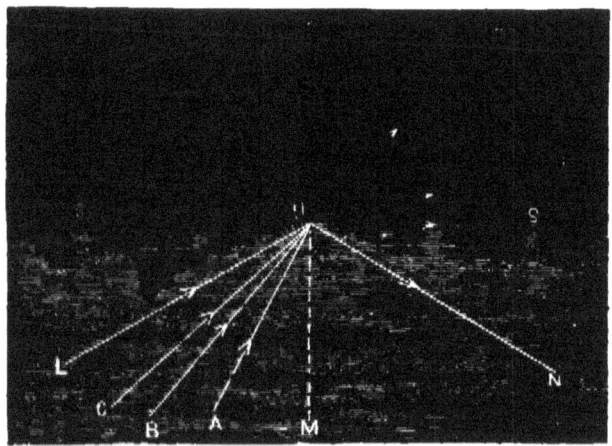

§ 324. Critical angle; total reflection. — Let SS' (Fig. 260), represent the boundary-surface between two media, and AO and BO incident rays in the more refractive medium (e.g., glass); then OD and OE may represent the same rays respectively after they enter the less refractive medium (e.g., air). It will be seen that, as the angle of incidence is increased, the refracted ray rapidly approaches the surface OS. Now, there must be an angle of incidence (e.g., COM) such that the angle

REFRACTION AND TOTAL REFLECTION. 355

of refraction will be 90°; in this case the incident ray C O, after refraction, will just graze the surface O S. This is called *the critical or limiting angle*. Any incident ray, as L O, making a larger angle with the normal than the critical angle, cannot emerge from the medium, and consequently is not refracted. Experiment shows that all such rays undergo internal reflection, *e.g.*, the ray LO is reflected in the direction O N. Reflection in this case is perfect, and hence is called *total reflection*. *Total reflection occurs when rays in the more refractive medium are incident at an angle greater than the critical angle*. Surfaces of transparent media, under these circumstances, constitute the best mirrors possible. The critical angle diminishes as the refractive index increases. For water it is about $48\frac{1}{2}°$; for flint glass, 38° 41'; and for diamond, 23° 41'. Light cannot, therefore, pass out of water into air with a greater angle of incidence than $48\frac{1}{2}°$. The brilliancy of gems, particularly the diamond, is due in part to their extraordinary power of internal reflection. It is evident that all incident light embraced in the angular space K O S, not reflected at the surface, is condensed by refraction into the angular space C O M of $48\frac{1}{2}°$, or that the whole light that passes into the water is condensed into an angular space of 97°. A diver, looking upward, can see external objects, as it were, only through a circular aperture overhead of limited diameter; while beyond this circle he sees, as the effect of total reflection, the various objects on the bottom.

§ 325. **Illustrations of refraction and total reflection.** —
Experiment 1. Place a bright coin in a tumbler of water, and tilt the glass till the light from the coin strikes the surface of the water above with sufficient obliquity, so that, looking upward toward that surface, you can see there a distinct image of the coin.
Experiment 2. Thrust the closed end of a glass test-tube into water, and incline the tube. Look down upon the immersed part of the tube, and its upper surface will look like burnished silver, or as if the tube contained mercury. Fill the test-tube with water, and immerse as before; the total reflection which before occurred at the surface of the air in the submerged tube now disappears. Explain.

Experiment 3. Place uncolored glass beads, or glass broken into quite small pieces, in a test-tube. They appear not only white, due to diffused reflection, but quite opaque, due to refraction and internal reflection. Pour some water into the tube, and it becomes somewhat translucent. Substitute spirits of turpentine for the water, and the translucency is increased. By mixing a small quantity of carbon bisulphide with the turpentine, or olive oil with oil of cassia, a liquid can be obtained whose refractive index is about the same as that of glass, when the light will pass through the liquid without obstruction, and the beads become transparent and nearly invisible. The last illustration shows that *one transparent body within another can be seen only when their refractive powers differ.* Place your eye on a level with the surface of a hot stove, and you may observe a wavy motion in the air, due to the mingling of currents of heated and less refractive air, with cooler and more refractive air.

Fig. 261.

A ray of light from a heavenly body A (Fig. 261) undergoes a series of refractions as it reaches successive strata of the atmosphere of constantly increasing density, and to an eye at the earth's surface appears to come from a point A' in the heavens. The general effect of the atmosphere on the path of light that traverses it is such as to increase the apparent altitude of the heavenly bodies. It enables us to see a body (B) which is actually below the horizon, and prolongs the apparent stay of the sun, moon, and other heavenly bodies above the horizon. Twilight is due both to refraction and reflection of light by the atmosphere.

LV. PRISMS AND LENSES.

§ 326. Optical prisms. — An optical prism is usually a transparent wedge-shaped body. Figure 262 represents a transverse section of such a prism. Let A B be a ray of light incident upon one of its surfaces. On entering the prism it is refracted *toward* the normal, and takes the direction B C. On emerging from the prism, it is again refracted, but

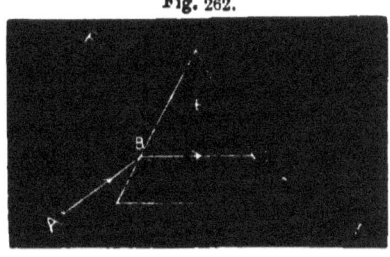

Fig. 262.

now *from* the normal in the direction C D. The object that emits the ray will appear to be at F. Observe that the ray A B, at both refractions, is bent toward the thicker part, or base, of the prism.

§ 327. Lenses. — Any transparent medium bounded by two curved surfaces, or one plane and the other curved, is a lens.

Experiment 1. Procure a couple of lenses thicker in the middle than at the edge; strong spectacle glasses, or the large lenses in an opera glass, will answer. Hold one of the lenses in the sun's rays, and notice the path of the beam in dusty air (made so by striking together two blackboard rubbers) after it passes through the lens; also, that on a paper screen all the rays may be brought to a small circle, or even a point, not far from the lens. This point is called the *focus*, and its distance from the lens, the *focal length* of the lens.

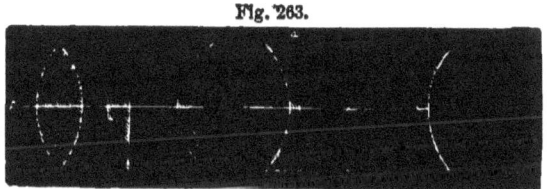

Fig. 263.

Find the focal length of this lens, and of the second, and then of the two together. You find the focal length of the two combined is less than of either alone, and learn that the more powerful a lens or combination of them is, the shorter the focal length; that is, the more quickly are the parallel rays that enter different parts of the lens brought to cross one another.

Experiment 2. Procure a lens thinner in the middle than at its edge. One of the small lenses or eye-glasses of an opera glass will answer. Repeat the above experiment with this lens, and notice that the light emerging from the lens, instead of coming to a point, becomes spread out.

Lenses are of two classes, converging and diverging, according as they collect or scatter beams of light. Each class comprises three kinds (Fig. 263) : —

CLASS I.			CLASS II.	
1. Double-convex	} Converging or convex lenses, thicker in the middle than at the edges.		4. Double-concave	} Diverging, or concave lenses, thinner in the middle than at the edges.
2. Plano-convex			5. Plano-concave	
3. Concavo-convex (or meniscus)			6. Convexo-concave	

A straight line, as A B, normal to both surfaces of a lens, and passing through its center of curvature, is called its *principal axis*. In every lens there is a point in the principal axis called the *optical center*. Every ray of light that passes through it has parallel directions at incidence and emergence, *i.e.*, can suffer at most only a slight lateral displacement. In lenses 1 and 4 it is half-way between their respective curved surfaces. A ray, drawn through the optical center from any point of an object, as A a (Fig. 269, p. 362), is called the secondary axis of this point.

§ 328. Effect of lenses. — We may, for convenience of illustration, regard a convex lens as composed, approximately, of two prisms placed base to base, as A (Fig. 264), and a concave lens as composed of two prisms with their edges in contact, as B. Inasmuch as a beam or pencil of light ordinarily strikes a lens in such a manner that the rays will be bent toward the thicker parts or bases of these approximate prisms, it is obvious that the lens A would tend to bend the transmitted rays toward one another, while the lens B would tend to separate them. *The general effect of all*

Fig. 264.

EFFECT OF LENSES. 359

convex lenses is to converge transmitted rays; and of concave lenses, to cause them to diverge. Incident rays parallel with the principal axis of a convex lens are brought to a focus F (Fig. 265) at a point in the principal axis. This point is called the *principal focus, i.e.,* it is the focus of incident rays parallel with the principal axis. It may be found by holding the lens so that the rays of the sun may fall perpendicularly upon it, and then moving a sheet of paper back and forth behind it until the image of the

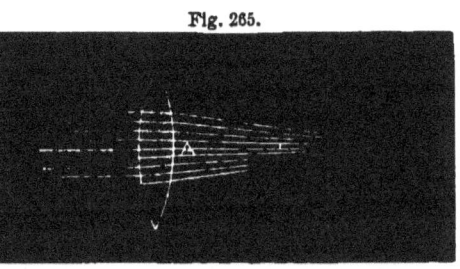

Fig. 265.

sun formed on the paper is brightest and smallest. Or in a room it may be found approximately by holding a lens at a considerable distance from a window, and regulating the distance of the paper so that a distinct image of the window will be projected upon it. The focal length is the distance of the optical center of the lens to the center of the image on the paper. The shorter this distance the greater is the power of the lens.

If the paper is kept at the principal focus for a short time it will take fire. Hence, this is the focus of *heat* as well as of *light.* The reason is apparent why convex lenses are sometimes called "burning glasses." A pencil of rays

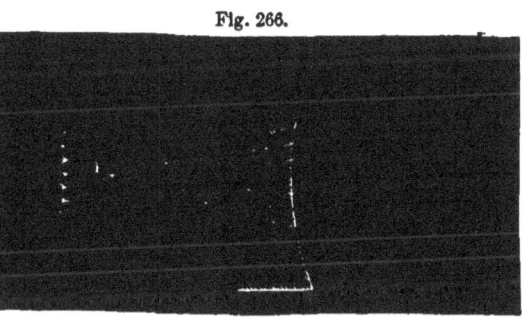

Fig. 266.

emitted from the principal focus F (Fig. 265), as a luminous point, becomes parallel on emerging from a convex lens. If the rays emanate from a point nearer the lens, they diverge after egress, but the divergence is less than before; if from a point beyond the principal focus, the rays are rendered convergent. A concave lens causes parallel incident rays to diverge as if they came from a point, as F (Fig. 266). This point is therefore its principal focus. It is, of course, a *virtual focus.*

§ 329. Conjugate foci. — When a luminous point S (Fig. 267) sends rays to a convex lens, the emergent rays converge to another point S'; rays sent from S' to the lens would converge to S. Two points thus related are called *conjugate foci.* The fact, that rays which emanate from one point are caused by convex lenses to collect at one point, gives rise to real images, as in the case of concave mirrors.

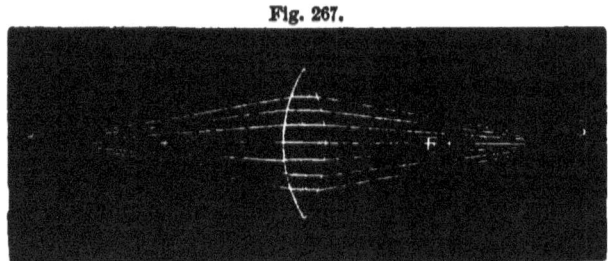
Fig. 267.

§ 330. Images formed. — Fairly distinct images of objects may be formed through *very small* apertures (page 330); but owing to the small amount of light that passes through the aperture, the images are very deficient in brilliancy. If the aperture is enlarged, brilliancy is increased at the expense of distinctness. (Why?) *A convex lens enables us to obtain both brilliancy and distinctness at the same time.*

Experiment 1. By means of a *porte lumière* A (Fig. 268) introduce a horizontal beam of light into a darkened room. In its path place some object, as B, painted in transparent colors or photographed on glass. (Transparent pictures are cheaply prepared by photographers for sunlight and lime-light projections.) Beyond the object place a convex lens L, and beyond the lens a screen S. The object being illuminated by the beam of light, all the rays diverging from any point a are bent by the lens so as to come together at the point a'. In like manner, all the rays proceeding from c are brought to the same point c'; and so also for all intermediate points. Thus, out of the billions of rays emanating from

each of the millions of points on the object, those that reach the lens
are guided by it, each to its own appropriate point in the image. It
is evident that there must result an image, both bright and distinct,
provided the screen is suitably placed, *i.e.*, at the place where the
rays meet. But if the screen is placed at S' or S'', it is evident
that a blurred image will be formed. Instead of moving the screen
back and forth, in order to "focus" the rays properly, it is cus-
tomary to move the lens.

Experiment 2. Fill some globular-shaped glass vessel (*e.g.*, a flask,
decanter, or fish-aquarium) with water, and place it 1^m in front of a
white wall of a darkened room. A little beyond the vessel place a
candle flame, and move it back and forth till a distinct image of the
flame is projected upon the wall by the water lens. Move the vessel
farther from the wall, and, on again focusing the flame, its image will
be larger than before. Repeat the same with a glass lens.

Fig. 268.

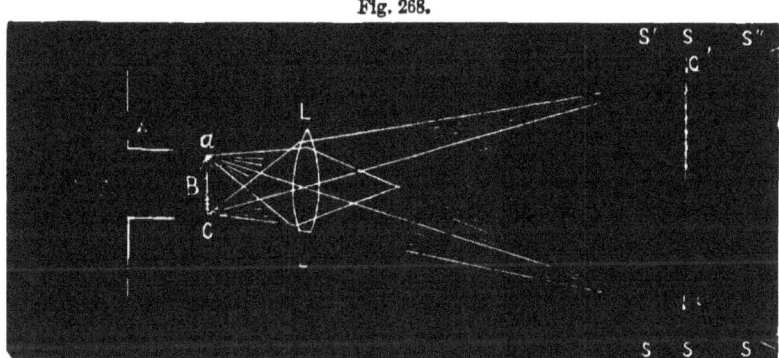

By properly varying the distances of the lens and flame from
the wall, in the last experiment, you may learn that when the
distance of the object is twice that of the principal focus, the
object and image are of equal size. When the image is within
twice the focal distance it is less, and when beyond this same
distance it is greater, than the object. *In all cases the corresponding linear dimensions of an object and its image are to one
another directly as their respective distances from the optical center.*

§ **331. To construct the image formed by a convex lens.**
— Given the lens L (Fig. 269), whose principal focus is at F (or F',

362   RADIANT ENERGY. — LIGHT.

for rays coming from the other direction), and object A B in front of it; any two of the many rays from A will determine where its image a is formed. The only two that can be traced easily are, the one along the secondary axis A O a, and the one parallel to the principal axis A A':

Fig. 269.

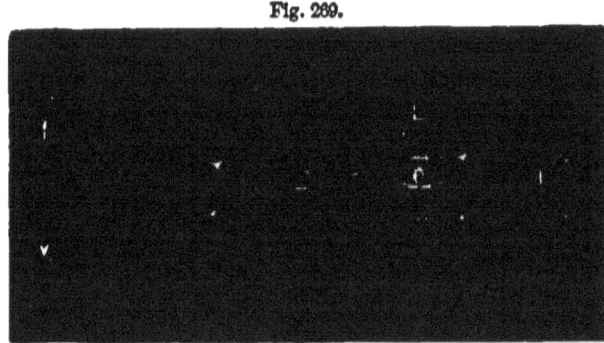

the latter will be deviated so as to pass through the principal focus F, and will afterward intersect the principal axis at some point a; so this is the conjugate focus of A; similarly for B, and all intermediate points along the arrow. Thus, a *real, inverted image* is formed at ab.

Fig. 270.

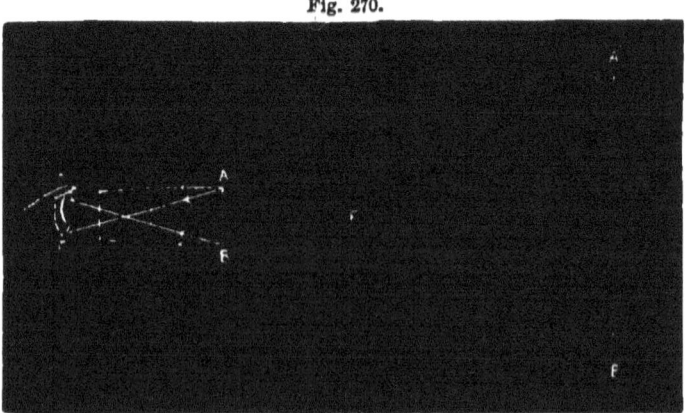

§ **332. Virtual images.** — Since rays that emanate from a point nearer the lens than the principal focus diverge after egress, it is evident that their focus must be virtual and on the same side of the lens as the object. Hence, *the image of an*

object placed nearer the lens than the principal focus is virtual, magnified, and erect, as shown in Figure 270. A convex lens used in this manner is called a *simple microscope.*

Since the effect of concave lenses is to scatter transmitted rays, pencils of rays emitted from A and B (Fig. 271), after

Fig. 271.

refraction, diverge as if they came from A' and B', and the image will appear to be at A'B'. Hence, *images formed by concave lenses are virtual, erect, and smaller than the object.*

§ **333. Spherical aberration.** — In all ordinary convex lenses the curved surfaces are spherical, and the angles which incident rays make with the little plane surfaces, of which we may imagine the spherical surface to be made up, increase

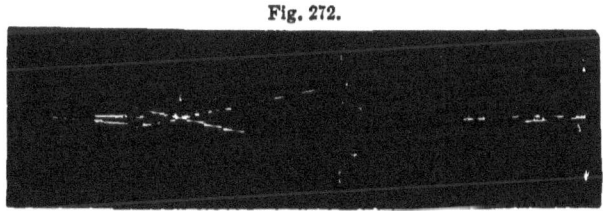

Fig. 272.

rapidly toward the edge of the lens. Hence, while those rays from a given point of an object, as A (Fig. 272), which pass through the central portion, meet approximately at the same point F, those which pass through the marginal portion are deviated so much that they cross the axis at nearer points, *e.g.*,

364   RADIANT ENERGY. — LIGHT.

at F'; so a blurred image results. This wandering of the rays from a single focus is called *spherical aberration*. The evil may be largely corrected by interposing a diaphragm D D' (Fig. 272), provided with a central aperture, smaller than the lens, so as to obstruct those rays that pass through the marginal part of the lens.

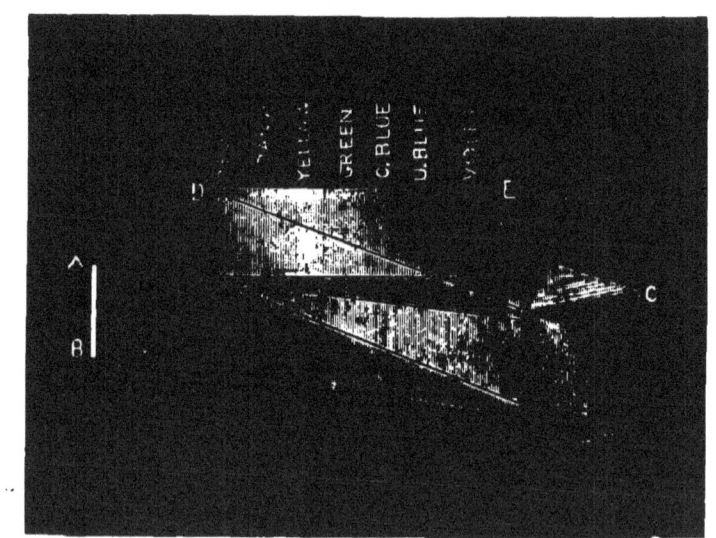

Fig. 273.

LVI. PRISMATIC ANALYSIS OF LIGHT. — SPECTRA.

§ 334. **Analysis of white light.** — **Experiment 1.** Paste tinfoil smoothly over one side of a glass plate about 5^{cm} square. In the center of the foil cut a slit 3^{cm} long by 1^{mm} wide, leaving smooth and parallel edges. Place the plate with the slit in the aperture of a *porte lumière* so as to exclude all light from a darkened room except that which passess through the slit. Near the slit interpose a double convex lens of (say) 10-inch focus. A narrow sheet of light will traverse the room and produce an image A B of the slit on a white screen placed in its path. Now place a glass prism C in the path of the beam with its axis (the straight line connecting the centers of the triangular faces) vertical. (1) The light now is not only turned from its former path, but that which before was a narrow sheet, is, after emerging from the prism, spread out fan-like into a wedge-shaped body, with its thickest part resting on the screen. (2) The image, before only a narrow vertical band, is now drawn out into a long horizontal ribbon of light D E. (3) The image, before white, now contains all the colors of the rainbow, from red at one end to violet at the other; it passes gradually through all the gradations of orange, yellow, green, blue, and violet. (The difference in deviation between the red and the violet is purposely much exaggerated in the figure.)

From this experiment we learn (1) that *white light is not simple in its composition, but the result of a mixture.* (2) *The colors of which white light is composed may be separated by refraction.* (3) *The cause of the separation is due to the different degrees of deviation which they undergo by refraction.* Red, which is always least turned aside from a straight path, is the least refrangible color. Then follow orange, yellow, green, blue, and violet in the order of their refrangibility. The many-colored ribbon of light D E is called the *solar spectrum.* This separation of white light into its constituents is called *dispersion.* The number of colors of which white light is composed is really infinite, but we have names for only seven of them; viz., *red, orange, yellow, green, cyan-blue,*[1] *ultramarine-blue, and violet;* and these are called the *primary* or *prismatic colors.* The names of the blues are derived from the names of the pigments which most closely resemble them. The rainbow is an illustration of a solar spectrum on a grand scale. It is the result of the dispersion of sunlight by rain drops.

The spectrum may be projected upon a screen, or it may be received directly by the eye, as in the two following experiments: —

Experiment 2. Upon a black card-board A (Fig. 274) paste a strip of white paper 3^{cm} long and 2^{mm} wide; and place the prism and the eye as in the figure. Now a beam of white light from the strip is

---

[1] See Rood's Modern Chromatics.

refracted and dispersed by the prism, and, falling upon the retina of the eye, you see, not the narrow white strip in its true position, but a spectrum in the position A'. This experiment is performed in a lighted room.

Experiment 3. Instead of a continuous white strip, paste short strips of red, white, and blue, end to end, on the black card, as represented in Figure 275. The spectrum of each color is given on the right, the light portions representing the illuminated parts.

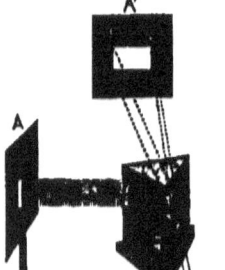

Fig. 274.

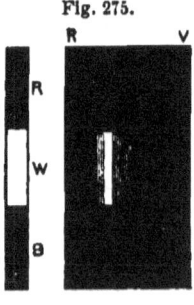

Fig. 275.

It will be seen that in the spectrum of the red, the green, blue, and violet portions are almost completely dark, but there is a faint trace of orange; in the spectrum of the blue, the red, orange, and yellow are wanting, blue and violet are present, and a small quantity of green. (What lessons does this experiment teach?)

Experiment 4. In place of the white strip of paper used in Exp. 2, admit light into a dark room through a narrow slit, and examine its spectrum.

§ **335. Synthesis of white light.** — The composition of white light has been ascertained by the process of analysis; can it be verified by *synthesis?* — *i.e.*, can the colors after dispersion be reunited? and, if so, will the result of the reunion be white light?

Experiment 1. Place a second prism (2) in such a position that light which has passed through one prism (1), and been refracted and decomposed, may be refracted back, and the colors will be reblended, and a white image of the slit will be restored on the screen.

Experiment 2. Place a large convex lens, or a concave mirror, so as to receive the colors after dispersion by a prism, and bring the rays to a focus on a screen. The image produced will be white.

Experiment 3. Receive the spectrum on a common plane mirror, and rapidly tip the mirror back and forth in small arcs at right angles to the path of the light, and the light reflected by the mirror upon a screen will produce a white image on the screen.

§ 336. **Cause of color and dispersion.** — *The color of light is determined solely by the number of waves emitted by a luminous body in a second of time, or by the corresponding wave-length.* In a dense medium, the short waves are more retarded than the longer ones; hence they are more refracted. This is the cause of dispersion. The ether waves diminish in length from the red to the violet. As pitch depends on the number of aerial waves which strike the ear in a second, so color depends on the number of ethereal waves which strike the eye in a second. From well-established data, determined by a variety of methods (see larger works), physicists have calculated the number of waves that succeed one another for each of the several prismatic colors, and the corresponding wave-lengths; the following table contains the results. The letters A, C, D, etc., refer to Fraunhofer's lines (see page 370).

		Length of waves in millimeters.	No. of waves per second.
Dark red	A	.000760	395,000,000,000,000
Orange	C	.000656	458,000,000,000,000
Yellow	D	.000589	510,000,000,000,000
Green	E	.000527	570,000,000,000,000
C. Blue	F	.000486	618,000,000,000,000
U. Blue	G	.000431	697,000,000,000,000
Violet	H	.000397	760,000,000,000,000

There is a limit to the sensibility of the eye as well as of the ear. The limit in the number of vibrations appreciable by the eye lies approximately within the range of numbers given in the above table; *i.e.*, if the succession of waves is much more or less rapid than indicated by these numbers, they do not produce the sensation of sight. It is evident that *the frequency of the waves emitted by a luminous body, and consequently the color of the light emitted, must depend on the rapidity of the vibratory motions of the molecules of that body*, i.e., *upon its temperature.* This has been shown in a convincing manner as follows: The temperature of a platinum wire is slowly raised by passing a gradually increasing current of electricity through it. At a temperature of about 540° C. it begins to emit light; and the light, analyzed by a prism, shows that it emits only red light. As the temperature rises, there will be added to the red of the spectrum, first yellow, then green, blue, and violet successively. When it reaches a white heat, it emits all the prismatic colors. It is significant that a white-hot body emits more red light than a red-hot body, and likewise more light of every color than at any lower temperature. The conclusion is, that *a body which emits white light sends forth simultaneously waves of a variety of lengths.*

§ 337. **Continuous spectra.** — The spectrum produced by the platinum is continuous; that is, the band of light is unbroken. If the spectrum is not complete, as when the temperature is too low, it will begin with red, and be continuous as far as it goes. *All luminous solids and liquids give continuous spectra.*

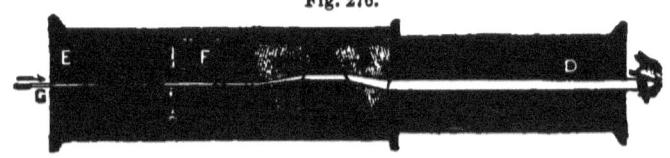

Fig. 276.

§ 338. **Spectroscope.** — A small instrument called a pocket spectroscope[1] will answer for all experiments given in this book. More elaborate experiments require more elaborate apparatus, a description of which must be sought for in larger works on this subject. This instrument contains three or more prisms, A, B, and C (Fig. 276). The prisms are enclosed in a brass tube D, and this tube in another tube E. F is a convex lens, and G is an adjustable slit. By moving the inner tube back and forth, the instrument may be so focused that parallel rays will fall upon prism A. By varying the kind of glass used in the different prisms,[2] as well as their structure, the deviation of light from a straight path, in passing through them, is overcome, while the dispersion is preserved. On account of the directness of the path of light through it, this instrument is called a *direct-vision spectroscope.*

[1] It is expected that the pupil will be provided with a pocket spectroscope, the cost of which need not exceed ten dollars.

[2] A and C are crown-glass, and B is flint-glass. See foot-note, p. 395.

§ 339. Bright line, absorption, or reversed spectra. —

Experiment 1. Open the slit a little less than 1^{mm} wide, and look through the spectroscope at the sky (not at the sun, for its light is too intense for the eye), and you will see a continuous spectrum.

Experiment 2. Repeat the last experiment with a candle, kerosene, or ordinary gas flame, and you will obtain similar results.

Experiment 3. Take a piece of platinum wire 10^{cm} long, seal one end of it by fusion to a short glass tube for a handle, and make a loop at the other end about 1^{mm} in diameter. Wet the loop in clean water, dip it into pulverized common salt, and introduce it into the almost invisible and colorless flame of a Bunsen burner. Instantly the flame becomes luminous and colored a deep yellow. Examine the light with a spectroscope, and you will find, instead of a continuous spectrum beginning with red, only a bright, narrow line of yellow in the yellow part of the spectrum, next the orange. Your spectrum consists essentially of a single bright yellow line on a comparatively dark ground (see Sodium, Fig. 277).

Experiment 4. Heat the platinum loop until it ceases to color the flame, then wet it and dip it into chloride of lithium, and repeat the last experiment. You obtain a carmine-tinted flame, and see through the spectroscope a bright red line and a faint orange line (see Lithium, Fig. 277).

Experiment 5. Use potassium hydrate, and you obtain a violet-colored flame, and a spectrum consisting of a red line and a violet line (the latter quickly disappears). Use strontium nitrate, and obtain a crimson flame, and a spectrum consisting of several lines in the red and the orange, and a blue line. (See Potass. and Stron., Fig. 277.)

Experiment 6. Use a mixture of several of the above chemicals, and you will obtain a spectrum containing all the lines that characterize the several substances.

Every chemical compound used in the above experiments contains a different metal, *e.g.*, common salt contains the metal sodium; the other substances used successively contain respectively the metals lithium, potassium, and strontium. These metals, when introduced into the flame, are vaporized, and we get their spectra when in a gaseous state. *All gases give discontinuous, or bright line, spectra, and no two gases give the same spectra.* The fact that in the second experiment we obtained continuous and similar spectra, appears to contradict the last two statements. But it should be remembered that all that gives light in those flames is small particles of *solid* carbon floating in the burning gas. We see, then, *that the spectroscope furnishes us with a reliable means of determining, at any time, whether light proceeds from a luminous solid or a luminous gas.*

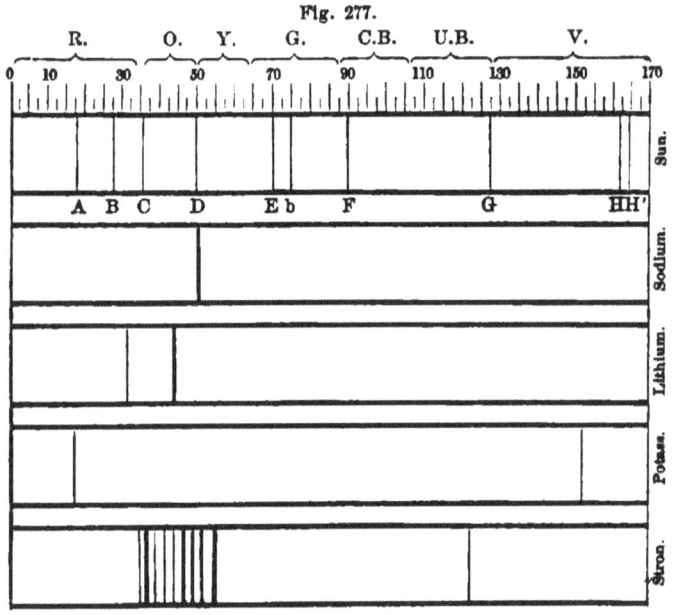

Fig. 277.

§ **340. Dark-line spectra.**—**Experiment 1.** Close the slit of the spectroscope so that the aperture will be very narrow; direct it once more to the sky, and slowly move the inner tube back and forth, and you will find, with a certain suitable adjustment which may be obtained by patient trial, that the solar spectrum is not in reality continuous, but is crossed by several *dark lines* (see Fig. 277).

Experiment 2. The electric light is now in so common use that it may be possible to perform this experiment. Between the electric light and the spectroscope introduce the flame of a Bunsen burner, and color it yellow with salt. Examine the electric light transmitted through this yellow flame.

In the last experiment you will naturally expect to find the yellow part of the spectrum uncommonly bright, for there would apparently be added to the yellow of the electric light the yellow of the salted flame. But precisely where you would look for the brightest yellow, there you discover that the spectrum is crossed by a dark line. If you use salts of lithium, potassium, and strontium in a similar manner, you will find in every case your spectrum crossed by dark lines where you would expect to find bright lines. Remove the Bunsen flame, and the dark lines disappear. It thus appears that *the vapors of different substances absorb or quench the very same rays that they are capable of emitting*; very much, it would seem, as a given tuning-fork selects from various sounds only those of a definite wave-length corresponding to its own vibration-period. The dark places of the spectrum receive light in full force from the salted flame; but the light is so feeble, in comparison with those places illuminated by the electric light, that the former appear dark by contrast. Light transmitted through certain liquids (as sulphate of quinine and blood) and certain solids (as some colored glasses) produces *dark-line* spectra. These spectra are obtained only when light passes through media capable of absorbing rays of certain wave-length; hence, they are commonly called *absorption spectra*. Since a given vapor causes dark lines precisely where, if it were itself the only radiator of light, it would cause bright lines, dark-line spectra are frequently called *reversed spectra*. There are then three kinds of spectra: *continuous spectra*, produced by luminous solids, liquids, or, as has been found in a few instances, gases under great pressure; *bright-line spectra*, produced by luminous vapors; and *absorption spectra*, produced by light that has been sifted by certain media.

§ 341. **Spectrum analysis.** — More elaborate spectroscopes contain many prisms, by which we greatly increase the *purity* of the spectrum. (By purity is meant a freedom from the overlapping of images of the slit, by which many lines of the spectra are concealed.) They also contain an illuminated scale which may be seen adjacent to the spectrum, by which the exact position of the lines, and their relative distances from one another, can be accurately determined, and a telescope by which the spectrum and scale may be magnified. The positions of some of the prominent lines of the solar spectrum were first determined, mapped, and distinguished from one another by certain letters of the alphabet by Fraunhofer; hence, the dark lines of the solar spectrum are commonly called *Fraunhofer's lines.* So far as discovered, no two substances have a spectrum consisting of the same combination of lines; and, in general, different substances but very rarely possess lines appearing to be common to both. Hence, when we have once observed and mapped the spectrum of any substance, we may ever after be able to recognize the presence of that substance, when emitting light, whether it is in our laboratory or in a distant heavenly body. The spectroscope, therefore, furnishes us a most efficient means of detecting the presence (or absence) of any elementary substance, even when it is combined or mixed with other substances. It is not necessary that the given substance should exist in large quantities; for example, the fourteen-millionth part of a milligram of sodium can be detected by the spectroscope. Substances that are not easily converted into vapors at low temperatures may be placed between the poles of an electric battery or an induction coil. The heat generated by electricity will vaporize all substances. After maps of the spectra of all known substances have been made out, if, on examination of a complex substance, any new lines should at any time appear in the spectrum, it would indicate the presence of a substance hitherto undiscovered. It was thus that the elements, caesium, rubidium, thallium, and indium were discovered.

§ 342. **Celestial chemistry and physics.** — The spectrum of iron has been mapped to the extent of 460 bright lines. The solar spectrum furnishes dark lines corresponding to nearly all these bright lines. Can there be any doubt of the existence of iron in the sun? By examination of the reversed spectrum of the sun, we are able to determine with certainty the existence there of sodium, calcium, copper, zinc, magnesium, hydrogen, and many other known substances. Again, from our knowledge of the way in which a reversed spectrum can be produced, we may conclude that the sun consists of a luminous solid, liquid, or an intensely heated and greatly condensed gas (called a *photosphere*), and that this nucleus is surrounded by an atmosphere of cooler vapor, in which exist at least all the substances just named. The moon and other heavenly bodies that are visible only by reflected sun-light give the same spectra as the sun, while those that are self-luminous give spectra which differ from the solar spectrum.

§ 343. **Heat and chemical spectra.**—If a sensitive thermometer is placed in different parts of the solar spectrum, it will indicate heat in all parts; but the heat generally increases from the violet toward the red. It does not cease, however, with the limit of the visible spectrum; indeed, if the prism is made of flint glass, the greatest heat is just beyond the red. A strip of paper wet with a solution of chloride of silver suffers no change in the dark; in the light it quickly turns black; exposed to the light of the solar spectrum, it turns dark, but quite unevenly. The change is slowest in the red, and constantly increases, till about the region indicated by G (Fig. 277), when it attains its maximum; from this point it falls off, and ceases at a point considerably beyond the limit of the violet. It thus appears that the solar spectrum is not limited to the visible spectrum, but extends beyond at each extremity. Those rays that lie beyond the red are usually called the *ultra-red* rays, while those that lie beyond the violet are called the *ultra-violet* rays. The ultra-red rays are of longer vibration-period, and the ultra-violet of shorter period, than the luminous rays.

§ 344. **Only one kind of radiation.**—The fact that radiant energy produces three distinct effects,—viz., luminous, heating, and chemical,—has given rise to a quite prevalent idea that there are three distinct kinds of radiation. There is, however, absolutely no proof that these different effects are produced by different kinds of radiation. *The same radiation that produces vision can generate heat and chemical action.* The fact that the ultra-red and ultra-violet rays do not affect the eye does not argue that they are of a different nature from those that do, but it does show that there is a limit to the susceptibility of the eye to receive impressions from radiation. Just as there are sound-waves of too long, and others of too short, period to affect the ear, so there are etherial waves, some of too long, and others of too short, period to affect the eye. It is true, however, that waves of long period from the sun are more energetic in producing heating effects than those of short period; and those of short period are more effective in generating chemical action in certain substances than those of long period; while only those which lie between the extremes affect the eye.

LVII. COLOR.

§ 345. Color produced by absorption. — "All objects are black in the dark;" this is equivalent to saying that *without light there is no color.* Is color a quality of an object, or is it a quality of the light which illuminates the object?

Experiment 1. We have found that common salt introduced into a Bunsen flame renders it luminous, and that the light when analyzed with a prism is found to contain only yellow. Expose papers or fabrics of various colors to this light in a darkened room. *No one of them exhibits its natural color except yellow.*

Experiment 2. Hold a narrow strip of red paper or ribbon in the red portion of the solar spectrum; it appears red. Slowly move it toward the other end of the spectrum; on leaving the red it becomes darker, and when it reaches the green it is quite black or colorless, and remains so as it passes the other colors of the spectrum. Repeat the experiment, using other colors, and notice that only in light of its own color does each strip of paper appear of its natural color; while in all other colors it is dark.

These experiments show that (1) *color is a quality of the light which illuminates, and not of the object illuminated;* (2) *in order that an object may appear of a certain color, it must receive light of that color; and of course if it receives other colors at the same time, it must be capable of absorbing them.* The energy of the waves absorbed is converted into heat, and warms the object. When white light strikes an object, it appears white if it reflects all the colors. If red light falls upon the same object, it appears red, for it is capable of reflecting red; or it appears green, if green light alone falls on it. If white light falls upon an object, and all the colors are absorbed except the blue, the object appears blue. When we paint our houses we do not apply color to them. We apply substances called *pigments*, that have a property of absorbing all the colors except those which we would have our houses appear.

Experiment 3. By means of a *porte lumière* introduce a beam of light into a dark room. Cover the orifice with a deep red (copper) glass. The white light, in passing through the glass, appears to be colored red. *Does the glass color the light red ?*

Experiment 4. With the slit and prism form a solar spectrum, and between the prism and screen interpose the red glass. All the colors of the spectrum instantly disappear except the red.

It thus appears that a red transparent body transmits only red, and absorbs all other colors. No body gives color to light that it reflects or transmits.

§ **346. Sky colors.** — **Experiment 1.** Dissolve a little white castile soap in a tumbler of water; or, better, stir into the water a few drops of an alcoholic solution of mastic, enough to render the water slightly turbid. Place a black screen behind the tumbler, and examine the liquid by reflected sunlight, — the liquid appears to be blue; examine the liquid by transmitted sunshine, — it now appears yellowish red.

Skylight is reflected light. The particles of atmospheric dust (of water, probably) that pervade the atmosphere, like the fine particles of mastic suspended in the water, reflect blue light; while, beyond the atmosphere, is a black background of darkness.

But we must not, from this, conclude that the atmosphere is blue; for, unlike blue glass, but like the turbid liquid, it transmits yellow and red rays freely, so that, seen by reflected light it is blue, but seen by transmitted light it is yellowish red.

Experiment 2. Pour some of the turbid liquid into a small test-tube, and examine it and the tumbler of liquid by transmitted light; the former appears almost colorless, while the latter is quite deeply colored.

When the sun is near the horizon, its rays travel a greater distance in the air to reach the earth than when it is in the zenith (see Fig. 261, p. 356); consequently, there is a greater loss by absorption and reflection in the former case than in the latter. But the yellow and red rays suffer less destruction, proportionally, than the other colors; consequently, these colors predominate in the morning and evening.

§ **347. Mixing colors.**—A mixture of all the prismatic colors, in the proportion found in sunlight, produces white. Can white be produced in any other way?

Experiment 1. On a black surface A (Fig. 278), about 4^{cm} apart, lay two small rectangular pieces of paper, one yellow and the other blue. In a vertical position between, and from 4^{cm} to 8^{cm} above these papers, hold a slip of plate glass C. Looking obliquely down through the glass you may see the blue paper by transmitted light and the yellow paper by reflection. That is, you see the object itself in the former case and the image of the object in the latter case. By a little manipulation, the image and the object may be made to overlap one another, when both colors will apparently disappear, and in their place the color which is the result of the mixture will appear. In this case it will be white, or, rather, gray, which is white of a low degree of luminosity. If the color is yellowish, lower the glass; if bluish, raise it.

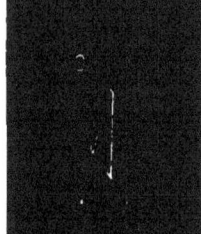

Fig. 278.

Experiment 2. Cut out of stiff drawing-paper two circular disks, each 16^{cm} in diameter. Paint one with chrome yellow, and the other with ultramarine blue. Cut a radial slit in each, and pass an edge of

MIXING COLORS. 377

one slit through the slit of the other, and so arrange them that one shall partly conceal the other, leaving rather more blue exposed than of the yellow, as in Figure 279. Attach the disks so combined to some apparatus by which they may be rapidly rotated; for example, to a "color top," such as are sold at toy stores. Rotate the disks, and the colors will be so blended in the eye as to appear gray; or, if either color predominates, arrange the disks so that less of that color will be exposed. Figure 280 represents "Newton's disk," which contains the seven prismatic colors arranged in a proper proportion to produce gray when rotated.

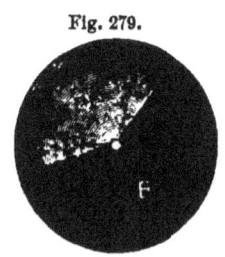

Fig. 279.   Fig. 280.   Fig. 281.

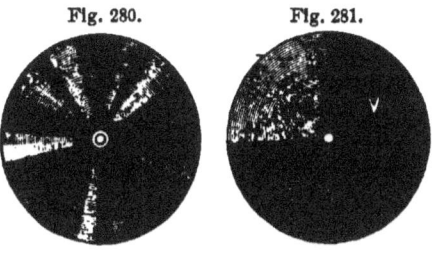

In a like manner, you may produce white by mixing purple and green; or, if any color on the circumference of the circle (Fig. 282) is mixed with the color exactly opposite, the resulting color will be white. Again, the three colors, red, green, and violet, arranged as in Figure 281, with rather less surface of the green exposed than of the other colors, will give gray. Green mixed with red, in varying proportions, will produce any of the colors between these two colors in the diagram (Fig. 282); green mixed with violet will produce any of

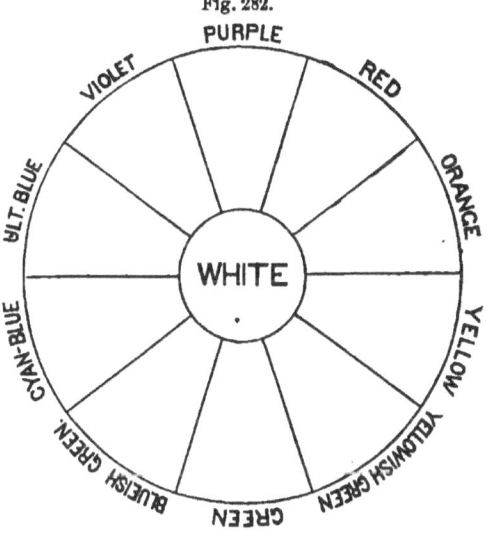

Fig. 282.

the colors between them; and violet mixed with red gives purple; but no two colors mixed will produce any of these three colors. Hence, a very widely accepted theory is adopted by many, that *red, green, and violet* are the three *primary color sensations*, and that the other colors of the spectrum are simply the products of mixtures, in varying proportions, of these three.

§ 348. Mixing pigments. — Experiment 1. Mix a little of the two pigments, chrome yellow and ultramarine blue, and you obtain a green pigment.

The last three experiments show that mixing certain colors, and mixing pigments of the same name, may produce very different results. In the first experiments you actually mixed colors; in the last experiment you did not mix colors, and we must seek an explanation of the result obtained. If a glass vessel with parallel sides containing a blue solution of sulphate of copper is interposed in the path of light which forms a solar spectrum, it will be found that the red, orange, and yellow rays are cut out of the spectrum, *i.e.*, the liquid absorbs these rays. And if a yellow solution of *bichromate* of potash is interposed, the blue and violet rays will be absorbed. It is evident that, if both solutions are interposed, all the colors will be destroyed except the green, which alone will be transmitted; thus: —

Cancelled by the blue solution, R̸ Ø̸ Y̸ G B V.
Cancelled by the yellow solution, R O Y G B̸ V̸.
Cancelled by both solutions, R̸ Ø̸ Y̸ G B̸ V̸.

In a similar manner, when white light strikes a mixture of yellow and blue pigments on the palette, it penetrates to some depth into the mixture; and, during its passage in and out, all the colors are destroyed except the green; so the mixed pigments necessarily appear green. But, when a mixture of yellow and blue lights enters the eye, we get, as the result of the *combined* sensations produced by the two colors, the sensation of white; hence. a mixture of yellow and blue gives white.

§ 349. **Complementary colors.** — **Experiment.** On a piece of white, or better, gray paper, lay a circular piece of blue paper 15^{mm} in diameter. Attach one end of a piece of thread to the colored paper, and hold the other end in the hand. Place the eyes within about 15^{cm} of the colored paper, and look steadily at the center of the paper for about fifteen seconds; then, without moving the eyes, suddenly pull the colored paper away, and instantly there will appear on the gray paper an image of the colored paper, — but the image will appear to be yellow. This is usually called an *after-image*. If yellow paper is used, the color of the after-image will be blue; and if any other color given in the diagram, Figure 282, the color of its after-image will be the color that stands opposite to it.

This phenomenon is explained as follows: When we look steadily at blue for a time, the eyes become fatigued by this color, and less susceptible to its influence, while they are fully susceptible to the influence of other colors; so that when they are suddenly brought to look at white, which is a compound of yellow and blue, they receive a vivid impression from the former, and a feeble impression from the latter; hence, the predominant sensation is yellow. Any two colors which together produce white are said to be *complementary* to each other. The opposite colors in the diagram, Figure 282, are complementary to one another.

§ 350. **Effect of contrast.** — When any two colors given in the circle, Figure 282, are brought in contrast, as when they are placed next one another, the effect is to move them farther apart. For example, if red and orange are brought in contrast, the orange assumes more of a yellowish hue, and the red more of a purplish hue. Colors that are already as far apart as possible, *e.g.*, yellow and blue, do not change their hue, but merely cause one another to appear more brilliant.

§ 351. **Color produced by interference.** — **Experiment 1.** In a vise or other convenient instrument, press two clean pieces of thick plate glass firmly together. A number of colors will be seen arranged in a certain order, and forming curves more or less regular around the point of pressure.

380   RADIANT ENERGY. — LIGHT.

Experiment 2. Paint one side of a piece of window glass with India ink so as to render it quite opaque; then, when dry, with the point of a needle rule fifteen to twenty parallel lines in the ink, about 2^{mm} apart, cutting quite through the ink, so that light may pass through the scratches. Now stand at a distance of ten feet or more from a kerosine or gas flame, and look through the glass with one eye at the flame, edge on; move the glass to and from the eye slowly, so as to properly focus it, and you will see many spectra of the flame on each side of it, separated by dark intervals.

Experiment 3. Place the ruled glass in the path of a beam of light thrown into a dark room by a *porte lumière*, and project an image of the glass on a screen by means of a convex lens of two to five inches focal length, and you will obtain a series of beautiful spectra.

Fig. 283.

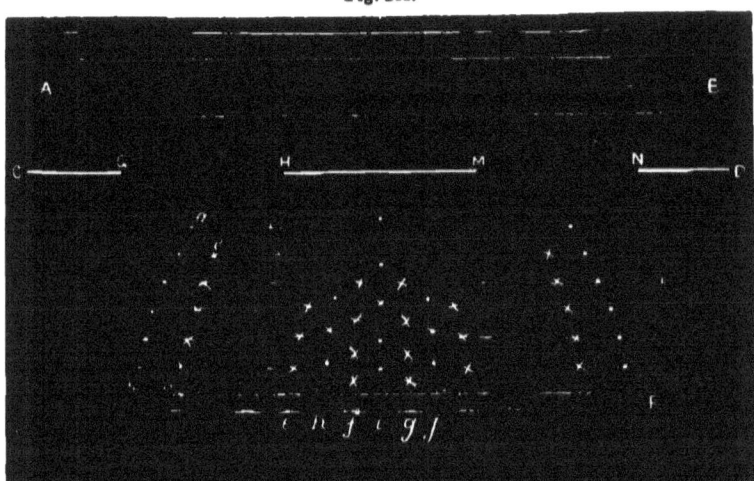

If in the path of the beam a red glass is interposed, a large number of alternating red and dark lines may be obtained, though the experiment is a difficult one. Let us study the last result. Let the series of parallel lines A B (Fig. 283) represent the series of waves constituting the beam of light before it strikes the ruled glass C D; and E F, the portions of the same waves that succeed in passing through the scratches, G H and M N. The wave-lengths and the width of the scratches, etc., are immensely exaggerated in the diagram. Now, if you watch waves of water as they beat against an obstacle rising above its surface, you will see that part of their energy is expended in forming a new set of waves, which we will call *secondary waves*, radiating from the obstacle and winding around behind it. In a similar manner, secondary waves of light are generated at the edges of obstacles against which light grazes. This apparent bending of the waves of light round the edges of opaque bodies receives the name of *diffraction*. Sections of such waves are represented in the diagram as crossing the original or primary waves at certain points, and also one another, behind the obstacle M. The continuous lines represent one phase of the waves, and the dotted lines the opposite phase, as crest and trough. Now, it will be seen that at certain points (denoted by heavy dots) which lie in the same line as ab, the primary and secondary waves meet in similar phases; and the consequence is, that the point b of the screen OP is illuminated by the combined action of the two sets of waves. But at other points (denoted by small crosses), as cd, the opposite phases of the two sets of waves coincide with one another, and the result is that they tend to neutralize one another; and consequently the point d of the screen is deprived of light, and a dark line occurs at this place. In a similar manner the points h, i, j, etc., are illuminated by the joint action of the two sets of waves, while the points e, f, g, etc., are deprived of light by their mutual destruction. Such will be the result when monochromatic light, or light of one wave-length is used, as is approximately the case when we interpose the red glass. But if white light, or light of various wave-lengths is used, it will happen that those places which are deprived of red light will receive light of other colors; hence the color effects produced when white light passes through the ruled glass. Of course waves are generated at the points G and N, as well as at the points H and M, but they are omitted for the sake of simplicity. This figure illustrates only in a very incomplete way the complex phenomena.

Such experiments as the above furnish a very strong argument for the wave theory of light, since two lights produce darkness apparently in a manner analogous to that in which two sounds produce silence.

Thin, transparent films of varying thickness, such as the film of a soap bubble, are well suited to show the effects of interference of light. Some of the light which strikes the anterior surface of the film is reflected; another portion enters the film, and is reflected from the posterior surface; but, by travelling twice through the film, the wave loses ground, so that, on emergence, its phases may or may not correspond with the phases of the former portion: this will depend evidently upon the thickness of the film at a given point, and the length of the waves striking that point. In this manner the phenomena obtained in the first experiment are explained; the film in this case is the layer of air between the two surfaces of glass.

Colors are produced by reflection from the surfaces of thin transparent films of all kinds; for example, the colors of the soap bubble, of oil on water, of the thin coating of metallic oxide formed in tempering steel, the changeable colors of the peacock's feathers and of certain insects' wings, the colors seen in cracks in glass and ice, are all colors of thin films. The halos seen around the moon or a street lamp on a misty evening, and the rainbow tints seen bordering the eyelashes when, with eyes partially closed, you look at a strong light, are examples of colors produced by diffraction.

Waves of light which emanate from the points H and M, Figure 283, travel equal distances to reach the point i on the screen; but to reach the point g, the waves from H must travel just one-half of a wave-length farther than the waves from M; and to reach the point j, they must travel just one wave-length farther. Hence, if we can ascertain the difference between the two distances, Hg and Mg, we obtain the wave-length for that color. In this manner the wave-lengths given in § 336 were ascertained.

LVIII. DOUBLE REFRACTION AND POLARIZATION OF LIGHT.

§ 352. Double refraction.—Experiment. Through a card make a pin-hole, and hold the card so that you may see skylight through the hole. Now bring a crystal of Iceland spar, Figure 284, between the eye and the card, and look at the hole through two parallel surfaces of the crystal. There will appear to be two holes, with light shining through each. Cause the crystal to rotate in a plane parallel with the card, and one of the holes

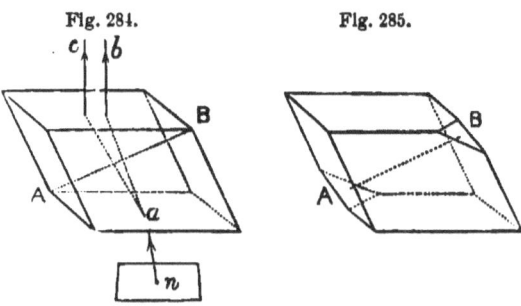

Fig. 284. Fig. 285.

will appear to remain nearly at rest, while the other rotates around the first. A ray of light na immediately on entering the crystal is divided into two parts, one of which obeys the regular law of refraction; the other does not. The former is called the *ordinary ray;* the latter, the *extraordinary ray.* The rays issue from the crystal parallel with one another. In all crystals which produce this phenomenon there is one direction, and in some two directions, in which, if an object is looked at through the crystal, it does not appear double. If all the edges of a crystal of Iceland spar are equal, and it is cut by two planes near each extremity of the line AB, which connects the two opposite solid obtuse angles, and at right angles to it, as shown in Figure 285, objects viewed in this line, or in any line parallel with it, do not appear double.

In every direction in which one looks through the crystal, except parallel to AB, objects seen through it appear double. (See Fig. 286.) The line AB is called the *optic axis* of the crystal, and is a line around which the molecules of the crystal appear to be arranged symmetrically. A crystal is called *uniaxial* when it has only one optic axis, and *biaxial* when it has two such axes. *Crystals* of many other substances possess the property of causing objects

Fig. 286.

seen through them to appear double. This phenomenon is called *double refraction*.

§ 353. **Polarization.** — Slices of crystals of the mineral tourmaline, cut in planes parallel with their axes, are prepared and sold for optical experiments. If two of these slices similarly situated, as in Figure 287, are placed between the eye and a card pierced by a hole, the hole will be plainly visible. But if one of the slices is slowly rotated in a plane at right angles with the beam of light, the hole will grow dimmer until the slice has passed through a quarter of a revolution (as represented in Figure 288), when it disappears. If the rotation is continued, the hole reappears, faint at first, but at the end of another quarter-revolution it reaches its maximum brightness. Thus, at each quarter-revolution it is alternately extinguished and restored.

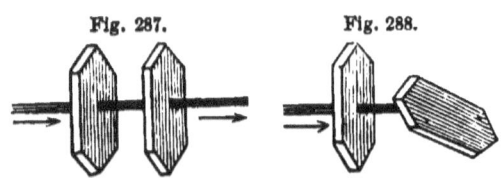

Fig. 287. Fig. 288.

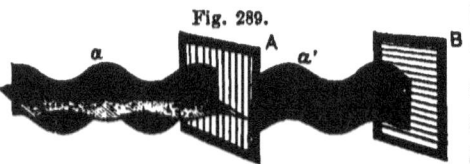

Fig. 289.

It appears, then, that light which has passed through one transparent slice of tourmaline differs so much from common light, that a second similar slice may act like an opaque body, and stop it altogether. The action of the tourmaline may be compared to that of a grating (A, Fig. 289) formed of parallel vertical rods, which will allow all vertical planes (as aa') to pass, but stops the planes (as cc') that are at right angles to these rods. Any plane that has succeeded in passing one grating will readily pass a second similarly placed. But if the second grating B is turned so that its rods are at right angles to the first, the plane that has succeeded in getting through the first grating will be stopped by the second. Light, in this condition, is said to be *polarized; polarization* is either the act of producing the change in the light, or the result of the change, and the instrument used is a *polarizer*.

In order to understand this phenomenon, it is necessary to know more of the undulatory theory of light. This theory supposes that the undulations in ether which constitute

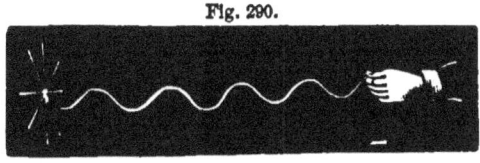

Fig. 290.

light are much like undulations in a cord when one end is shaken by a hand, as seen in Figure 290. If the hand moves vertically, all the undulations will lie in a vertical plane; if the movements of the hand are horizontal or oblique, the undulations lie in corresponding planes. So we can produce these waves on the rope in any plane passing through the rope, and can change rapidly from one plane to another. These waves appear differently when viewed from different sides. If we could look endwise at a ray of light for an instant, it is believed that we should see the ether particles vibrating, as in the figure of the rope, in

Fig. 291.

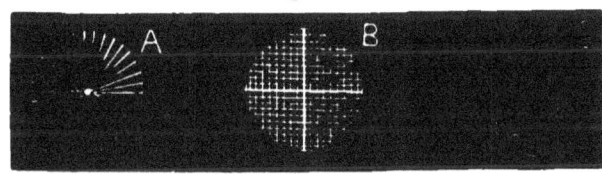

one plane; but in only a thousandth of a second so many million waves reach the eye, that there is time for the vibrating particles, which, like the hand, start the waves, to vibrate in many planes. In an ordinary beam of light, as it reaches the eye, there are therefore undulations in all possible planes, as is partially represented by the cross section A, Figure 291. But any motion may be considered as the effect of two forces that would produce motions in directions at right angles to one another. So here, for many practical purposes, the vibrations may be regarded as taking place in only two sets of planes at right angles to one another, as represented by B of the same figure. Now, when a ray of light, consisting, according to supposition, of undulations in planes at right angles to one another, strikes a slice of tourmaline, its molecular structure allows those undulations which are in planes parallel with its axis to pass through, but it absorbs those undulations that are in planes at right angles to its axis. By this means the undulations are reduced to those in parallel planes only, as represented in C. The unaided eye cannot usually detect any difference between common and polarized light. An instrument which will enable the eye to detect polarized light is called an *analyzer;* thus the first slice of tourmaline serves as a polarizer, and the second slice as an analyzer. A complete polarizing apparatus, called a *polariscope,* used for observing the phenomena of polarized light, consists essentially of a polarizer and an analyzer.

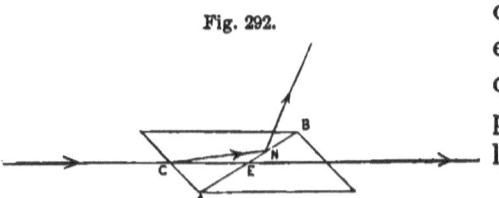

Fig. 292.

The favorite analyzer is the Nicol prism, which consists of a crystal of Iceland spar divided diagonally, as A B, Figure 292, and the two surfaces cemented together again with Canada balsam. The extraordinary ray C E, falling upon the transparent balsam, passes through it; but the ordinary ray C N strikes the balsam at a greater than its critical angle, and is therefore reflected out of the crystal, and thus got rid of. Now, when polarized light enters this prism in one position, it will pass freely through it, but if the prism is turned 90°, none will pass through. In the example given above, light is polarized by *absorption.*

§ **354. Polarization by double refraction and by reflection.** — If light which has undergone double refraction, as in passing through a crystal of Iceland spar, is examined with an analyzer, it is found that both the ordinary and the extraordinary rays are completely polarized in planes at right angles to each other. Again, light reflected obliquely from smooth surfaces, such as water, glass, and polished furniture, etc., is found on examination to be partially polarized. There is a definite angle of incidence at which the maximum polarizing effects are produced. This angle varies with different substances. With glass it is 55°; with water, 53°.

§ 355. **Description of a simple polariscope.** — D (Fig. 293) is a plate of glass, about 15^{cm} square, used as a polarizer. A is the analyzer, — preferably a Nicol prism, — so placed as to view the center of the glass at the proper polarizing angle (about 55°). The prism may be mounted in a cork, and the whole should be free to rotate in its support. S is a piece of ground glass used to cut off the images of outside objects. G is a glass shelf, on which objects to be examined are placed. The instrument, covered with a black cloth, is placed on a table with S toward a window. The prism can be obtained of any dealer in optical apparatus.

Fig. 293.

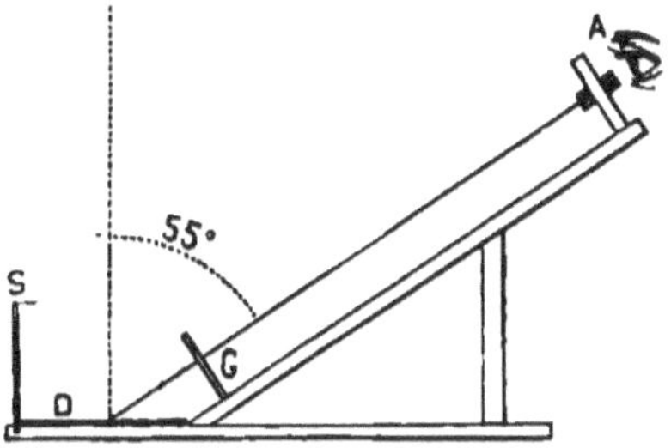

§ 356. **Colors by polarization.** — **Experiment.** Place on the support G a thin film of selenite or mica, and slowly rotate the analyzer. A beautiful display of colors will appear. At a certain point they will appear of maximum brilliancy, then they will gradually fade away and change into their complementaries.

This is really a phenomenon of interference, brought about through the combined agency of the object examined and the polariscope. If a piece of plate glass, subjected to pressure by means of a screw-clamp, or a piece of unannealed or poorly annealed glass, — a glass stopper, for example, — is examined, it will exhibit analogous phenomena.

LIX. THERMAL EFFECTS OF RADIATION.

§ 357. Diathermancy and athermancy. — What becomes of radiations that strike a body depends largely upon the character of the body. If the nature of the body is such that its molecules can accept the motion of the ether, the undulations of ether are said to be absorbed by the body, and the body is thereby heated; that is, the undulations of ether are transformed into molecular motion or *heat*. A good illustration of this is the experiment with blackened glass, page 325. On the other hand, the unblackened glass allows the radiations to pass freely through it, and very little is transformed into heat. Notice how cold window-glass may remain, while radiations pour through it and heat objects within the room. It must be constantly borne in mind, that *only those radiations that a body absorbs heat it; those that pass through it do not affect its temperature.* Bodies that transmit radiant heat freely are said to be *diathermanous*, while those that absorb it largely are called *athermanous*. The most diathermanous solid is rock salt. Among the most athermanous solids are lamp-black and alum. Carbon bisulphide, among liquids, is exceptionally transparent to all forms of radiation; while water, transparent to short waves, absorbs the longer waves, and is thus quite athermanous.

Experiment 1. Bring the bulb of an air thermometer into the focus of a burning-glass exposed to the sun's rays. The radiation concentrated on the enclosed air scarcely affects this delicate instrument.

Experiment 2. Cover the outside of the bulb of the air thermometer with lamp-black and repeat the last experiment. The lamp-black absorbs the radiant heat, and the heat conducted through the glass to the enclosed air raises its temperature and causes it to expand and rapidly push the liquid out of the stem.

Dry air is almost perfectly diathermanous. All of the sun's radiations that reach the earth pass through a layer of air, from fifty to two hundred miles in depth, which contains a vast amount of aqueous vapor. This vapor, like water, is comparatively opaque to long waves; hence it modifies very much the character of the radiations which reach the earth. This fact, together with what we have learned from Exp. 2, enable us to understand the method by which our atmosphere becomes heated. First, a very considerable portion of the radiant energy which comes to us from the sun, in the form of relatively long waves, is stopped by the watery vapor in the air, which is, in consequence, heated. Most of that which escapes this absorption heats the earth by falling upon it. The warmed earth loses its heat, — partly by conduction to the air, still more largely by radiation outward. The form of radiation, however, has been greatly changed; for now, coming from a body at a low temperature, it is chiefly in long waves that the energy is transmitted; while, as we have seen, it was largely in the form of short waves that the earth received its heat. But it is exactly these long waves which are most readily stopped by the atmosphere; hence, the atmosphere, or rather the aqueous vapor of the atmosphere, acts as a sort of trap for the energy which comes to us from the sun. Remove the watery vapor (which serves as a "blanket" to the earth) from our atmosphere, and the chill resulting from the rapid escape of heat by radiation would put an end to all animal and vegetable life. Glass does not screen us from the sun's heat, but it can very effectually screen us from the heat radiated from a stove or any other terrestrial object. Glass is diathermanous to the sun's radiations (simply because they have already lost most of the very long waves by atmospheric absorption), but quite athermanous to other radiations. This is well illustrated in the case of hot-beds and green-houses. The sun's heat passes through the glass of these enclosures, almost unobstructed, and heats the earth; but the radiations given out in turn by the earth are such as cannot pass out through the glass, hence the heat is retained within the enclosures.

RADIANT ENERGY. — LIGHT.

§ 358. All bodies radiate heat. — Hot bodies *usually* part with their heat much more rapidly by radiation than by all other processes combined. But cold bodies, like ice, radiate heat even when surrounded by warm bodies. This must be so from the nature of the case, for the molecules of the coldest bodies possess some motion, and being surrounded by ether, they cannot move without imparting some of their motion to the ether, and to that extent become themselves colder.

§ 359. Theory of Exchanges. — Let us suppose that we have two bodies, A and B, at different temperatures, — A warmer than B. Radiation takes place not only from A to B, but from B to A; but, in consequence of A's excess of temperature, more heat passes from A to B than from B to A, and this continues until both bodies acquire the same temperature. At this point radiation by no means ceases, but each now gives as much as it receives, and thus equilibrium is kept up. This is known as the "Theory of Exchanges."

§ 360. Good absorbers, good radiators. — **Experiment.** Select two small tin boxes of equal capacity, — one should be bright outside, while the other should be covered thinly with soot from a candle flame. Cut a hole in the cover of each box large enough to admit the bulb of a thermometer. Fill both boxes with hot water, and introduce into each a thermometer. They will register the same temperature at first. Set both in a cool place, and in half an hour you will find that the thermometer in the blackened box registers several degrees lower than the other. Then fill both with cold water, and set them in front of a fire or in the sunshine, and it will be found that the temperature in the blackened box rises fastest.

As bodies differ widely in their absorbing power, so they do in their radiating power, and it is found to be universally true that *good absorbers are good radiators, and bad absorbers are bad radiators*. Much, in both cases, depends upon the character of the surface as well as the substance. Bright, polished surfaces are poor absorbers and poor radiators; while tarnished, dark, and roughened surfaces absorb and radiate heat rapidly.

COMPOUND MICROSCOPE. 391

Dark clothing absorbs and radiates heat more rapidly than light. (Which is better to wear at all seasons? Why? Why are certain parts of steam engines kept scrupulously bright?)

§ 361. Dew. — It requires no elaborate experiments to show that some bodies radiate heat more rapidly than others. All nature testifies to this every still, cloudless summer night. During the day objects on the earth's surface receive more heat by radiation than they lose, but as soon as the sun has set this is reversed. Then everything begins to cool as its heat is radiated into space. Objects becoming cool, the air in contact with them becomes chilled; its watery vapor condenses, and collects in tiny liquid drops on their surfaces. But these dew-drops collect much more abundantly on certain things, such as grasses and leaves, than on others, such as stones and earth. The reason that it does not collect on the latter so freely, is because of their poor radiating power; they do not get cool as rapidly.

Fig. 294.

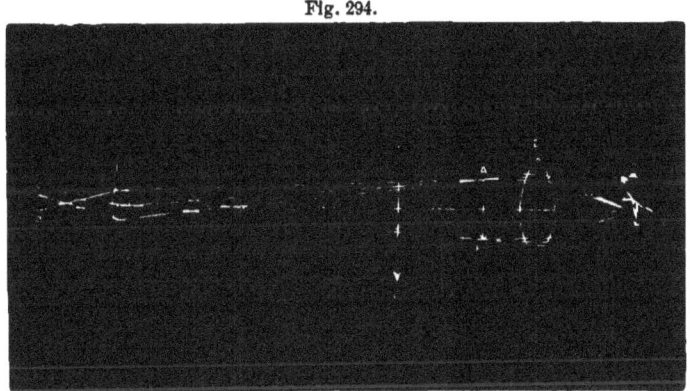

LX. SOME OPTICAL INSTRUMENTS.

§ 362. Compound microscope. — The simple microscope was described on page 362. When it is desired to magnify an object more than can be done conveniently and with distinctness by a single lens, two convex lenses are used, — one (O, Fig. 294) called the *object-glass*, to form a magnified real image

A'B' of the object AB; and the other (E) called the *eye-glass*, to magnify this image so that the image A'B' appears of the size A"B". In the same sense as we look at the object with one lens when we use a simple microscope, here we look at A'B'.

§ 363. **Astronomical telescope.** — The astronomical refracting telescope consists essentially, like the compound microscope, of two lenses. The object-glass (O, Fig. 295) forms a real diminished image ab of the object AB; this image, seen through the eye-glass E, appears magnified and of the size cd. The object-glass is of large diameter, in order to collect as much light as possible from a distant object for a better illumination of the image. Some idea of the power of some of our

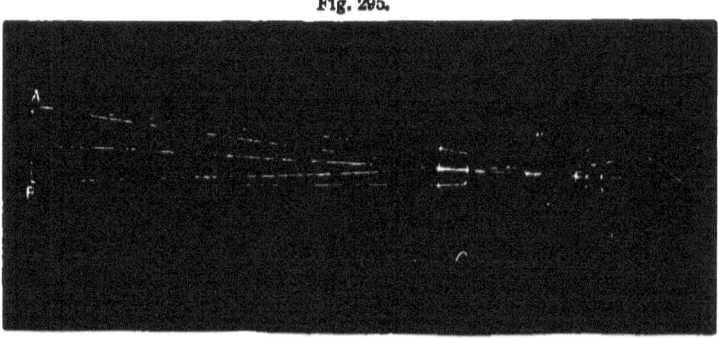

Fig. 295.

best telescopes may be obtained from the fact that Mr. Clark of Cambridgeport has made a telescope of such magnifying power, and possessing such distinctness of definition, that a ball two inches in diameter, and two hundred and fifty miles distant (about the distance between Boston and New York), would be distinctly visible as a body of perceptible dimensions, if properly illuminated.

§ 364. **Photographer's camera.** — The *photographer's camera*, or *camera obscura*, of which AB, Figure 296, represents a vertical section, consists of a dark box painted black on the interior. A screen of ground glass S forms a partition in the box. A sliding tube T contains a convex lens L. If an object D is placed some distance in front, and the distance of the lens from the screen is suitably adjusted, a distinct, real, and inverted image can be seen upon the screen by looking through the aperture C. When the image is properly focused, the photographer replaces the ground glass plate by a sensitized plate, and the chemical power of the sun's rays paints a true picture of the object on this plate.

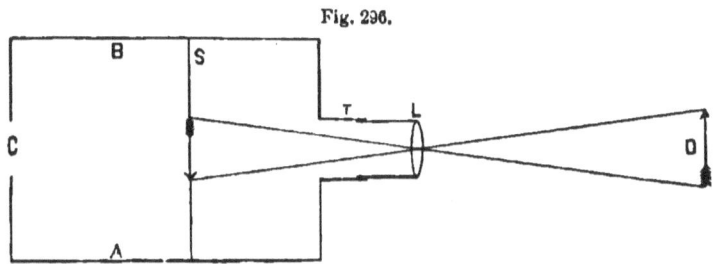

Fig. 296.

§ 365. **The human eye.**—Figure 297 represents a horizontal section of this wonderful organ. Covering the front of the eye, like a watch-crystal, is a transparent coat 1, called the *cornea*. A tough membrane 2, of which the *cornea* is a continuation, forms the outer wall of the eye, and is called the *sclerotic coat*, or "white of the eye." This coat is lined on the interior with a delicate membrane 3, called the *choroid coat*; the latter is covered with a black pigment, which prevents internal reflection. The inmost coat 4, called the *retina*, is formed by expansion of the optic nerve O. The front of the choroid coat *it* is called the *iris*; its color constitutes the so-called

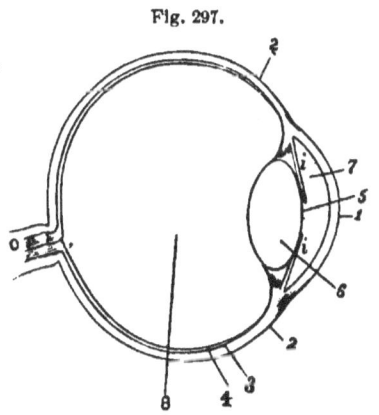

Fig. 297.

"color of the eye." In the center of the iris is a circular opening 5, called the *pupil*, whose function is to regulate, by involuntary enlargement and contraction, the quantity of light admitted to the interior chamber of the eye. Just back of the iris is a tough, elastic, and transparent body 6, called the *crystalline lens*. This lens divides the eye into two chambers; the anterior chamber 7 is filled with a limpid liquid, called the *aqueous humor;* the posterior chamber 8 is filled with a jelly-like substance, called the *vitreous humor.*

The eye is a camera obscura, in which the retina serves as a screen. Images of outside objects are projected by means of the crystalline lens, assisted by the refractive powers of the humors, upon this screen, and the impressions thereby made on this delicate network of nerve filaments are conveyed by the optic nerve to the brain. If the two outer coatings are removed from the back part of the eye of an ox, recently killed, so as to render it somewhat transparent, true images of whole landscapes may be seen formed upon the retina of the eye, when it is held in front of your eye. With the ordinary camera, the distance of the lens from the screen must be regulated to adapt itself to the varying distances of outside objects, in order that the images may be properly focused on the screen. In the eye this is accomplished by changing the convexity of the lens. We can almost instantly and involuntarily change the lens of the eye, so as to form on the retina a distinct image of an object miles away or only a few inches distant. The nearest limit at which an object can be placed, and form a distinct image on the retina, is about five inches. On the other hand, the normal eye in a passive state is adjusted for objects at an infinite distance. Curious enough, the retina on careful examination is found to be covered with little projections which have received, from their appearance, the names of *rods* and *cones*. These project from the nerve fibres very much like nap from the threads of velvet. It is thought that these rods and cones receive and respond to the vibrations of light; in other words, that they co-vibrate with the undulations of the ether, and thereby we get our sensation of light.

§ **366. Chromatic aberration.** — There is a serious defect in ordinary convex lenses, to which we have not before alluded, called *chromatic aberration*, which has required the highest skill of man to correct. The convex lens both *refracts* and *disperses* the light that passes through it. The tendency, of course, is to bring the more refrangible rays, as the violet, to a focus much sooner than the less refrangible rays, such as the red. The result is a disagreeable coloration of the images that are formed by the lens, especially by that portion of the light that passes through the lens near its edges. This evil has been overcome very effectually by combining with the convex lens a plano-concave lens. Now, if a crown-glass convex lens is taken, Fig. 298. a flint-glass concave lens may be prepared that will correct the dispersion of the former without neutralizing all its refraction.[1] A compound lens, composed of these two lenses (Fig. 298) cemented together, constitutes what is called an *achromatic lens*.

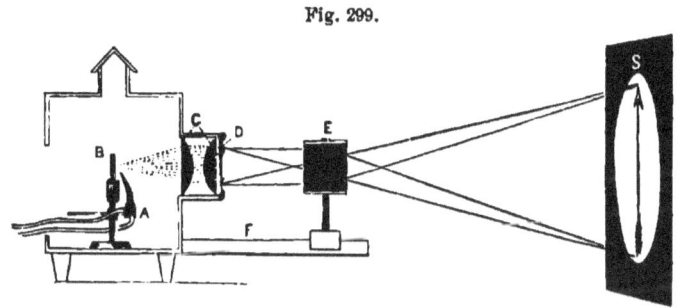

Fig. 299.

§ 367. Stereopticon. — This instrument is extensively employed in the lecture-room for producing on a screen magnified images of small transparent pictures on glass, called *slides;* also for rendering a certain class of experiments visible to a large audience by projecting them on a screen. The light most commonly used is the *lime light*, though the electric light is preferred for a certain class of projections. The flame of an oxyhydrogen blow-pipe A, Figure 299, is directed against a stick of lime B, and raises it to a white heat. The light of the lime is converged — by means of a convex lens c, called the *condensing lens* (usually two plano-convex lenses are used) — upon the slide

[1] The refractive and dispersive powers of the two lenses are not proportional.

D, and strongly illuminates it. In front of it is placed another convex lens E (or a system of lenses), called the *projecting lens*. The latter lens produces (or projects) a real, inverted, and magnified image of the picture on the screen S. The mounted lens E may be slid back and forth on the bar F, so as properly to focus the image. (For useful information relating to the operation of projection, see Dolbear's *Art of Projection*.)

APPENDIX.

Inches.

Millimeters   Centimeters

The area of this figure is a square decimeter. A cube of water, one of whose sides is this area, is a cubic decimeter or a liter of water, and at the temperature of 4° C. weighs a kilogram. The same volume of air at 0° C., and under a pressure of one atmosphere, weighs 1.293 grams. The gram is the weight of 1^{cc} of pure water at 4° C.

Square Inch.

Square Centimeter

SECTION B.

Cutting glass. — Bottoms of glass bottles may be cut off, and plate glass may be easily cut in any pattern desirable, by observing the following directions. Procure an iron rod B, Fig. 300, 25^{cm} long and 7^{mm} in diameter, and insert one end in a wooden handle C, and let the exposed end be filed to a smooth surface. With a pointed piece of soap, trace a line on the glass where you would cut; and, if it is a bottle that is to be cut, file a short gash A (to a depth varying with the thickness of the glass) in the bottle in the direction of the line drawn. Heat, in a Bunsen flame, the free end of the rod to a bright red heat (the hotter the better), and apply the heated end to the glass, as

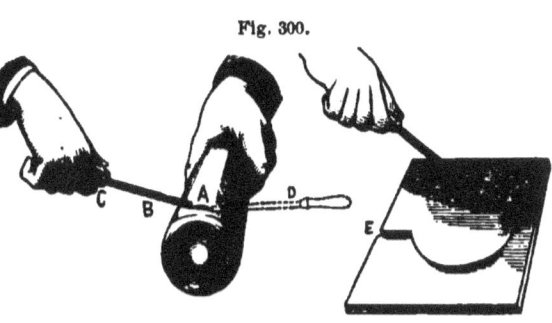

Fig. 300.

in the figure, about 1^{mm} from one extremity of the gash for (say) about five seconds (longer if the glass is very thick; not, however, long enough to crack the glass), and then quickly apply it in the same manner to the other extremity of the gash, as D, and hold it firmly till you see a fine crack creeping toward the rod; then slowly move the rod along the traced line, and the crack will follow faithfully the movements of the rod. If plate glass is to be cut, file a small gash E in one edge; and, commencing with this gash as before, you may cut in the glass a circle, or any design you desire. To bore holes in glass, make a thick paste by partially dissolving gum camphor in spirits of turpentine; nip off a short piece from the end of a small rat-tail file, and, keeping the ragged end wet with the paste, you can readily bore a hole by employing strong pressure, and by a twisting movement as in boring.

SECTION C.

TABLES OF SPECIFIC GRAVITIES OF BODIES.

[The standard employed in the tables of solids and liquids is distilled water at 4° C.]

I. Solids.

Antimony	6.712	Diamond	3.530
Bismuth	9.822	Glass, flint	3.400
Brass	8.380	Human body	0.890
Copper, cast	8.790	Ice	0.920
Iridium	23.000	Quartz	2.650
Iron, cast	7.210	Rock salt	2.257
Iron, bar	7.780	Saltpetre	1.900
Gold	19.360	Sulphur, native	2.033
Lead, cast	11.350	Tallow	0.942
Platinum	22.069	Wax	0.969
Silver, cast	10.470	Cork	0.240
Tin, cast	7.290	Pine	0.650
Zinc, cast	6.860	Oak	0.845
Anthracite coal	1.800	Beech	0.852
Bituminous coal	1.250	Ebony	1.187

II. Liquids.

Alcohol, absolute	0.800	Nitric acid	1.420
Bisulphide of carbon	1.293	Oil of turpentine	0.870
Ether	0.723	Olive oil	0.915
Hydrochloric acid	1.240	Sea water	1.026
Mercury	13.598	Sulphuric acid	1.841
Milk	1.032	Water, 4° C., distilled	1.000
Naphtha	0.847	Water, 0° C., distilled	0.999

III. Gases.

[Standard: air at 0° C.; barometer, 76^{cm}.]

Air	1.0000	Hydrogen	0.0693
Ammonia	0.5367	Nitrogen	0.9714
Carbonic acid	1.5290	Oxygen	1.1057
Chlorine	3.4400	Sulphuretted hydrogen	1.1912
Hydrochloric acid	1.2540	Sulphurous acid	2.2474

APPENDIX. 403

SECTION D.

TABLE OF NATURAL TANGENTS.

Deg.	Tangent.	Deg.	Tangent.	Deg.	Tangent.	Deg.	Tangent.
1	.017	24	.445	47	1.07	70	2.75
2	.035	25	.466	48	1.11	71	2.90
3	.052	26	.488	49	1.15	72	3.08
4	.070	27	.510	50	1.19	73	3.27
5	.087	28	.532	51	1.23	74	3.49
6	.105	29	.554	52	1.28	75	3.73
7	.123	30	.577	53	1.33	76	4.01
8	.141	31	.601	54	1.38	77	4.33
9	.158	32	.625	55	1.43	78	4.70
10	.176	33	.649	56	1.48	79	5.14
11	.194	34	.675	57	1.54	80	5.67
12	.213	35	.700	58	1.60	81	6.31
13	.231	36	.727	59	1.66	82	7.12
14	.249	37	.754	60	1.73	83	8.14
15	.268	38	.781	61	1.80	84	9.51
16	.287	39	.810	62	1.88	85	11.43
17	.306	40	.839	63	1.96	86	14.30
18	.325	41	.869	64	2.05	87	19.08
19	.344	42	.900	65	2.14	88	28.64
20	.364	43	.933	66	2.25	89	57.29
21	.384	44	.966	67	2.36	90	Infinite.
22	.404	45	1.000	68	2.48		
23	.424	46	1.036	69	2.61		

SECTION E.

Galvanometer. — A galvanometer that will answer sufficiently well all the purposes of this book can be easily and cheaply prepared as follows: Make a wooden frame A (Fig. 301), 10^{cm} square and 2.5^{cm} thick, joined by wooden or brass pins in grooves; on it wind 50 to 60 turns ($\frac{1}{4}$ lb.) insulated No. 16 wire in three layers, leaving 1^{cm} space in the center (in the figure this space is exaggerated in order to show the position of the needles), and insert the extremities in the brass screw-cups L and K. In this frame insert a copper or brass wire D, carrying a cork E, which supports a silk fibre F and a strip of

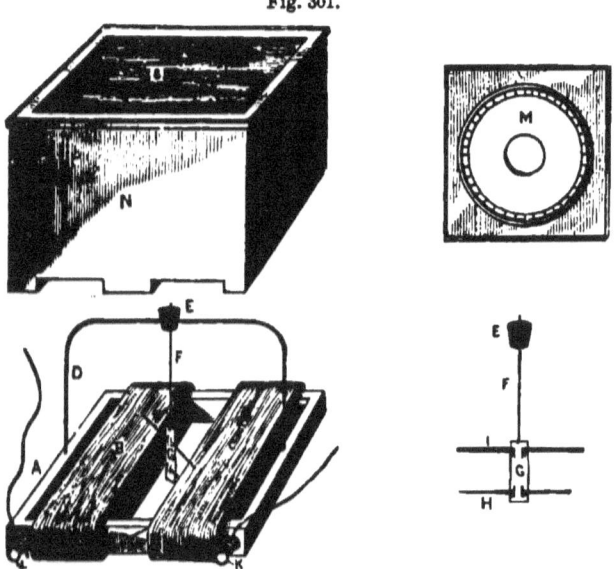

Fig. 301.

paper G. Magnetize a large sewing-needle H, and insert in the paper, as in the figure; also insert a small copper wire I in the paper for a pointer, and suspend the whole so that the needle will swing freely between the upper and lower windings of wire, and the pointer will be just above the coils. Prepare a graduated circle on a card M, having a hole in the center through which to pass the needle, and lay it on the coil. To prevent disturbance from currents of air, cover the whole with a frame N, having a glass plate O laid over its top. Connect the battery wires with the screw-cups L and K. The cost of material need not exceed 75 cents.

SECTION F.

Kind of battery to use. — Several things must be considered in the selection of a battery for a particular use. Among the most important of these are the intensity of current required, and the service required; i.e., whether continuous, temporary, or occasional currents are wanted. The cost is of consequence, but that must be governed mainly by the preceding considerations. In the following arrangement, preferences are given to the several batteries by numbers, in the order in which they occur against the several uses specified: —

NAMES OF BATTERIES, ETC.

1. Smee.
2. Leclanche.
3. Gravity.
4. Daniell.
5. Grenet.
6. Bunsen or Grove.
7. Magneto or dynamo machines.
8. Thermo-batteries.

USES CELLS ARE SUITED FOR.

Strong, Continuous Currents.

Electrotyping or Electro-plating.................. 7, 4, 1, 3.
Electro-magnets................................. 3, 4, 1.
Electric light................................... 7, 6.
Telegraph (closed circuit)....................... 3, 4.

Temporary.

Induction coils .... 5, 6, 4, 3.
Medical coils .... 5, 1.

Occasional.

Annunciators, domestic bells..................... 2, 1, 3, 4.
Exploding fuses.................................. 2, 4.
Electrical measurements (constant current)....... 8, 4, 3.

SECTION G.

Apparatus to illustrate wave-motion. — The most efficient apparatus for this purpose that we have seen may be constructed as follows. Procure forty wooden return-balls (sold at toy stores); suspend them by strings (better, fine wires) about 1^m long, as in Figure 302, and about 7^{cm} apart. Connect all the balls horizontally by small

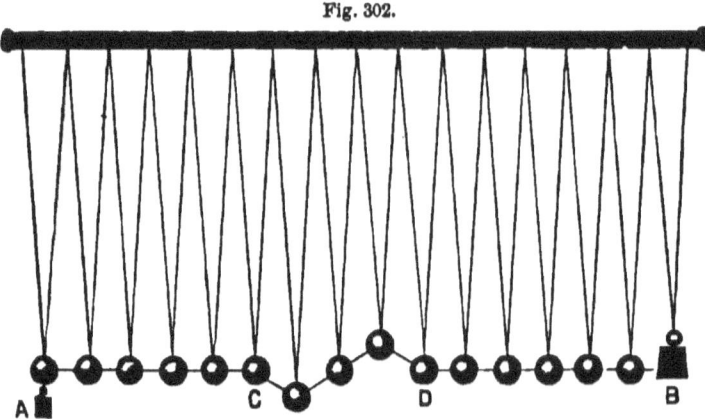

Fig. 302.

elastic cord (better, small spiral wire coil), and connect the ball at one extremity of the series with a suspended weight B (weighing about 1^k), and from the ball at the other extremity suspend a small weight A, which may be easily removed when desirable. By a simple vibration given with the hand to A, a wave, as CD, will be projected through the series, and on reaching B will be reflected; though when reflection is wanted, B had better be replaced by a hook attached to a wall.

APPENDIX. 407

SECTION H.

Porte Lumière. — Two half-sections of a tube A and B (Fig. 303) may easily be sawn from a block of pine wood. These glued together at their edges make the tube C. This tube is 20^{cm} long and 15^{cm} in diameter, with a bore of 11^{cm} diameter. Raise a window-sash about 50^{cm}, and fit a board D just to fill the opening. In the middle of this board cut a hole just large enough to receive the tube, and allow it to turn in the hole freely. Attach a bolt E to the board D, and about 12^{cm} from one end of the tube bore a row of holes around the tube, 1^{cm} in depth and about 1^{cm} apart, to receive the bolt. By means of a hinge, attach to the outer edge of the tube a board G, 30^{cm} long, 14^{cm} wide, and 1.5^{cm} thick. A mirror F, 26^{cm} long and 12^{cm} wide, is fastened by tacks with large heads to the upper surface of this board. A stout string attached to one of the long edges of the board is carried through the tube and fastened to a binding screw H. When the mirror is to be adjusted so as to receive the sun's rays and reflect them through the tube, rotate the tube, and raise or depress the mirror by means of the string, so as to adapt it to the position and élevation of the sun in the heavens, and then fasten by means of the bolt and string. A window on the *south* side of a building should be selected for experiments with this apparatus. The portion of the window not occupied with the board D, as well as other windows not in use, may be darkened with curtains of black enamelled cloth. The whole cost of the above apparatus need not exceed $1.00.

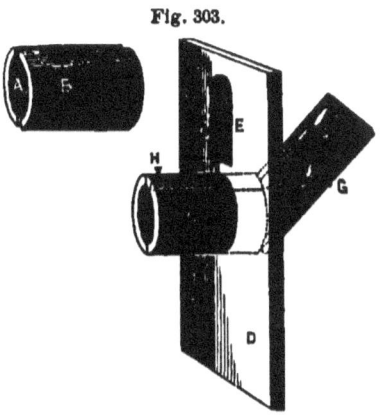

Fig. 303.

SYLLABUS.

CHAPTER I.

MATTER AND ITS PROPERTIES.

I. INTRODUCTION.

An experiment is a question put to Nature.

By experiment we learn that invisible air, like matter, and like nothing else with which we are acquainted, can displace matter (*e.g.*, a bubble), exert pressure, and has weight; hence, we conclude that air is matter, and that matter can exist in an invisible state.

No visible body of matter, however compact it may appear, ever *fills* the space enclosed by its surface; but every visible body is a collection of countless smaller bodies called *molecules*, separated by invisible spaces called *pores*. Every molecule is in motion, and this motion is such as to prevent permanent contact between one another.

By the *mass* of a body we understand the *quantity of matter* in it; and by its *density*, *the mass in a unit volume*.

Substances whose molecules can be separated into molecules differing in substance from the original molecules are called *compound*. Substances whose molecules have never been so separated are called *simple*. There are known only about 71 of the latter, and an innumerable number of the former.

Any change in a body that does not cause a change of substance is a *physical change*. A change of substance is a *chemical change*

Matter is nowhere created, nowhere annihilated. The mass of the universe is constant.

The tendency which matter possesses to *push* and to *pull* is called *force*. Whatever tends to produce or alter motion is force. It is manifested only in pushes and pulls.

Forces are classified as *molecular* or *molar*, according as they act between molecules or larger bodies; *attractive* or *repellent*, according as they are manifested in pulls or pushes.

When, by external force, the molecules of a body are brought nearer together, it is said to be *compressed* or *condensed;* if they are brought together by internal forces, the body is said to *contract*, and if they are separated by internal forces, it is said to *expand*.

II. THREE STATES OF MATTER.

Any substance may exist in any one of three states,—solid, liquid, or gaseous.

General characteristics of matter in the solid state: Immobility of the molecules, and permanence of shape.

Liquid state: Greater mobility of the molecules, easily poured, and shaped by the containing vessels.

Gaseous state: Almost perfect freedom of motion of the molecules; unlimited tendency to expand; great compressibility.

Owing to their tendency to flow, both liquids and gases are called *fluids*.

* The state which a body assumes depends on temperature and pressure.

III. PHENOMENA OF ATTRACTION.

The force of attraction which exists between all bodies at distances however great is called the force of *gravity*, and the phenomenon is called *gravitation*. *Weight* is a term applied to the measure of this force as exerted between the earth and terrestrial objects. Weight varies *as* the mass, and *with* the distance from the centre of the earth. At the same place, weight is proportional to the mass.

When attraction is molecular, and between like molecules, it is called *cohesion;* when between unlike molecules, it is called *adhesion.*

When a body of matter in a solid state exhibits method in the arrangement of its molecules, it is said to be *crystalline;* otherwise, *amorphous.* Matter crystallizes, usually, while passing from the liquid to the solid state.

A theory, suggested as a possible explanation of the cause of crystallization, is based upon the hypothesis that the molecules of crystalline matter possess something akin to *polarity*, in which case the molecules would tend to arrange themselves somewhat as iron filings do when they possess polarity.

Hardness is due to some peculiar action (not well understood) of cohesion, that enables a body to resist another body tending to scratch it.

When molecular forces tend to restore a body to its original shape and volume after having yielded to some force, they are called *elastic forces*, and the body is said to possess the property of *elasticity*.

Substances which tend to break rather than suffer a permanent alteration in form are said to be *brittle*.

Substances which, though brittle when a force is applied suddenly, will suffer a permanent change in form if subjected to a gradual and long-continued stress, are called *viscous*.

All substances in the solid state possess, to some extent, the property of *fluidity*, and hence are more or less *flexible, malleable*, and *ductile*. This implies, also, that they possess a power of preventing rupture when subjected to a pulling force. This power, due to cohesion, is called *tenacity*.

Liquids ascend or are depressed in capillary tubes according as the adhesion between the liquid and the tubes is greater or less than the cohesion in the liquids. For the *four laws of capillary action*, see page 36.

A liquid will dissolve a solid when the adhesion is greater than the cohesion.

Absorption of gases by solids is caused chiefly by molecular attraction, and is said to be *superficial* when the gases are taken into the *cavities* of solids, and *intermolecular* when taken into the pores.

Absorption of gases by liquids is intermolecular, and is caused both by attraction of the molecules and their incessant motion.

Diffusion of liquids is caused mainly by the motion of their molecules.

Osmose, or the diffusion of liquids through porous septa, is imperfectly understood, though it is supposed that adhesion between the liquids and the septa is the chief agent.

Diffusion of gases depends almost wholly on molecular motion. All gases diffuse regardless of the force of gravity.

Osmose depends on the size of molecules, size of pores, and on molecular motion; very complex.

CHAPTER II.

DYNAMICS.

IV. DYNAMICS OF FLUIDS.

Dynamics treats of force and motion. When several forces so act on a body as to neutralize one another's effect, both the forces and the body are said to be in *equilibrium*. A body in a state of rest or uniform motion is in equilibrium.

(Can absolute rest or uniform motion exist except there is absolute equilibrium? Is there any body known to be in a state of absolute equilibrium, *i.e.*, in equilibrium with reference to all external forces?)

When a pushing force is resisted, *i.e.*, when any portion of the force is not effective in producing motion, there results a *pressure*. Under similar conditions a pulling force causes *tension*. (The word *tension* is also applied to the expansive power of gases.)

At every point in a body of fluid, gravity causes pressure to be exerted equally in all directions. In gases the pressure increases *with* the depth; in liquids, *as* the depth.

The average sea-level atmospheric pressure (and consequently the tension of the air at this level) is 1033.3^g (about 1^k) per square centimeter, or 14.7 lbs. (about 15 lbs.) per square inch. An atmosphere (when the term is used to denote pressure) is the pressure of 1^k per square centimeter. Any instrument which will measure atmospheric pressure is a *barometer*.

MARIOTTE'S or BOYLE'S LAW: At the same temperature the volume of a body of gas varies inversely as the pressure, density, or elastic force.

In consequence of the mobility and perfect elasticity of fluids, any pressure exerted on a given area of a fluid enclosed in a vessel is transmitted undiminished to every equal area of the interior of the vessel. In the hydrostatic (or hydraulic) press we have a practical application of this principle.

The pressure on the bottom or sides of a vessel, due to the gravity of the liquid which it holds, depends on the depth and area of the surface pressed upon, and the density of the liquid, and is independent of the shape of the vessel and the quantity of liquid.

The pressure upon any portion of a vessel is the weight of a column of that liquid whose base is the area of the portion pressed upon, and whose hight is the average depth of that portion.

The free surface of a body of liquid, at rest, partakes of the sphericity of the earth, but for most practical purposes may be regarded as level.

V. BUOYANT FORCE OF LIQUIDS.

A solid immersed in a fluid is buoyed up by it in consequence of the unequal pressures upon the top and bottom at their different depths, and the amount of buoyancy (or the apparent loss of weight of the solid) is the weight of a body of that fluid whose volume is equal to the volume of the solid. A floating solid displaces its own weight of fluid. The absolute weight of a body is its weight in a vacuum.

VI. DENSITY AND SPECIFIC GRAVITY.

The specific gravity of a substance is the ratio of the density of that substance to the density of another substance taken as a standard. It is found by dividing the weight of a given volume of the substance by the weight of an equal volume of the standard. (In finding the specific gravity of solids and liquids, state various methods of ascertaining the weight of an equal volume of the standard.)

VII. MOTION.

Motion and rest are wholly relative terms, *i.e.*, they are applicable to an object only when considered in connection with some other object. There is no such thing as absolute rest in the universe.

Velocity is given in units of space and time.

Motion is uniform or varied. Varied motion is accelerated or retarded. We may conceive of uniform motion though it nowhere exists.

VIII. FIRST LAW OF MOTION. — INERTIA.

Motion always arises from *mutual* action between at least two bodies, and cannot originate in an object isolated from all others.

Motion in a body is caused only by another body's parting with some of its power of producing motion.

Bodies receive motion gradually and part with it gradually.

No body possesses any innate power to change its condition with reference to motion or rest. It is sometimes convenient to speak of this complete absence of power as a property of matter, under the name of *inertia*.

First Law of Motion. — A body at rest remains at rest (why?), and a body in motion moves with uniform velocity (why?) in a straight line (why?), unless acted upon by some external force to change its condition.

IX. SECOND LAW OF MOTION, AND APPLICATIONS.

Second Law of Motion. — A given force has the same effect in producing motion, whether the body on which it acts is in motion or at rest; whether it is acted upon by that force alone, or by others at the same time.

It is usually possible to substitute for two or more forces a single force which will produce the same result as the combined forces. Such a force is called a *resultant*.

The resultant of two forces acting at an angle to each other may be represented by a diagonal of a parallelogram, of which the components form two adjacent sides.

Any single force may be resolved into two or more components.

The resultant of parallel forces having the same direction is their sum; the resultant of two parallel forces acting in opposite directions is motion in the direction of the greater force proportionate to their difference.

When two parallel forces having the same direction act upon a body at different points, the distances of their points of application from the points of application of their resultant are inversely as their intensities.

A pair of forces, equal, parallel, opposite, and applied at opposite extremities of an object, produces only rotation, and is called a *couple*. A couple has no resultant and no equilibrant.

X. CENTER OF GRAVITY.

The center of gravity of a body is the point of application of the resultant of the forces of gravity acting on its molecules. To support any body, it is only necessary to provide an equilibrant for this resultant, and to apply it at some point in the line of direction. Whether a body will stand or fall depends upon whether its line of direction falls within its base, *i.e.*, whether its support is applied in the line of direction.

A body tends to assume a position such that its c.g. will be as low as possible, and when in such position it is said to be in *stable equilibrium*. When a disturbance would lower its c.g., it is said to be in *unstable equilibrium;* and when disturbance would not lower or raise its c.g., it is in *neutral equilibrium*. A broad base and low c.g. give stability to a body.

XI. CURVILINEAR MOTION.

A curved line is one whose direction changes at every point. To produce curvilinear motion, a continuous (why continuous?) force must be applied at an angle (why?) to its otherwise straight path. (See First Law of Motion.) *Centrifugal force* is the result of the tendency of a body to move in a straight line; *centripetal force* is the force which compels it to depart from a straight line.

Centrifugal force increases as the mass and the square of the velocity; hence, to produce circular motion, the centripetal force must increase as the mass and the square of the velocity.

XII. ACCELERATED AND RETARDED MOTION.

A body impelled by a single constant force, and encountering no resistance, always has a *uniformly accelerated motion*. A moving body, encountering constant resistances, has *uniformly retarded motion*. The acceleration or retardation per unit of time is represented by k (or g, when the force is gravity).

Formulas for uniformly accelerated motion:—

(1) $V = (\frac{1}{2} k \times 2T) = kT$
(2) $s = \frac{1}{2} k (2T - 1)$
(3) $S = \frac{1}{2} k T^2$

Hence, *velocity varies as the time, and the entire distance traversed as the square of the time.*

The acceleration due to gravity in the Northern States, near the sea level, when there are no resistances, is (g) 9.8^m (or 32.2 ft.) per second. This is the measure of the force of gravity at these places.

All bodies, of whatever mass, density, or substance, fall with equal velocities in a vacuum. (Why?)

The horizontal distance attained by a projectile is its *range* or *random*. The greatest range is obtained at an angle a little less than 40°.

A body projected horizontally, with any velocity, will reach the ground in precisely the same time that it would if dropped vertically. (Why?)

XIII. THE PENDULUM.

The time of vibration of a pendulum varies inversely as the number of vibrations.

The time of vibration and the number of vibrations are independent of both the mass and the length of arc.

The time of vibration and the number of vibrations depend upon both the length of the pendulum and the force of gravity.

The time of vibration varies as the square root of the length of the pendulum.

The number of vibrations varies inversely as the square root of the length of the pendulum.

The time of vibration varies inversely as the square root of the force of gravity.

The number of vibrations varies as the square root of the force of gravity.

The length of a pendulum is the distance from the point of suspension to its center of oscillation. These two points are interchangeable.

The center of percussion is coincident with the center of oscillation.

XIV. MOMENTUM.—THIRD LAW OF MOTION.

Momentum is the product of mass, multiplied by velocity; or, it is the product of force, multiplied by the time during which it acts.

Third Law of Motion. To every action there is an equal and opposite reaction.

The momentum of the reaction is equal to the momentum of the action.

Law of Reflection. — When the striking body and the body struck are perfectly elastic, the angle of reflection is equal to the angle of incidence.

XV. WORK.—ENERGY.

Work is done whenever force acts through space. It is estimated by multiplying resistance by the space, or force by the space through which it acts. It is commonly expressed in kilogrammeters or foot pounds.

In estimating the rate of doing work, or the power of an agent to do work, time is taken into consideration. The unit employed is a horse-power, which is a power capable of doing 33,000 ft. lbs. = $4,570^{kgm}$ per minute.

Power to do work is called *energy*. Every moving body possesses energy due to its motion; energy due to motion is called *kinetic energy*. A body may possess energy due to an advantage of position given it by work done upon it. Energy due to position is called *potential*. Potential energy becomes kinetic on the return of bodies to their original positions.

Formulas for energy:

(1) $\text{Energy} = \dfrac{WV^2}{2g}$;

or,

(2) $\text{Energy} = \dfrac{MV^2}{2}$.

But, since $W = Mg$,

$M = \dfrac{W}{g}$;

hence, in using the second formula, the value of M must be found by dividing W by g.

Force may be measured by the change of momentum it produces in a second.

In the C.G.S. system, the *centimeter, gram,* and *second* are taken as the units of distance, mass, and time respectively, and in it the *dyne* is the unit of force. A dyne is a force which, acting for a second, will give to a gram of matter a velocity of one centimeter per second.

In the gravitation system the term *gram* (or pound, etc.) is applied to both mass and force.

In the C.G.S. system the unit of work is the *erg*. The erg and kilogrammeter measure both work and energy, or power to do work. An erg is the work done, or the energy imparted, by a force of one dyne working through a distance of one centimeter.

Energy may be transformed from one condition to another, as from kinetic to potential; or, from one variety to another, as from heat to mechanical or molar motion.

Physics treats of transferences and transformations of energy.

XVI. MACHINES.

Advantages of machines: (1) They enable us to exchange power for velocity, or velocity for power. (2) They enable us to employ a force in a direction which is more convenient than the direction in which the resistance is to be moved. (3) They enable us to employ other forces than our own in doing work.

LAW OF MACHINES: The work applied to a machine is equal to the effective work plus the internal work done by the machine. The useful work done by a machine is less than the work done upon the machine. In a perfect machine they would be equal. None exists.

In every machine $P : W :: W : p$; *i.e.*, The power and resistance vary inversely as the distances they respectively travel.

CHAPTER III.

MOLECULAR ENERGY. — HEAT.

XVII. WHAT HEAT IS. — SOME SOURCES OF HEAT.

Two theories of heat have successively prevailed, viz.: (1) that heat is matter; (2) that it is motion. Molar motion is convertible into heat, and heat is convertible into molar motion. It is scarcely conceivable that motion can be converted into matter, or matter into motion. But it is a matter of everyday observation, that when motion of one kind (or thing) ceases, motion of another kind (or thing) takes its place. Conclusion: heat is motion, *i.e.*, molecular motion; this implies the existence of kinetic energy.

Molecules may possess potential energy, which becomes kinetic during chemical action, *i.e.*, the clashing together of molecules in consequence of affinity, thereby generating heat, as in combustion. This is the origin of animal heat and muscular motion. The sun is the ultimate source of nearly all the energy at man's command.

Temperature is determined by the average kinetic energy of the individual molecule; quantity of heat, by the average kinetic energy of the individual molecule multiplied by the number of molecules.

XIX. DIFFUSION OF HEAT.

Heat is diffused by *conduction*, *convection*, and *radiation*. (Explain the first two. How is ventilation usually accomplished? By which method do we receive heat from the sun? Why not by either one of the other two methods?)

XX. EFFECTS OF HEAT.—EXPANSION.

Effects of heat: Expansion, liquefaction, vaporization, change of temperature, and specific heat in part.

(In what state is matter least expansive? Why? In what state is the coefficient of expansion the same for all substances? What is the coefficient? State an exception to the general rule that matter expands with a rise of temperature.)

XXI. THERMOMETRY.

Change of temperature is measured by expansion. A thermometer measures expansion, hence it measures temperature.

(State the method of construction and graduation of a mercury thermometer. What kind of thermometer is more sensitive than a mercury thermometer? Why?)

Formulas for conversion of thermometer readings:

$$\tfrac{5}{9}(F - 32) = C\,;\ \tfrac{9}{5}C + 32 = F.$$

Absolute temperature is reckoned from an absolute zero, or state of no heat. It may be found by adding 273 to the C. reading, or 460 to the F. reading.

LAW OF CHARLES: The volume of a given body of gas at a constant pressure varies as its absolute temperature.

Conversely, the tension of a given body of gas, whose volume is constant, varies as its absolute temperature.

MARIOTTE'S LAW: At a constant temperature, the volume of a given body of gas varies inversely as the external pressure. At a constant temperature, the product of the volume and tension of a given body of gas is constant. The product of the volume and tension of a body of gas varies as its absolute temperature.

The tension of a body of gas is due to the kinetic energy of its molecules.

MOLECULAR ENERGY. — HEAT. 423

XXII. LIQUEFACTION AND VAPORIZATION.

See laws of fusion and boiling, page 161.

Distillation, or the separation of mixed liquids by vaporization, is conducted on the principle that the temperature of the boiling points of different substances differs.

The rapidity of evaporation varies with the temperature (why?), amount of surface exposed, and dryness of the atmosphere, and inversely with the pressure upon the liquid. Dewpoint is the temperature of the atmosphere when saturated with watery vapor. The atmosphere is dry when the difference between its temperature and dewpoint is great. The term *dryness*, when applied to the atmosphere, signifies capacity for receiving more moisture, and does not necessarily imply deficiency of moisture.

[The molecules of every body of liquid are in motion. The distances traversed by the molecules in the interior of a body are limited by the proximity of neighboring molecules on all sides. When they reach the free surface of the body, they are not subject to this restraint, and more or less of them depending upon the temperature (*i.e.*, the energy of their movements), become released from the force of cohesion and pass off as a vapor. *This is evaporation.* On the other hand, molecules of the vapor, resting upon the liquid surface, beat against it, and, it is supposed, become entangled in it and thus return to the liquid state. When the number of molecules which thus return to the liquid state equals those which escape, the space above the liquid is said to be *saturated.*]

XXIII. HEAT CONVERTIBLE INTO POTENTIAL ENERGY, AND VICE VERSA.

Heat is measured in calories; temperature, in degrees. A calorie is the quantity of heat required to raise the temperature of 1^k of water from 0° to 1° C.

Eighty calories are consumed in converting one kilogram of ice into water. Five hundred and thirty-seven calories are consumed in converting 1^k of water at 100° C. into steam. In the first case, the heat is consumed in doing interior work, such as neutralizing, in part, the force of cohesion. In the second case, the larger portion (about $\frac{12}{13}$) of the heat is consumed in the interior work of completely overcoming cohesion, and the remaining portion ($\frac{1}{13}$) in the exterior work of overcoming atmospheric pressure.

The temperature of a body is reduced, either by imparting heat to a colder body, or by the consumption of its heat in doing work. By the latter method artificial cold is produced. The work done is usually that of melting or dissolving some solid, vaporizing a solid or liquid, or producing expansion in a gas against resistance. (Give illustrations of each.)

Heat which is consumed in melting, dissolving, and vaporizing, is restored when the opposite changes occur. (Explain.)

XXIV. SPECIFIC HEAT.

Equal quantities of heat applied to equal weights of different substances raise their temperatures unequally. In the case of solids and liquids, this is explained by the fact that a portion of the heat applied is always consumed in doing internal work; and since in different substances the amount consumed in doing work varies, consequently the amount of heat left to raise the temperature of different substances must vary.

The number of heat units required to raise a body 1° C. is called its capacity for heat.

The specific heat of a body is the ratio of its capacity for heat to that of an equal weight of water.

Hydrogen gas has the greatest capacity for heat. Water ranks next.

XXV. THERMO-DYNAMICS.

A definite quantity of mechanical work can produce a definite quantity of heat; and conversely, this heat can perform the original amount of work. One calorie is equivalent to 424^{kgm} of work. The quantity, 424^{kgm}, is called the *mechanical equivalent of heat.*

Doctrine of correlation of energy: Any kind of energy can be converted into any other kind.

Doctrine of conservation of energy: When one form or kind of energy disappears, an exact equivalent of another form or kind always takes its place, so that the sum total of energy in the universe is constant.

XXVI. STEAM ENGINE.

A steam engine is a machine by means of which a portion of the motion of the molecules of steam (*i.e.*, its heat) is transformed into molar or mechanical motion.

In a non-condensing engine, a large amount of energy is wasted in producing motion against the resistance of atmospheric pressure.

CHAPTER IV.

ELECTRICITY.

XXVII. CURRENT ELECTRICITY.

Just as a difference of level is necessary to produce a current of water, and a difference of temperature to cause a flow of heat, so a difference of electrical condition, called a difference of *potential*, is necessary to cause a flow of electricity. To establish the necessary conditions in each case (*i.e.*, difference of level, etc.) energy must be expended. On the other hand, the return of each to its normal condition or state of equilibrium is attended with the development of energy. The constant expenditure of chemical potential energy in a voltaic cell causes a constant inequality of potential, and this in turn causes a constant tendency to equalization of potential throughout the circuit; in other words, a continuous current.

As the stress (called *force of gravity*) between an elevated body of water and the earth is the cause of the so-called waterpower, so it is probable that a stress between two parts of a body having different potential is the cause of a power usually called *electro-motive force*.

The greater the disparity between the two solid elements of a voltaic cell with reference to the action of the liquid on them, the greater the difference of potential or electro-motive force of the cell; hence, the stronger the current.

The office of a voltaic battery is to create inequality of potential, i.e., to generate electro-motive force.

The zinc element generally needs to be amalgamated to prevent local action and consequently a waste of energy.

Hydrogen ought not to be allowed to collect on the electronegative plate, as it offers a resistance to the current, but chiefly because it tends to reduce the difference of potential

ELECTRICITY. 427

between the two plates, and thereby reduce the current. This may be prevented by either mechanical or chemical action.

The effects of electricity are heating, luminous, electrolytic, physiological, and magnetic.

XXX. ELECTRICAL MEASUREMENTS.

Upon the strength of the current depends the magnitude of these effects. By the strength of the current is meant the quantity of electricity which flows through a circuit in a unit of time. The voltameter and galvanometer measure the strength of the current. In the tangent galvanometer the strength of current is proportional to the tangent of the angle of deflection.

OHM'S LAW: The strength of the current varies as the electromotive force, and inversely as the resistance in the entire circuit; *i.e., an effect is directly proportional to that which tends to produce it, and inversely proportional to that which tends to oppose it.*

Resistance varies as the length, and inversely as the squares of the diameters of cylindrical conductors.

When the external resistance is much greater than the internal, it is best to connect cells "tandem" in order to increase the electro-motive force. When the internal resistance is greater than the external, the cells should be connected "abreast" in order to diminish the internal resistance. When the external and internal resistances are nearly equal, it may be best to connect them in both ways.

In a circuit having a given external resistance the greatest possible efficiency is obtained from a given battery when the external and internal resistances are about equal.

XXXI. MAGNETS AND MAGNETISM.

LAW OF MAGNETS: Like poles repel, unlike poles attract one another.

THE LAWS OF CURRENTS on which Ampère's theory of magnets is based are as follows: Parallel currents in the same direction attract one another; parallel currents in opposite directions repel one another. Angular currents tend to become parallel and flow in the same direction.

Ampère's theory assumes that there are constant currents circulating around the molecules of every magnetizable substance, and that the resultant of these forces in a magnet is the same as though "a sheet of currents" circulated around the magnet as a whole. The deflection of the magnetic needle is due to the tendency of angular currents to become parallel.

The attractions or repulsions of the poles of magnets are due to the attractions or repulsions of parallel currents according as they flow in the same or opposite directions.

The earth's magnetism is probably due to the circulation of currents around the earth, from east to west, in planes at right angles to its axis. Substances which are attracted by a magnet are called *paramagnetic;* those which are repelled are called *diamagnetic.*

XXXII. MAGNETO-ELECTRIC AND CURRENT INDUCTION.

When Ampèrian currents (*i.e.*, a magnet) or a battery current approaches a neighboring closed circuit, a momentary current is induced in this circuit opposite in direction to the inducing current; when carried away, a current is induced in the same direction as the primary or inducing current. The same happens whenever in one of two neighboring circuits a current is started or stopped by making or breaking the circuit. At these instants not only are secondary or induced currents sent through the neighboring circuit, but, likewise, corresponding currents are induced in its own circuit. The latter are called *extra currents.*

Briefly, *any magnetic or electrical disturbance in the neighborhood of a circuit will induce currents in that circuit.*

The currents established by magneto and dynamo machines are induced currents. Induced currents have high *tension*, *i.e.*, great power of overcoming resistances.

XXXIII. THERMO-ELECTRICITY.

When two different metals so connected as to form a circuit are brought in contact, and heated or cooled at the junction, a thermo-electric current is established.

The electro-motive force, and consequently the strength of the current, depends upon the elevation or depression of temperature at their junction, and upon the metals employed.

XXXIV. FRICTIONAL ELECTRICITY.

Friction between two bodies, especially if they are composed of unlike substances, tends to produce a difference of potential in the bodies. As long as the bodies remain in this condition they are said to be electrified or charged with electricity; one with positive, the other with negative electricity. Electricity in this condition is said to be *static*. On the return of either to its normal potential a discharge is said to occur, a current is established, and the electricity for the time being is said to be *dynamic*. As commonly understood, an electrified body is one that has a different potential from that of the earth, and is said to be positively electrified when its potential is higher than that of the earth, and negatively electrified when lower than that of the earth. Similarly-electrified bodies repel one another; dissimilarly-electrified bodies attract one another. [Phenomena of electric attraction and repulsion are thought by many to be phenomena of *ether-stress*, or of "action at a distance" through the medium of ether.]

An electrified body brought near a body whose potential is zero tends to electrify it by induction, causing the part of the body nearest itself to be of a different potential from itself, while the remote part of the body, if it is insulated, becomes of like potential. The electricity in the former case is said to be *bound*, while the latter is *free*, since, if the insulated body is connected with the earth, a discharge of the latter occurs; in other words, the location of this potential is transferred to the earth.

Electrification is confined to the surface of a body.

XXXV. ELECTRICAL MACHINES.

By means of the so-called electrical machines mechanical energy is transformed into electric energy. These machines are capable of producing a great change in potential, consequently their electro-motive force is great, but their internal resistance is so great that the strength of current they are capable of yielding is extremely small, and consequently they are of little practical value.

The phenomena of electricity in a statical state are limited to those of attraction and repulsion. All other effects are produced by electricity in the dynamical state, and the magnitude of the effects generally varies as the square of the current.

No one knows what electricity is. [It is not "a form of energy."] For practical purposes, it suffices to regard it as a medium for communication of energy, and to know the laws to which it is subject.

XXXVI. USEFUL APPLICATIONS OF ELECTRICITY.

These are well-nigh innumerable. Some of the more important are those pertaining to medical and surgical operations, electric lighting, electrotyping, electroplating, telegraphy, telephony, and the production of mechanical motion through the instrumentality of electro-motors in great variety. [One of the most recent applications is that of storing energy, as in the so-called *storage-batteries*. Energy of any kind, *e.g.* water-power of any of the great water-falls, may be transformed into electric energy, and the latter may be transformed into potential energy and stored in these batteries. These batteries may be transported to distant places, and the potential energy restored to electric energy, and made to do any species of work.]

CHAPTER V.

SOUND.

XXXVIII. VIBRATION AND WAVES.

A vibration is a recurrent change of position. The propagation of a vibration through a series of particles produces wave motion. A succession of such propagations produces a wave-line. Only the wave-form advances. The distance between any point on one wave and a similarly situated point on its successor or predecessor is a wave-length. The greatest distance which a particle reaches from its median position is the amplitude of the vibration or wave. The distance traversed by a given wave in one second is the velocity of propagation. If v be the velocity, l the wave-length, and n the number of waves per second,

$$v = nl,\ n = \frac{v}{l},\ \text{and}\ l = \frac{v}{n}.$$

When a given particle is subjected simultaneously to two or more impulses, due to two or more trains of waves, the motion of the particle is the resultant of the several impulses, and the phenomenon is called *interference*. Interference may intensify or nullify motion. In cords, membranes, etc., it may result in vibration in segments, or stationary vibrations, the points of least vibration being called *nodes*, the points of greatest motion, *antinodes*, and the portion between two nodes, a *ventral segment* or *loop*.

Waves are longitudinal or transverse, according as the particles vibrate in the same plane with the path of the wave, or across it.

Wave motion is one of the most common and most important

methods of transmission of energy. Elasticity is essential in a medium, that it may transmit waves made up of condensations and rarefactions ; and the greater the elasticity, the greater the facility and the rapidity with which a medium transmits waves.

XXXIX. SOUND-WAVES.

Sound is vibration that may be appreciated by the ear, and originates in a vibrating body. It is transmitted by wave motion, and therefore cannot travel in a vacuum.

XL. VELOCITY OF SOUND.

The velocity of sound in gases varies as the square root of their elasticity, and inversely as the square root of their densities.

XLI. REFLECTION AND REFRACTION OF SOUND.

Sound-waves are reflected in accordance with the general law of reflection. Echoes are the result of reflected sound-waves.

When the form or direction of a wave is altered in consequence of changing media of different densities the phenomenon is called *refraction*.

XLII. LOUDNESS OF SOUND.

Loudness, which is a measure of the intensity of a sensation, varies with the intensity of sound, but there are many reasons why they are not exactly proportional.

The intensity of sound varies as the square of the amplitude of the vibrations of the sounding body. It is also affected by the density of the medium in which it is produced, being weaker when originating in a rare medium. The intensity of sound varies inversely as the square of the distance from the source. When sound is re-enforced or intensified by interference the phenomenon is called *resonance*. Re-enforcement is produced by means of sounding-boards, columns of air, etc.

SOUND. 433

When one vibrating body causes another body having the same vibration period to vibrate, the vibrations of the latter are called *sympathetic vibrations*. They represent the accumulated effect of a series of impulses transmitted from the sounding body through the sound medium to the body thus caused to vibrate.

When a vibratile body is compelled to surrender its own vibration period and to vibrate in an arbitrary manner imposed upon it by another, the phenomenon is known as *forced vibrations*.

The sensation of noise is caused by irregularity in the impulses received by the ear and by its inability to distinguish pitch. Regularity and simplicity are characteristics of musical sound.

XLIII. PITCH OF SOUND.

Pitch depends upon vibration-frequency or wave-length; the greater the number of vibrations per second, or the shorter the wave-length, the higher the pitch. The siren is one of many instruments used to determine vibration-frequency.

A wavy sound caused by interference of sound waves is known by the name of *beats*. The number of beats per second due to two simple tones is equal to the difference of their respective vibration-numbers. Beats are one of the chief causes of discord.

XLIV. VIBRATION OF STRINGS.

The vibration-frequency of strings of the same material varies inversely as their lengths and the square roots of their weights, and directly as the square roots of their tension.

XLV. OVERTONES AND HARMONICS.

Sounds proceeding from instruments vibrating in parts are called *overtones*. If the vibration-number of the overtone bears some simple ratio to the vibration-number of the fundamental, the overtone is called a *harmonic*.

Generally, only those notes harmonize whose fundamental tones bear to one another ratios expressed by small numbers; and the smaller the numbers which express the ratios of the rates of vibration, the more perfect is the harmony.

XLVI. QUALITY IN SOUND.

Quality, or that property of sound which enables us to distinguish different sounds having the same pitch and intensity, is shown both by analysis and synthesis to depend upon what overtones combine with its fundamental, and on their relative intensities; or, briefly, upon the form of vibration. Sounds differ only in intensity, pitch, and quality. The manometric flame apparatus is well suited to illustrate these three properties.

XLVIII. SOME SOUND-RECEIVING INSTRUMENTS.

The ear is a mechanical contrivance for the transmission of vibration to the organs of sensation. For example, the aërial waves cause forced vibrations in the outer membrane or tympanum; these are communicated by a chain of bones to the membranous walls of the vestibule, and thereby to the liquid contained in the cavity. The bristles suspended in this liquid take up and analyze the vibrations, much as when we sing into a piano with the dampers down, only those strings respond which are in unison with the sound produced by the voice. The bristles stir the nerve filaments connected with them, and the nerve transmits to the brain the impressions received.

XLIX. MUSICAL INSTRUMENTS.

In pipes the vibration-frequency varies inversely as its length. When the fundamental of an open pipe is sounded its air-column divides itself into two equal vibrating sections with a node in the center, hence an open pipe should be twice as long as a closed pipe to produce the same pitch.

CHAPTER VI.

RADIANT ENERGY.—LIGHT.

L. INTRODUCTION.

That we receive energy from the sun, and that nearly all the energy employed by man came from the sun, is certain. The average amount of energy received by the earth is estimated to be 83 ft.-lbs. per square foot of surface per second of time (Daniell). This energy appears to be transmitted in the form of wave-motion. This method of transmission, or indeed any method, would seem to require a medium for its transmission. The hypothetical medium has received the name of *ether*, and this method of transmission is called *radiation*. Radiant energy manifests itself as heat, light, or chemism according to the object upon which it acts. Light is the vibration of ether that may be appreciated by the organ of sight.

Luminous bodies are seen by the light which they emit; illuminated bodies, by the light which they reflect. Every point of a luminous body emits light in every direction.

The intensity of light diminishes as the square of the distance from the source increases. (Why?)

The apparent size of an object diminishes as its distance from the eye increases. (Why?)

The velocity with which light traverses interplanetary space is about 186,000 miles per second.

LIII. REFLECTION OF LIGHT.

Light is so reflected that the angle of reflection is equal to the angle of incidence. Owing to the greater or less roughness of the surfaces of all objects light becomes by reflection more or less scattered or *diffused*.

The amount of light reflected from a smooth surface increases rapidly with the angle of incidence.

Concave mirrors tend to produce a convergence of rays; convex mirrors cause divergence; plane mirrors do not alter the relation of rays.

Images formed by plane and convex mirrors are virtual images. Images formed of objects situated between a concave mirror and its principal focus are virtual; in all other situations the images are real.

(Describe the variety of images that may be formed by a concave mirror. Describe an image formed by a convex mirror; also one formed by a plane mirror.)

LIV. REFRACTION.

When light passes obliquely from a rarer into a denser medium it is refracted toward a perpendicular to the boundary plane; if from a denser into a rarer medium, it is refracted from the perpendicular.

The ratio of the sine of the angle of incidence to the sine of the angle of refraction is called the *index of refraction*, and is the same as the ratio of the velocity of the incident to that of the refracted light.

When a ray passes obliquely from a vacuum into a medium, the index of refraction is greater than unity, and is called the *absolute index of refraction*. The relative index of refraction, from any medium A, into another B, is found by dividing the absolute index of B by the absolute index of A.

When the angle of incidence is such that the angle of refraction is 90°, *i.e.*, the reflected ray moves in the plane of the refracting surface, the angle is called the *critical angle*. Total reflection occurs when rays in the more refractive medium are incident at an angle greater than the critical angle.

LV. LENSES.

The general effect of convex lenses is to converge transmitted rays; and of concave lenses, to cause them to diverge.

The corresponding linear dimensions of an object and its image formed by a convex lens are to one another as the respective distances from the optical center of the lens.

(Describe the variety of images that may be formed by convex and concave lenses.)

LVI. PRISMATIC ANALYSIS OF LIGHT.—SPECTRA.

When a beam of white light passes through an optical prism, the colors of which it is composed are separated by refraction, owing to their different degrees of refrangibility. If the different colors are again brought together, white light is reproduced.

Difference of color is a difference of vibration-rate or wavelength. In a dense medium the shorter waves are more retarded than the longer ones; hence, they are more refracted. A body which emits white light sends forth simultaneously waves of a variety of lengths.

Luminous solids and liquids give continuous spectra, while gases usually give discontinuous or bright-line spectra. Hence, the spectrum reveals the state of the substance emitting light.

A vapor of any substance is opaque to those rays which it would itself emit if luminous. Hence, when white light traverses vapors capable of absorbing certain rays, the spectrum formed by the transmitted light will be crossed by dark lines. These dark lines occur where bright lines would be formed if the same vapors were rendered luminous, hence the former are sometimes called *reversed spectra*.

No two substances give spectra consisting of the same combination of lines. Spectrum analysis consists in determining the presence or absence of given substances in a luminous vapor by the presence or absence of their characteristic lines in the spectrum. Likewise the substances which are present in the solar atmosphere and the photosphere can be determined by the reversed lines of the solar spectrum.

The solar spectrum is not limited in either direction by the visible spectrum. Although the eye is not susceptible to impressions from the ultra-red and ultra-violet rays, yet the former are quite energetic in producing heat, and the latter in generating chemical action.

LVII. COLOR.

Color is a quality of the light which illuminates, and not of the object illuminated. No body gives color to light which it reflects or transmits.

The tendency of atmospheric dust is to absorb the colors at the violet end of the spectrum, and to transmit the colors at the red end. On the other hand, it tends to reflect the colors of the violet end, and absorb those of the red. Hence the redness of the light at sunrise and sunset, when the light passes long distances through this dust; also, the blueness of sky-light, which is reflected light.

Red, green, and violet are thought to be the three primary color-sensations, and all other colors are supposed to be the product of mixed sensations of these three in varying proportions.

A color resulting from a mixture of pigments is the color that is left after the two pigments have absorbed all the other colors, and is not the result of a combination of colors.

When any two colors combined will produce white, they are said to be complementary to each other.

Waves of light, like sound-waves, may interfere so as to produce mutual destruction. If the light is monochromatic, darkness is the result. If the light is white, and only waves of certain length interfere, then a color is produced which is the result of the subtraction of the annihilated color from white light. Interference may be caused by reflection from thin films, and by the bending of rays of light around the edges of opaque objects.

LVIII. DOUBLE REFRACTION AND POLARIZATION OF LIGHT.

Light transmitted through the crystals of certain substances, notably Iceland spar, suffers a double refraction, *i.e.*, it becomes divided, and pursues two different paths.

Ordinary light is supposed to consist of vibrations in ether, in every possible plane at right angles to the path of the light. When, by reflection or transmission through certain substances, it is reduced to vibrations of one plane, it is said to be *polarized*.

LIX. THERMAL EFFECTS OF RADIATION.

When a body absorbs largely the radiations which strike it, *i.e.*, when the undulations of the ether are largely transformed into molecular motion, the body becomes heated thereby, and is said to be *athermanous*. But if the nature of the molecules of a body is such that their motions are not readily quickened by the undulations, but the body allows a large portion of the undulations to pass through it unabsorbed, then is it slightly heated thereby, and is said to be *diathermanous*.

All bodies emit radiations, and, in common parlance, are said to radiate heat. Good absorbers are good radiators; bad absorbers are bad radiators. The absorbing and radiating power of a body of the same substance depends largely upon the character of its surface, *i.e.*, whether it be bright and smooth, or tarnished and rough. Dew, which is the result of condensation of the watery vapor of the air, collects most abundantly on good radiators, inasmuch as they part with their heat rapidly, and, consequently, become cooler than poor radiators.

INDEX.

[NUMBERS REFER TO PAGES.]

A.

Aberration, Chromatic, 394.
　Spherical, 363.
Absorption, 38, 39.
Acceleration, Unit of, 128.
Achromatic lens, 395.
Action and reaction, 116.
Adhesion, 33.
Air, a medium of wave-motion, 277.
　Weight of, 3.
Air-pump, 54.
Air-waves, 282.
Alphabet, Telegraphic, 266.
Amalgamating zinc, 187.
Ampère, a unit of current, 206.
Ampère's rule, 183; theory, 218.
Analysis of light, 364.
Angles of incidence and reflection, 118.
Antinodes, 276.
Armature, 197.
Artesian wells, 70.
Athermancy, 388.
Atmosphere, a unit of pressure, 49.
Attraction, Phenomena of, 20.
　mutual, 13, 20, 21.

B.

Ballistic curve, 109.
Barometer, 50.
Battery, Bunsen or Grove, 189.
　Gravity, 191.
　Grenet or bottle, 189.
　Smee, 188.
　Kind of to use, 405.
　Qualities of good, 210.
Batteries, Arrangement of, 207.
　Thermo, 235.
　Various, 188.
Beam of light, 328.
Beats, 302.

Bells, 323.
Blake transmitter, 271.
Boiling, Laws of, 161.
Brittleness, 31.
Buoyant force of fluids, 76.

C.

Camera Obscura, 392.
　Photographer's, 392.
Candle-power, 334.
Capillarity, 34.
Celestial chemistry and physics, 372
Center of gravity, 96.
Centrifugal force, 102.
Centripetal force, 102.
C.G.S. system, 125.
Chemical changes, 9.
Cohesion, 23.
Coil, Rhumkorff's, 233.
Coils, Induction, 232.
Cold, Methods of producing, 167
Color by absorption, 374.
　by interference, 379.
　by polarization, 387.
　Cause of, 367.
　Effect of contrast of, 379.
Colors, Complementary, 379.
　Mixing, 376.
　Primary, 365.
　Sky, 375.
Compound substances, 8.
Compressibility of gases, 52.
Condenser, 249.
Conservation of energy, 174.
Constitution of matter, 6.
Correlation of energy, 174.
Couple, Mechanical, 95.
Critical angle, 354.
Current attraction, 215.
　Extra, 231.

INDEX.

Current induction, 230.
  Strength of, 197, 201.
Currents, Earth, 224.
  Laws of, 215, 217.
Curvilinear motion, 101.
Cutting glass, 401.

D.

Density, 7, 79.
Dew, 391.
  point, 164.
Dialysis, 40.
Diamagnetism, 225.
Diathermancy, 388.
Diffraction, 381.
Diffusion, 39.
Discharge, Electrical, 242.
Discord, Cause of, 307.
Dispersion of light, 365.
Distillation, 162.
Ductility, 32.
Dynamics defined, 44.
  of fluids, 44.
Dynamo machines, 227.
Dyne, 125, 128.

E.

Ear, 315.
Earth, a magnet, 220, 223.
Elasticity, 29.
Electric candle, 260.
  lamp, 260.
  light, 259.
Electrical attractions, etc., 238, 252.
  machines, 245.
  measurements, 197.
Electricity, Chemical effect of, 192.
  Current, 179.
  Frictional, 237.
  Heating effect of, 191.
  how it originates, 184.
  Luminous effect of, 192, 253.
  Magnetic effect of, 196.
  Physiological effect of, 195.
  Thermo, 234.
  Two states of, 238.
  Useful applications of, 258.
  What is, 257.
Electrification, 237.
  on surface, 244.
  Two kinds of, 239.

Electro-chemical series, 186.
Electrodes, 183.
Electrolysis, 193.
Electro-magnet, 196.
Electro-magnetic machines, 262.
Electro-motive force, 204.
Electrophorus, 246.
Electroplating, 262.
Electroscope, 237.
Electrotyping, 261.
Energy, Conservation of, 174.
  contrasted with momentum, 123.
  Correlation of, 174.
  defined, 121.
  Formula for, 124.
  Potential and kinetic, 121.
  Radiant, 327.
  Transformation of, 128, 129, 257.
  Unit of, 128.
Engine, Steam, 175.
Engines, Kinds of steam, 177.
Equilibrant force, 95.
Equilibrium, 44.
  Three states of, 98.
Erg, 126, 128.
Ether, a medium of motion, 326.
Evaporation, 163.
Expansibility of gases, 52.
Expansion, Abnormal, 150.
  by heat, 148.
  Coefficients of, 149.
  Power of, 150.
Experiment defined, 1.
Eye, Human, 393.

F.

Falling bodies, 104.
Fire-alarm, Electric, 267.
Flexibility, 29.
Foci, Conjugate, 360.
Focus, Principal, 346, 359.
  Virtual, 360.
Foot-pound, 120.
Force, Absolute unit of, 125.
  Centrifugal, 102.
  Centripetal, 102.
  defined, 12, 13.
  Equilibrant, 95.
  Gravity unit of, 125.
  Measure of a, 124.

INDEX.

Force, Measure of the effect of, 126.
  Resultant, 91.
Forces, Composition of, 91, 94, 95.
  Graphic representation of, 90.
  Molar, 13.
  Molecular, 13.
  Resolution of, 92.
Fraunhofer's lines, 372.
Fusion, Laws of, 161.

G.

Galvanometer, 198, 404.
  Tangent, 199.
Galvanoscope, 184.
Gaseous bodies, Laws of, 156.
Gases, Kinetic theory of, 157.
Gravitation, 14, 20.
Gravity, Acceleration of, 106.
  Center of, 96.
  Force of, 14, 21.

H.

Hardness, 28.
Harmonics, 305.
Harmony, Cause of, 307.
Hearing, Limits of, 301.
Heat, Capacity for, 171.
  Conduction of, 142.
  Convection of, 143.
  convertible, 138, 165.
  defined, 139.
  Diffusion of, 142.
  Expansion by, 148.
  from chemical action, 140.
  Mechanical equivalent of, 175.
  Origin of animal, 140.
  Reference tables for specific, 172.
  Some sources of, 138.
  Specific, 170.
  units, 165.
Helix, 196.
Horse-power, 121.
Hydrogen at the copper plate, 185.
Hydrometers, 83.
Hydrostatic bellows, 64.
  press, 64.

I.

Images, After, 379.
  Formation of, 346, 360.
  Real, 347.
  through apertures, 330.
  To construct, 347, 361.
  Virtual, 342, 362.
Impenetrability, 1, 6.
Induction, 241.
  coils, 232.
Inertia, 90.
Interference of light, 379.
  of sound-waves, 274, 322.
Insulation, 243.

J.

Joule's equivalent, 175.

K.

Kilogrammeter, 120.
Kinetic energy, 121.
  theory of gases, 157.

L.

Law, Mariotte's, 156.
  of Charles, 156.
Laws of fusion and boiling, 161.
  of gaseous bodies, 156.
Lenses, 357.
  Effects of, 358.
Leyden jar, 250.
Light, a form of energy, 325.
  Analysis of, 364.
  Diffused, 340.
  Electric, 259.
  invisible, 327.
  Reflection of, 339.
  Synthesis of, 366.
Lightning, 255.
  rods, 255.
Liquid surface level, 69.
Luminous and illuminated bodies, 328

M.

Machines, 131.
  Law of, 133.
  Uses of, 132.
Magnets and magnetism, 212, 224.
  Law of, 213.
  Natural, 223.
  not sources of energy, 225.
Magnetic transparency, 213.

INDEX.

Magnetism, Cause of the earth's, 223.
Magneto machines, 227.
 electric induction, 226.
Malleability, 32.
Manometric flames, 312.
Mariotte's law, 57, 156.
Mass, 7, 20.
Matter a constant quantity, 10, 11.
 Conditions of, 24.
 Crystalline and amorphous, 24.
 Three states of, 15.
Metric system, 399.
Microphone, 270.
Microscope, Simple, 362.
 Compound, 391.
Minuteness of particles of matter, 3.
Mirrors, Reflection from, 341.
Molecule, 4.
Momentum, 115.
Motion, Accelerated, 104.
 Curvilinear, 101.
 First law of, 89.
 Formulas for uniformly accelerated, 106.
 Kinds of, 87.
 Retarded, 107.
 Second law of, 91.
 Third law of, 117.
 versus rest, 86, 87.
Multiple reflection, 343.
Musical instruments, 319.
 Scale, 300.

N.
Nodes, 275.
Noise, 297.

O.
Ohm, 202.
Ohm's law, 205.
Opacity, 328.
Oscillation, Center of, 111.
Osmose, 40.
Overtones, 305.

P.
Parabolic curve, 109.
Paramagnetism, 225.
Pencil of light, 328.
Pendulum, 110.

Pendulum, Center of oscillation of, 111.
 Center of percussion of, 113.
Phenomenon, 1.
Phonograph, 317.
Photometry, 333.
Physical changes, 9.
Physics defined, 129.
Pigments, 375.
 Mixing, 378.
Pitch, 298.
Points, Effects of, 252.
Polarity, 28, 214.
Polariscope, 387.
Polarization, 384.
 of plates, 188.
Poles of battery, 183.
Porosity, 7.
Potential, Electric, 183, 244.
 energy, 121, 168.
Porte lumière, 339, 407.
Press, Hydrostatic, 64.
Pressure in fluids, 44–79.
Primary colors, 365.
Prisms, Optical, 357.
Projectiles, 108.
Pump, Air, 54–57.
 Force, 76.
 Lifting, 74, 75.

Q.
Quality of sound, 309.
Qualities of perfect battery, 210.

R.
Radiation, 327.
 Thermal effects of, 388.
Radiator, 327.
Radiometer, 325.
Random of projectiles, 108
Ray, 328.
Reaction, 116.
Reflection, Angle of, 118.
 Law of, 118.
 Multiple, 343.
 Total, 355.
Refraction, 350.
 Cause of, 351.
 Double, 383.
 Index of, 352.

INDEX.

Relay and repeater, 264.
Repulsion mutual, 13.
Resonance, 290.
Resonators, 291.
Resistance, Formula for, 202.
  Internal, 203.
  External, 204.
Rest, 86, 87.
Resultant force, 91.

S.

Shadows, 331.
Simple substances, 8.
Siphon, 72.
Siren, 299.
Solution of solids, 37.
Sonometer, 303.
Sound, Analysis of, 309.
  how it originates, 280.
  how it travels, 281.
  Loudness of, 288.
  media, 283.
  Musical, 297.
  Pitch of, 298.
  Quality of, 309.
  Reënforcement of, 290.
  Reflection of, 285.
  Refraction of, 287.
  Synthesis of, 310.
  Velocity of, 284.
  what it is, 283.
Sounder, 264.
Sounding air-columns, 319.
  plates, 321.
Sound-waves, 272, 274, 280.
Speaking tubes, 289.
Specific gravity, 80.
Spectra, Bright-line, 369.
  Continuous, 368.
  Dark-line, 370.
  Heat and chemical, 373.
Spectrum analysis, 371.
  Solar, 365.
Spectroscope, 368.
Stability of bodies, 99.
Steam engine, 175.
Stereopticon, 395.
Summary of elec. measurements, 206.
  of mechanical units and formulas, 127.
Sun as a source of energy, 141.

T.

Table of boiling points, 161.
  of E.M.F., 205.
  of indices of refraction, 353.
  of melting points, 161.
  of metric system, 399.
  of natural tangents, 403.
  of specific gravities, 402.
  of specific heat, 172.
Telegraph, 263.
  Fac-simile, 266.
Telegraphic alphabet. 266.
Telephone, Bell, 269, 318.
Telephone, Dolbear, 271.
  String, 318.
Telescope, Astronomical, 392.
Temperature, Absolute, 155.
  defined, 141.
  measured by expansion, 151.
Tenacity, 32.
Tension, 44.
Theory of exchanges, 390.
Thermo batteries, 235.
Thermopile, 236.
Thermo-dynamics, 174.
Thermometer, Air, 154.
  Construction of, 151.
  Graduation of, 152.
Thermometry, 151.
Transformation of energy, 128, 129, 257.
Translucency, 328.
Transparency, 328.
Tubes, Speaking, 289.

U.

Undulatory theory, 327.

V.

Vacuum, Absolute, 56.
Variation of needle, 222.
Velocity, Accelerated, 104.
  defined, 87.
  of electric discharge, 256.
  of light, 337.
  of sound, 284, 292.
  Unit of, 128.
Ventilation, 145.

INDEX.

Ventral segment, 276.
Vibration, Direction of, 273.
  of strings, 303.
  Propagation of, 274.
  Sound, 272.
Vibrations, Complex, 273, 305.
  Composition of, 311.
  Forced, 295.
  Stationary, 275.
  Sympathetic, 295.
Viscosity, 31.
Visual angle, 335.
Vocal organs, 323.
Volt, 205.
Voltaic arc, 259.
Voltameter, 198.

W.

Waves, Air, 282.
  Interference of, 274, 294.
  Longitudinal, 276.
  Reflection of, 274.
  Sound, 272, 274, 280.
  Water, 276.
Wave-length, 274, 292.
  Measuring, 292.
Wave-lengths of light, 367.
Wave-line, 274, 279.
Wave-motion, Apparatus to illustrate, 406.
Wave-propagation, 278.
Weber, 206.

APPARATUS ADAPTED TO GAGE'S PHYSICS.

Immediately following the first appearance of the book, in November, 1882, the Publishers received many calls for apparatus especially adapted to the carrying out of the plan of the book. It appearing almost a necessity, Mr. Gage reluctantly consented to give some attention to the furnishing of schools with cheap and efficient apparatus, thereby rendering it possible for every school in the land, however limited its means, to teach this branch in a rational manner. In future, he will devote a portion of his time to the study of (1) new forms of apparatus, and (2) methods of making the same pieces, with slight modifications, answer a variety of purposes. His popular "little marvels," the New Porte-Lumière, Seven in One Apparatus, Eight in One Apparatus, improved Pascal's Vases, Bunsen Batteries, Apparatus for making electrical measurements, etc., are a sufficient testimony to his success thus far.

Only a minimum profit is charged on this apparatus, so that no discounts are possible, and the school which has but a dollar to expend can purchase on terms which will compare favorably with the lowest net prices ever offered. A set of this apparatus will be kept on constant exhibition at our office, 13 Tremont Place, Boston.

For price lists, and other information respecting the apparatus, address

A. P. GAGE, English High School, Boston, Mass.

Unsolicited testimonial from **L. B. Charbonnier**, *Professor of Physics in the University of Georgia.*

The apparatus ordered from you has been received to-day. Like all previously bought from you, it gives entire satisfaction. You are really doing an excellent work for our schools in furnishing such apparatus as you do, and at the most reasonable cost. I have had excellent opportunity to judge of the quality of your work, as I have under my charge an extensive collection of apparatus bought from different makers here and in Europe. The apparatus bought of you is used by the students of the lower class in the laboratory; and hence I have been able to compare your work with that of other makers. I feel it due you to testify to the excellence of your work. There is no reason why physical science should not be now fully illustrated in our schools, when the inexpensiveness of your apparatus brings it within the reach of the most moderate means.

Athens, Ga., October 13, 1887.

GINN & COMPANY, Publishers, Boston, New York, and Chicago.

Introduction to Physical Science.

By A. P. GAGE, Instructor in Physics in the English High School, Boston, Mass., and Author of *Elements of Physics*, etc. 12mo. Cloth. viii + 353 pages. With a chart of colors and spectra. Mailing Price, $1.10; for introduction, $1.00; allowance for an old book in exchange, 30 cents.

THE great and constantly increasing popularity of Gage's *Elements of Physics* has created a demand for an equally good but easier book, on the same plan, suitable for schools that can give but a limited time to the study. The *Introduction to Physical Science* has been prepared to supply this demand.

Accuracy is the prime requisite in scientific text-books. A false statement is not less false because it is plausible, nor au inconclusive experiment more satisfactory because it is diverting. In books of entertainment, such things may be permissible; but in a text-book, the first essentials are correctness and accuracy. It is believed that the *Introduction* will stand the closest expert scrutiny. Especial care has been taken to restrict the use of scientific terms, such as *force, energy, power*, etc., to their proper significations. Terms like *sound, light, color*, etc., which have commonly been applied to both the effect and the agent producing the effect have been rescued from this ambiguity.

Recent Advances in physics have been faithfully recorded, and the relative practical importance of the various topics has been taken into account. Among the new features are a full treatment of electric lighting, and descriptions of storage batteries, methods of transmitting electric energy, simple and easy methods of making electrical measurements with inexpensive apparatus, the compound steam-engine, etc. Static electricity, which is now generally regarded as of comparatively little importance, is treated briefly; while dynamic electricity, the most potent and promising physical element of our modern civilization, is placed in the clearest light of our present knowledge.

In **Interest and Availability** the *Introduction* will, it is believed, be found no less satisfactory. The wide use of the *Elements* under the most varied conditions, and, in particular, the author's own experience in teaching it, have shown how to improve where improvement was possible. The style will be found suited to the grades that will use the book. The experiments are varied, interesting, clear, and of practical significance, as well as simple in manipulation and ample in number. Certain subjects that are justly considered difficult and obscure have been omitted; as, for instance, certain laws relating to the pressure of gases and the polarization of light. The *Introduction* is even more fully illustrated than the *Elements*.

In General. The *Introduction*, like the *Elements*, has this distinct and distinctive aim, — to elucidate science, instead of "popularizing" it; to make it liked for its own sake, rather than for its gilding and coating; and, while teaching the facts, to impart the spirit of science, — that is to say, the spirit of our civilization and progress.

George E. Gay, *Prin. of High School, Malden, Mass.:* With the matter, both the topics and their presentation, I am better pleased than with any other Physics I have seen.

R. H. Perkins, *Supt. of Schools, Chicopee, Mass.:* I have no doubt we can adopt it as early as next month, and use the same to great advantage in our schools. (*Feb.* 6, 1888.)

Mary E. Hill, *Teacher of Physics, Northfield Seminary, Mass.:* I like the truly scientific method and the clearness with which the subject is presented. It seems to me admirably adapted to the grade of work for which it is designed. (*Mar.* 5, '88.)

John Pickard, *Prin. of Portsmouth High School, N.H.:* I like it exceedingly. It is clear, straightforward, practical, and not too heavy.

Ezra Brainerd, *Pres. and Prof. of Physics, Middlebury College, Vt.:* I have looked it over carefully, and regard it as a much better book for high schools than the former work. (*Feb.* 6, 1888.)

James A. De Boer, *Prin. of High School, Montpelier, Vt.:* I have not only examined, but studied it, and consider it superior as a text-book to any other I have seen. (*Feb.* 10, '88.)

E. B. Rosa, *Teacher of Physics, English and Classical School, Providence, R.I.:* I think it the best thing in that grade published, and intend to use it another year. (*Feb.* 23, '88.)

G. H. Patterson, *Prin. and Prof. of Physics, Berkeley Sch., Providence, R.I.:* A very practical book by a practical teacher. (*Feb.* 2, 1888.)

George E. Beers, *Prin. of Evening High School, Bridgeport, Conn.:* The more I see of Professor Gage's books, the better I like them. They are popular, and at the same time scientific, plain and simple, full and complete. (*Feb.* 18, 1888.)

Arthur B. Chaffee, *Prof. in Franklin College, Ind.:* I am very much pleased with the new book. It will suit the average class better than the old edition.

W. D. Kerlin, *Supt. of Public Schools, New Castle, Ind.:* I find that it is the best adapted to the work which we wish to do in our high school of any book brought to my notice.

C. A. Bryant, *Supt. of Schools, Paris, Tex.:* It is just the book for high schools. I shall use it next year.

Introduction to Chemical Science.

By R. P. WILLIAMS, Instructor in Chemistry in the English High School, Boston. 12mo. Cloth. 216 pages. Mailing Price, 90 cents; for introduction, 80 cents; Allowance for old book in exchange, 25 cents.

IN a word, this is a working chemistry — brief but adequate. Attention is invited to a few special features : —

1. This book is characterized by directness of treatment, by the selection, so far as possible, of the most interesting and practical matter, and by the omission of what is unessential.

2. Great care has been exercised to combine clearness with accuracy of statement, both of theories and of facts, and to make the explanations both lucid and concise.

3. The three great classes of chemical compounds — acids, bases, and salts — are given more than usual prominence, and the arrangement and treatment of the subject-matter relating to them is believed to be a feature of special merit.

4. The most important experiments and those best illustrating the subjects to which they relate, have been selected ; but the modes of experimentation are so simple that most of them can be performed by the average pupil without assistance from the teacher.

5. The necessary apparatus and chemicals are less expensive than those required for any other text-book equally comprehensive.

6. The special inductive feature of the work consists in calling attention, by query and suggestion, to the most important phenomena and inferences. This plan is consistently adhered to.

7. Though the method is an advanced one, it has been so simplified that pupils experience no difficulty, but rather an added interest, in following it ; the author himself has successfully employed it in classes so large that the simplest and most practical plan has been a necessity.

8. The book is thought to be comprehensive enough for high schools and academies, and for a preparatory course in colleges and professional schools.

9. Those teachers in particular who have little time to prepare experiments for pupils, or whose experience in the laboratory has been limited, will find the simplicity of treatment and of experimentation well worth their careful consideration.

For testimonials, see the special circular.

www.ingramcontent.com/pod-product-compliance
Lightning Source LLC
Chambersburg PA
CBHW022109300426
44117CB00007B/643